ENVIRONMENTAL REMEDIATION TECHNOLOGIES, REGULATIONS AND SAFETY

ESSAYS ON FUNDAMENTAL AND APPLIED ENVIRONMENTAL TOPICS

ENVIRONMENTAL REMEDIATION TECHNOLOGIES, REGULATIONS AND SAFETY

Additional books in this series can be found on Nova's website under the Series tab.

Additional E-books in this series can be found on Nova's website under the E-book tab.

ENVIRONMENTAL RESEARCH ADVANCES

Additional books in this series can be found on Nova's website under the Series tab.

Additional E-books in this series can be found on Nova's website under the E-book tab.

ENVIRONMENTAL REMEDIATION TECHNOLOGIES, REGULATIONS AND SAFETY

ESSAYS ON FUNDAMENTAL AND APPLIED ENVIRONMENTAL TOPICS

DRAGUTIN MIHAILOVIC
EDITOR

Nova Science Publishers, Inc.
New York

For permission to use material from this book please contact us:
Telephone 631-231-7269; Fax 631-231-8175
Web Site: http://www.novapublishers.com

Library of Congress Cataloging-in-Publication Data

Essays on fundamental and applied environmental topics / editor, Dragutin Mihailovic.
p. cm.
Includes index.
ISBN 978-1-61942-522-4 (hardcover)
1. Environmental engineering. 2. Environmental engineering--Mathematical models. 3. Environmental sciences--Mathematical models. 4. Environmental health. 5. Environmental policy. I. Mihailovic, Dragutin T.
TA170.E88 2011
628--dc23

2011048500

Published by Nova Science Publishers, Inc. †New York

This book is dedicated to Gordana, Ivan and Anja

CONTENTS

PREFACE

To preserve this world for future generations, we must strive for sustainable development that meets the needs of the present without compromising future options and resources. However, some scientists suspect that recent floods, heat waves and droughts in worldwide regions are indicators of severe hazardous events to come. Increased pollution in the atmosphere is causing a gradual increase in air temperature. The environment is becoming increasingly endangered, especially as humans encroach on and modify fragile environments. Environmental issues are a primary focus of the world scientific community in the 21st century.Perhaps the most comprehensive and detailed list of these important environmental topics is outlined by the Royal Society of London for Improving Natural Knowledge (RS), the oldest national scientific society in the World and the leading national organization for the promotion of scientific research in Great Britain. The RS has emphasized the following fundamental issues: (i) Are we alone in the Universe?; (ii) Emission of harmful greenhouse gases and climate change; (iii) What is the consciousness? What is the consciousness?; (iv) How to make decisions in a insecure world; (v) Extension of the average lifespan; (vi) Is culture the only characteristic of people; (vii) Managing the Earth's resources; (viii) The importance of the internet; (ix) Study of the master cells; (x) The importance of maintaining biodiversity; (xi) The role of geoengineering in climate change; and (xii) The importance of new vaccines. The aforementioned topics, which are not listed in hierarchical order, highlight issues that can be linked to the environment and are primarily affected by climate change and related issues.

A unique characteristic of these issues is that they are closely connected with the survival of individual human beings on the Earth. This is the first time in the history of science that environmental problems are at the top of the scientific priority list, from fundamental inquiry to the development of the practical applications. Environmental science can be defined from the point of view of many individual scientific disciplines. However, it is defined in the Random House Dictionary as "the branch of science concerned with the physical, chemical, and biological conditions of the environment and their effect on organisms." This *can be accepted as the broadest because it implicitly includes the main feature of environmental science, which is its multidisciplinary nature* (underlined by D.T. Mihailovic). The field of environmental science is abundant with various interfaces and is ready for the application of new fundamental approaches that can lead to a better understanding of environmental phenomena. We have slightly evolved the previous definition by focusing on the concept of an *environmental interface,* which is defined as an interface between two abioticor biotic environments that are in relative motion and exchange energy and substances through physical, biological and chemical processes, fluctuating temporally and spatially regardless of a space and time scale. We define environmental science as "*the branch of science concerned with interactions in environmental interfaces that are regarded as natural complex systems.*"

Environmental science encompasses issues such as climate change, conservation, biodiversity, water quality, groundwater contamination, soil contamination, use of natural resources, waste management, sustainable development, disaster reduction, air pollution, and noise pollution. The core components of environmental science are derived from the atmospheric sciences, ecology, environmental chemistry and the geosciences. Atmospheric science examines the Earth's gaseous outer layer and emphasizes interrelationships with other systems. It includes disciplines ranging from meteorological studies (e.g., atmospheric chemistry and atmospheric physics), greenhouse gases phenomena, atmospheric dispersion modeling of airborne contaminants, sound propagation phenomena as it relates to noise pollution, and light pollution. Studies in *Ecology* typically analyze the dynamics of biological populations and an aspect of their environment. These studies may address endangered species, predator/prey interactions, habitat integrity, effects of environmental contaminants on populations, or an analysis of the impact of proposed land development on species viability. Interdisciplinary analysis of ecological systems that are impacted by one or more stressors related to fields in environmental science, such as water pollution, may also be evaluated. Environmental chemistry involves chemical alterations in the environment. Principal areas of study include soil contamination and water pollution. Topics of analysis include chemical degradation in the environment, multi-phase transport of chemicals and chemical effects on biota. The *geosciences* include environmental geology, environmental soil science, volcanic phenomena and the evolution of the Earth's crust. Hydrology and oceanography are also included as geosciences in some classification systems.

This book is organized into two broad parts: Fundamental and Modeling Aspects (Part One) and Application Issues (Part Two). Part One has 7 chapters that cover various theoretical and modeling aspects of fundamental issues, such as the nonlinear dynamics of intercellular exchange of biochemical substances, use of formal concept analysis and the category theory in modeling living systems, generation of information in biological systems, calculation of the albedo for different geometries, turbulence above very rough surfaces, vertical mixing transport in the chemical transport model and the effects of permeability on hyporheic flows across a triangular dune. Part Two has 8 chapters and addresses a broad range of environmental application issues, such as dynamic downscaling in climate models, moss biomonitoring techniques used to study spatial and temporal atmospheric deposition, surveillance of mosquito populations, the rapid evolution of enteroviruses, monitoring of forest ecosystems, the impact of climate change on plant diseases and pests, the impact of climate change on water river flow and the social aspects of local environmental policy. The chapters mostly cover the aforementioned "century" scientific topics. This book is the result of the authors experiences over a long period of time and previous work on the following projects: (i) "Studying climate change and its influence on the environment: impacts, adaptation and mitigation" (No. 43007) and (ii) "Sustainable agriculture and rural development in achieving the strategic goals of the Republic of Serbia within the Danube Region" (No. 46006). The projects were financed by the Ministry of Education and Science of the Republic of Serbia within the framework of integrated and interdisciplinary research from 2011-2014. These two projects assembled scientists from different departments at the

University of Novi Sad and the University of Belgrade (Serbia) in close collaboration with scientists from the University of Sarajevo (Bosnia and Herzegovina) and the University of Naples "Federico II" (Italy).

Part One begins with a chapter by *D.T. Mihailović, M. Budinčević, I. Balaž* and *D. Perišić*that describes the exchange of biochemical substances to establishcommunication between cells. The authors propose a form of coupled difference logistic equations as a suitable approach for modeling the exchange of biochemical substances between two cells. The synchronization of exchange between cells and sensitivity to fluctuations of intra- and intercellular environmental parameters using methods of nonlinear dynamics, such as the largest Lyapunov exponent and cross sample entropy as measures, were investigated. Instead of developing an accurate quantitative model of substance exchange between cells, the authors were interested in formalizing the basic process and creating an appropriate strategy that allows further investigation of synchronization induced by fluctuations of intra- and intercellular environmental parameters. Chapter 2, written by *S. Crvenković, D.T. Mihailović* and *I. Balaž*, considers formal concept analysis and category theory in modeling interactions between living systems and their environment. After several theoretical considerations, the authors developed a formal representation of how organisms "observe" their environment using Formal Concept Analysis (FCA) and the category theory. FCA is a branch of the applied lattice theory that defines a concept as a unit of two parts: extension and intension. Extension covers all objects belonging to the particular concept and intension includes all of the attributes valid for those objects. To analyze the algebraic properties of changeable systems, the authors further considered the concept of lattices as objects in a category of lattices and morphisms that preserve lattice completeness. They elaborated on some of the functional consequences of the ability of biological systems to perform variable identification of environmental signals on different scales. Chapter 3, written by *I. Balaž* and *D.T. Mihailović*, is devoted to the generation of information in biological systems through an epistemological essay. Modern information theory deals with the transfer of information between the source and the receiver. The information and means of encoding, decoding and transfer are already established. In man-made systems, function is a direct result of the proper design of the information source and how information is received and processed. In natural systems, function is analyzed on short time scales. The authors investigated the functional aspects of processing information in biological systems and analyzed the prerequisites that enable biological systems to generate new information sources and channels. These findings were interpreted in terms of Shannon's information theory and show how current measures of information quantity can be used within a new context. In Chapter 4, *D. Kapor, D.T. Mihailović, A. Ćirišan* and *A. Firanj* introduce a new approach for calculating the aggregated albedo of the heterogeneous grid-cell for regions of different heights, given that different geometries ("trilateral prism", "slope" and "canyon") are commonly encountered in grid-cells in environmental models. In most situations, the analytical solution for the loss of a coefficient is not accessible, and the Monte Carlo ray tracing method (MCRT) was determined to be the most reliable approach for calculating the aggregated albedo. The authors found that the method is highly efficient and less time-consuming with increasing dimensions compared to deterministic methods. The differences in albedo obtained conventionally and with the MCRT method were emphasized by the calculation of the effective surface temperature of the grid-cell for all geometries considered using the land surface scheme LAPS. The chapter by *B. Lalić, D.T. Mihailović* and *D. Kapor* addresses the limitations and uncertainties in logarithmic wind profiles above very rough, forest and urban surfaces, which is an important issue in environmental modeling. The authors analyzed the impact of roughness and element size on the deviation of the vertical wind profile from the logarithmic profile using the results of micrometeorological measurements. Their study was devoted to the consequences of (i) different numerical values of von Karman's constant and (ii) arbitrary determinations of wind profile parameters, such as roughness length and zero plane displacement height, in theoretical and practical considerations in environmental models. It was emphasized that remarkable relative deviations in wind profile values exist compared to the chosen reference profile above very rough surfaces when different values of Karman's constant were used. In the next chapter, *D.T. Mihailović*, *Z. Podraščanin* and *I. Arsenić* present the results of numerical simulations of non-local eddy diffusion and vertical mixing schemes for use in chemical transport and air quality models. The easiest and most common way for parameterization of vertical turbulent mixing in an atmospheric boundary layer (ABL) is the use of K-schemes. In K-schemes, the mixing takes place between the nearest model layers, which is a reasonable approach only when the scale of turbulent motion is much smaller than the scale of mean motion, for example, during stable and neutral atmospheric conditions. Because these schemes cannot describe the effects of large scale eddies that are dominant in the convective boundary layer (CBL),they are not recommended for use in the CBL. To avoid K-schemes, a "non-local" turbulent kinetic energy (TKE) scheme can be used. However, even the TKE scheme can represent a turbulent scale in any stability condition in which the mixing is local and is only suitable to represent small scale turbulence, which reduces its reliability. The asymmetrical convective mixing scheme (ACM) is suitable for use during convective conditions. Because none of these schemes represent mixing at all scales, the authors comprehensively considered a linear combination of the non-local TKE and ACM schemes. The features of the considered schemes were evaluated using numerical tests and chemical transport and 1-D models. Part One ends with a chapter by *C. Gualtieri*that considers the effects of permeability on hyporheic flows across a triangular dune. Despite the traditional separation between the studies of surface water and groundwater flows, rivers and aquifers are strongly connected and their interaction gives rise to a continuous exchange of water and solutes, which exerts a significant influence on water quality. The *hyporheic zone* where groundwater and stream water mix. In this chapter, the author presents results of a numerical study that investigated the influence of the permeability of a porous medium on a coupled free fluid-porous medium flow. Laminar flow in the water column above triangular bedforms and Darcian flow in the underlying permeable sediments were considered. The results confirmed the close relationship between the characteristics of the separation region in the water column downstream of the bedform crest and those of the hyporheic zone in the porous medium. The numerical results also highlighted the influence of permeability on the development of the hyporheic zone.

Part Two begins with a chapter by *B. Rajković*, *K. Veljović* and *V. Djurdjević* describing monthly, seasonal and climate case studies using the dynamic downscaling method. Monthly forecasts are the shortest integrations that fall in the category of long-term integrations using current modeling skills and computer resources. The authors focus on the possible applications and general meanings of monthly forecasts relative to standard 5- to 10-day forecasts. Another interesting issue they address is the ability to conduct forecasts without resorting to the nudging of large scales during integration. The quantitative scores for such forecasts are also discussed. The other types of integrations/simulations are climate time scale integrations because there is a paramount need for assessment of their quality. According to the authors, the first step in that direction involves downscaling the ERA-interim data set, which can be achieved by performing integrations in bothnon-coupled and coupled modes. Chapter 9, written by *M. Krmar*, *D. Radnović* and *M.V. Frontasyeva*, presents a moss biomonitoring technique used to study the spatial and temporal atmospheric deposition of heavy metals and airborne radionuclides. This technique, first introduced in Scandinavia using terrestrial mosses, has been shown to be suitable for studying the atmospheric deposition of heavy metals and other trace elements, including airborne radionuclides. By measuring the concentration of heavy elements every 5 years since 1990 in most European countries, information regarding the spatial distribution of atmospheric deposition and temporal changes in depositional pattern has been collected. The authors point out that airborne radionuclides, such as ^{7}Be and ^{210}Pb, can be used as tracer elements for exploring the atmospheric transport and deposition of aerosols. They concluded that recent measurements show that mosses can be a superior sampling medium in spatial and temporal studies of the atmospheric deposition of airborne radionuclides. The chapter by *D. Petrić, M. Zgomba, R. Bellini* and *N. Becker* addresses the surveillance of mosquito populations as a key element in understanding the spread of invasive/vector species and vector-borne diseases in Europe. Increased mobility and international trade together with projected climate changes play an important role in the dissemination of vectors and the pathogens and parasites they can transmit. Since the beginning of Millennium, a number of pathogen introductions have been recorded in Europe. The latest introduction (Ravenna, Italy, 2007) is the tropical Chikungunya virus, which is transmitted by the "Asian tiger mosquito" species, introduced in Italy in 1990. Determining the actual cause of changes in distribution and behavior of the mosquito vectors is difficult and complex and largely depends on the availability of data obtained by monitoring. Important invasive vector species, relevance and sustainability of different monitoring techniques and surveillance programs are discussed by the authors in order to assist in preparedness regarding vector-borne diseases. The chapter by *I.B. Hrnjaković Cvjetković, J.P. Radovanov* and *V.S. Milošević* describes the predominant enterovirus serotypes in the Vojvodina region of Serbia. Enteroviruses are small, non-enveloped, single-stranded RNK viruses that belong to the family *Picornaviridae*. They exhibit a high degree of genetic variability, which is the basis of their rapid evolution. Enteroviruses infect millions of people worldwide. Clinical manifestation of enteroviral infections range from mild respiratory illness to severe diseases, such as myocarditis, neonatal sepsis, type I diabetes and aseptic meningitis. Enteroviruses are excreted in large quantities in the feces of infected individuals and remain viable for a long time in the environment. They are spread mainly by the fecal-oral route, direct contact with secretions from an infected person, and via contaminated water and food. Detection of enterovirus serotypes enables the identification of endemic circulating virus types that are responsible for an outbreak and can assist in monitoring changes in pathogenicity, virulence and the epidemic potential of the different enteroviral types. To identify the predominant enteroviruses in the etiology of aseptic meningitis in the Vojvodina region of Serbia, a total of 70 samples were obtained from pediatric patients suspected of having aseptic meningitis. The presence of enteroviruses was detected by isolation in cell culture and confirmed using an indirect immunofluorescence assay. The authors concluded that echoviruses (52.8 %) were the most prevalent, followed by coxsackie B viruses (27.8 %), and coxsackie A viruses (8.3 %). Chapter 12, written by *S. Orlović, Z. Galić, S. Stojnić* and *B. Klašnja* includes key results of long-term monitoring of forest ecosystems in Serbia through the National Forest Inventory. Systems for monitoring forest health at the national level and a system recommended by the IPCC are discussed as well as the impact of climate change on forestry. The authors comprehensively elaborated on the monitoring program and the following forest types: *Fraxineto - Querceto typicum*, *Querceto frainetto- cerris*, *Querceto petrea*, *Fagenion moesiace montanum*, *Blechneto - Piceto - Abieto - Fagetum sphagnetosum*, *Populus* x *euramericana*, and coniferous plantation of *Pinus nigra* and *Pseudotsuga* sp. Information about additional relevant parameters, including microclimate, crown condition, chemism of soil, wood increment, phenology, litter and oxidative drought stress, is presented. The chapter by *R. Jevtić, S. Maširević* and *D. Vajgand* considers the impact of climate change on diseases and pests that target small grains and sunflowers in the Vojvodina region of Serbia. The authors emphasized that the mean annual temperature in the Vojvodina region in the last ten-year period has risen to 12 °C compared to 11.1 °C from 1881 - 1990. Climate change has resulted in the dominance of pathogens that either require higher temperatures for their development or are better able to adapt to drought conditions. For example, small grain fungal diseases of the genus *Fusarium* and *Septoria* spp. have a dominant role and cause significant damage, and the causal agents of sunflower spots develop rapidly at high air temperatures. According to the authors, rust (*Puccinia* spp.) pathogens have been forming fewer generations and transitioning quickly from the uredo to the teleuto stage compared with the previous time period as a direct result of climate change. *Orobanche cumana* is a serious problem in the sunflower. Broomrape has caused significant damage in susceptible hybrids in Serbia since the 1990s. Climate change also influences pests. As insects, moths do not have a constant body temperature, and the speed of their development is directly correlated with air temperature. The cotton bollworm (*Helicoverpa armigera*) has become dominant because of the increase in air temperature. The chapter by *E. Hadžić* and *N. Drešković* is devoted to the impacts of climate change on river flow in the Sarajevo Valley of Bosnia and Herzegovina. The authors considered the impacts of air temperature and precipitation fluctuation on the flow of the Miljacka River and the Bosna River during the hydrological instrumental period (1926 - 2010). Climate variability was evaluated for a longer meteorological instrumental period (1888 - 2010) using data on a yearly basis from the Bjelave meteorological station, which is representative of the Sarajevo Valley. Although an increase in the air temperature and a slight increase in the amount of precipitation was shown, the authors concluded that there is still evident stagnation in the river flow. In the final chapter, *J. Čikić* and *Ž. Petrović* explain the roles of nongovernmental organizations (NGOs) and local environmental policy through a case study of the city of Novi Sad in Serbia. The status of these organizations in local environmental policy was analyzed according to data collected in a survey that included all environmental NGOs registered in the city of Novi Sad. Five major issues were analyzed: human resources, available equipment and facilities, areas of interest and activities of various NGOs, cooperation and information and funding. These issues were addressed with respect to the potential negotiating power of the environmental NGOs in the policy actors' network and the ability to influence the development of environmental awareness in the community.

The objective of this book is to bring together scientists and engineers working in research institutions, universities and academia who engage in the study of the theoretical, modeling, measuring and software aspects of the environmental sciences. It is intended to provide a forum for participants to obtain up-to-date information and exchange new ideas and expertise through the presentation of recent achievements in this field. A combination of professionals from a wide spectrum of different areas focused on theoretical, modeling and experimental approaches, including software design and measurements, in the environmental sciences.

The editor wishes to thank all of the authors for their continuous and dedicated effort that made the realization of this book possible. The editor also thanks the reviewers of the project for the thoughtful and detailed suggestions that have improved the content and presentation of this book. The reviewers are: Dr. Zoran Ivić, Institute for Nuclear Sciences "Vinča", Laboratory for Theoretical Physics and Condensed Matter, Belgrade, Serbia (Chapter 1); Dr. Taichi Haruna, Department of Earth and Planetary Sciences, Kobe University, Kobe, Japan (Chapters 2 and 3); Dr. Ljuba Budinski Petković, Faculty of Technical Sciences, University of Novi Sad, Novi Sad, Serbia (Chapter 4); Dr. Branko Grisogono, Department of Geophysics, Faculty of Sciences, University of Zagreb, Zagreb, Croatia (Chapter 5); Dr. Tamás Weidinger, Department of Meteorology, Eötvös Loránd University, Budapest, Hungary (Chapter 6); Dr. Fabian A. Bombardelli, Department of Civil and Environmental Engineering, University of California, Davis, United States (Chapter 7); Dr. Silvio Gualdi, National Institute of Geophysics and Vulcanology, Bologna, Italy (Chapter 8); Dr. Eiliv Steinnes, Department of Chemistry, Norwegian University of Science and Technology, Trondheim, Norway (Chapter 9); Dr. Frantisek Rettich, National Institute for Public Health, Prague, Czech Republic (Chapter 10); Dr. Marija Kulauzov, Center of Bacteriology, Institute of Public Health of Vojvodina, Faculty of Medicine, University of Novi Sad, Novi Sad, Serbia (Chapter 11); Dr. Carsten Lorz, Institute for Soil Science and Site Ecology, Dresden University of Technology, Dresden, Germany (Chapter 12); Dr. Mirko Ivanović, Department of Phytomedicine, Faculty of Agriculture, University of Belgrade, Belgrade, Serbia (Chapter 13); Dr. Samir Djug, Department of Biology, Faculty of Sciences, University of Sarajevo, Sarajevo, Bosnia and Herzegovina (Chapter 14); and Dr. Cosmin Salasan, Faculty of Agricultural Management at the Banat University of Agricultural Sciences and Veterinary Medicine Timisoara, Timisoara, Romania (Chapter 15).

This book was supported by PhD candidates Miss Ana Firanj and Miss Zorica Podraščanin, who invested time and effort in its development. The editoracknowledges the assistance of the Editorial Office of the Nova Science Publishers with gratitude.

PART ONE: FUNDAMENTAL AND MODELING ASPECTS

In: Essays on Fundamental and Applied Environmental Topics ISBN 978-1-61942-522-4
Editor: Dragutin Mihailovic

Chapter 1

TWO MAPS SERVING THE INTERCELULAR EXCHANGE OF BIOCHEMICAL SUBSTANCES AND THEIR BEHAVIOUR IN THE PRESENCE OF FLUCTUATIONS

D. T. Mihailović[1,*], M. Budinčević[2], I. Balaž[3] and D. Perišić[2]

[1]Department of Field and Vegetable Crops, Faculty of Agriculture, University of Novi Sad, Novi Sad, Serbia
[2]Department of Mathematics and Informatics, Faculty of Sciences, University of Novi Sad, Novi Sad, Serbia
[3]Department of Physics, Faculty of Sciences, University of Novi Sad, Novi Sad, Serbia

ABSTRACT

The exchange of biochemical substances is an essential way to establish the communication between cells. Despite great diversity among organisms in specific mechanisms where proteins and small molecules are involved, the general scheme of the communication process remains fairly universal. It is notable that all phases of the process are heavily influenced by perturbations of either internal or external parameters of such systems. We propose a form of coupled difference logistic equations as a suitable approach for modeling the exchange of biochemical substances between two cells. We then investigate the synchronization of exchanges between cells and their sensitivity to fluctuations of intra- and intercellular environmental parameters using methods of nonlinear dynamics, i.e., using the largest Lyapunov exponent and cross sample entropy as measures. Instead of developing an accurate quantitative model of substance exchanges between cells, we are instead interested in the formalization of the basic shape of the process and creating an appropriate strategy that allows further investigation of

[*] Corresponding author: E-mail address: *guto@uns.ac.rs;* Phone: +381 21 485 3203. Fax: +381 21 635 0552.

synchronization induced by fluctuations of intra- and intercellular environmental parameters.

1. Introduction

Communication between cells is ubiquitous in the biological world. From single cell bacteria to complex eukaryotic organisms, cellular communication is a way to create more complex structures through integration and maintaining of functionality. Organisms evolved various auxiliary ways to ensure that the transfer of signals can be performed timely and efficiently (e.g., development of vascular systems starting from early chordates). However, at the molecular level the basic scheme of signal exchange remains the same: signaling molecules should reach the cellular receptor, which in turn activates regulatory response, modulating production of the targeted molecular species. These species then either directly or indirectly influence production of arriving signals. In this general approach, several points should be noted. Since communication is established by exchange of biochemical substances (substances in the further text) through the surrounding environment, this process is heavily influenced by environmental factors. In single cell organisms, environmental fluctuations are even more prominent, since substances have to be released into the external environment, which is not included in the homeostasis created by the organism. Additionally, even in a clonal population and under a strongly controlled environment, a significant level of fluctuation of constituting parameters will remain due to protein disorder [1, 2] and the so-called intrinsic noise [3, 4]. Finally, due to thermal and conformational fluctuations, biochemical processes are inherently random [5]. If the described process of substance exchange between cells is considered through the optics of a complex system that should maintain its functionality under strong influence of both internal and external fluctuations, we are approaching the problem of robustness [6 - 8]. Some elaborated formal treatments of this problem are still in their infancy. One of the main reasons for this is the fact that its focus is beyond already developed tools of dynamical systems theory. It indicates that a new, more general approach for describing the system has to be developed.

In this chapter, our focus is only on questions about how the system, which is basically stochastic and inherently influenced by internal and external perturbations, can continue to function. In Section 2, we give a short overview of the general mechanism for exchanging substances between two cells, representing a cooperative communication process. Furthermore, we identify the main parameters of the process and derive a system of two coupled logistic equations as an appropriate model of the given process. In Section 3, we analyse the coupled maps which represent the intercellular exchange of substances using methods from nonlinear dynamics. Section 4 is devoted to investigating the synchronization of the model and its sensitivity to fluctuations in environmental parameters. Some concluding remarks are given in Section 5. It should be emphasized that our goal is not the development of an accurate quantitative model of exchange of substances between cells. Rather, we are interested in the formalization of the basic shape of the process, and creating the appropriate strategy that allows further investigation of the synchronization induced by fluctuations of intra- and intercellular environmental parameters.

2. A Simple Model of Intercellular Exchange

2.1. Empirical Background

Ranging from bacteria, in which quorum sensing [9] and colony formation [10] are efficient mechanisms for rapid switching between different phenotypes, to vertebrates, in which sophisticated humoral control ensures proper functioning of the organism as an integrated system, communication between cells is one of the main prerequisites for assembling cells into highly organized structures. Despite a great variety of specific mechanisms and an even greater number of molecules involved, the general scheme remains fairly universal (see, for example [11]), as is seen in Figure 1, which is adopted as a scheme of intercellular exchange model we propose in this chapter.

Signaling molecules are ones that are deliberately extracted by the cell into intracellular environments and that can affect behavior of other cells of the same or different type (species or phenotype) by means of active uptake and subsequent changes in genetic regulations. Signaling Molecules can be excreted as either a side product of other metabolic processes or as purposefully synthesized and transported from the cell. Once they appear in the intercellular environment, they can be transported to other cells, which can be affected themselves. Let us note that the term *environment*, in this chapter, includes both the (i) *intracellular environments* (inside the cell) and (ii) *intercellular environments* (which surrounds cells). Because active uptake is one of the milestones of the process, a very important factor in establishing communication is a current set of receptors and transporters in the cellular membrane during the communication process. They constitute the backbone of the whole process, but simultaneously are a very important source of perturbations of the process, due to protein disorders and intrinsic noise. As a result, the process of exchange is constantly under inherent fluctuation due to the aforementioned parameters. Another important factor is intercellular environments, which could interfere with the process of exchange. These factors include distance between cells, mechanical and dynamical properties of the fluid which serves as a channel for exchange and various abiotic and biotic factors influencing physiology of the involved cells. Finally, in order to define the exchange process as communication, signaling molecules should induce change in genetic regulations. Signaling molecules can influence production of a number of different genes, but the synthesis of molecules that are able to directly or indirectly affect production of arriving signals is necessary to call this process a communication. Therefore, the concentration of signaling molecules inside of the cell which are destined to be extracted can serve as an indicator of the dynamics of the whole process of communication. These signaling molecules can be either the same for all involved cells or they can be different, acting directly or indirectly on production of arriving signals.

Additionally, the influence of affinity in the functioning of living systems is also an important issue. It can be divided into following aspects: (a1) affinity of genetic regulators towards arriving signals, which determines the intensity of cellular response and (a2) affinity for uptake of signaling molecules. The first aspect is genetically determined and therefore species specific. The second aspect is more complex and is influenced by the affinity of the receptors for binding specific signaling molecules and the number of active receptors and their conformational fluctuations (protein disorder).

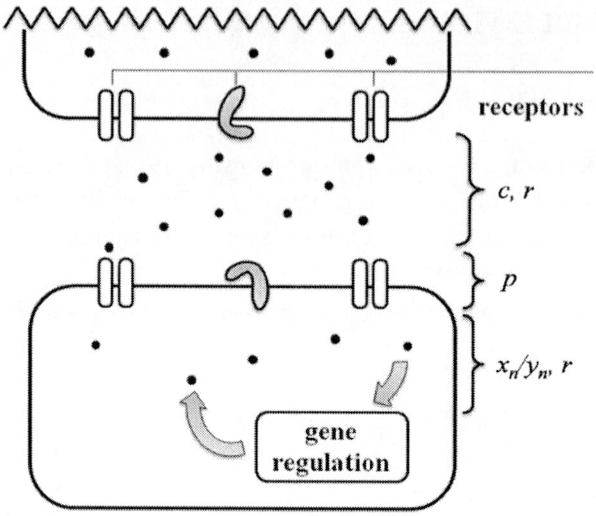

Figure 1. Schematic representation of intercellular communication. Here, c represents the concentration of the signaling molecule in the intercellular environment coupled with the intensity of the response they provoke, while r includes the collective influence of environmental factors which can interfere with the process of communication. x_n and y_n represent the concentration of signaling molecules in cell environment, while p denotes the cellular affinity to uptake the substances.

2.2. Model Philosophy

As is obvious from the empirical description, we can infer the successfulness of the communication process by monitoring: (i) the number of signaling molecules, both inside and outside of the cell and (ii) their mutual influence. The concentration of signaling molecules in an intercellular environment is subject to various environmental influences, and taken alone often can indicate more about state of the environment than about the communication itself. Therefore, we choose to follow the concentration of signaling molecules inside of the cell as the main indicator of the process. In this case, the parameters of the system are (i) the affinity p by which cells perform uptake of signaling molecules (a2), which depends on the number and the state of the appropriate receptors, (ii) the concentration c of the signaling molecules in the intercellular environment within the radius of interaction, (iii) the intensity of the cellular response (a1) x_n and y_n and (iv) the influence of other environmental factors, which can interfere with the process of communication. In this case we estimate parameter r, which can be taken collectively for intra- and intercellular environments inside of the one variable, indicating the overall disposition of the environment to the communication process.

The time development (n is the number of time steps) of the concentration in cells (x_n, y_n) can be expressed as

$$x_{n+1} = (1-c)\Psi(x_n) + h_2\,(y_n), \qquad (1a)$$

$$y_{n+1} = (1-c)\Psi(y_n) + h_1\,(x_n). \qquad (1b)$$

The map, h represents the flow of materials from cell to cell, and $h_1(x)$ and $h_2(y)$ are defined by a map that can be approximated by a power map,

$$h_1(x) \sim cx^p, \tag{2a}$$

$$h_2(y) \sim cy^q. \tag{2b}$$

If $h_1(x) \sim cx^p$ and $h_2(y) \sim cy^q$, the interaction is expressed as a nonlinear coupling between two cells. The dynamics of intracellular behavior is expressed as a logistic map (e.g., [12, 13]),

$$\Psi(x_n) = r\,x_n(1 - x_n), \tag{3a}$$

$$\Psi(y_n) = r\,y_n(1 - y_n). \tag{3b}$$

Because the concentration of the signaling molecules can be regarded as their population for a fixed volume, and because we are focused on the mutual influence of these populations, it is clear that we should use the coupled logistic equations. Instead of considering cell-to-cell coupling of two explicit n-gene oscillators [14], we consider the generalized case of intercellular communication. In this case, the investigation of conditions for which two equations are synchronized and how this synchronization behaves under changes of intra- and intercellular environments can give some answers to the question of how functionality in the system is maintained. Therefore, having in mind (i) that cellular events are discrete [6] and (ii) using the aforementioned reasoning, we consider systems of difference equations of the form

$$\mathrm{X}_{n+1} = \mathrm{F}(\mathrm{X}_n) \equiv \mathrm{L}(\mathrm{X}_n) + \mathrm{P}(\mathrm{X}_n), \tag{4}$$

with notation

$$\mathrm{L}(\mathrm{X}_n) = ((1-c)rx_n(1-x_n), (1-c)ry_n(1-y_n)), \qquad \mathrm{P}(\mathrm{X}_n) = (cx_n^p, cy_n^q), \tag{5}$$

where $\mathrm{X}_n = (x_n, y_n)$ is a vector representing the concentration of the signaling molecules inside of the cell, while $\mathrm{P}(\mathrm{X}_n)$ denotes the stimulative coupling influence of members of the system, which is here restricted only to positive numbers in the interval (0, 1). The starting point X_0 is determined such that $(x_0, y_0) \in (0,1)$. Parameter $r \in (0,4)$ is the so-called logistic parameter, which in logistic difference equations determines an overall disposition of the environment to the given population of signaling molecules and exchange processes. The affinity to uptake signaling molecules is indicated by p. Let us note that we require that the sum of all affinities of cells p_i exchanging substances has to satisfy the condition $\sum_i p_i = 1$ or in the case of two cells, $p + q = 1$, i.e., $q = 1 - p$. Because $\mathrm{X} = 0$ is a fixed point (4), in order to ensure that zero is not simultaneously the point of attraction, we define $p \in (0,1)$ as an exponent. Finally, c represents coupling of two factors: the concentration of the signaling molecules in the intracellular environment and the intensity of response they can provoke. This form is taken because the effect of the same intracellular concentration of signaling molecules can vary greatly with variation of affinity of genetic regulators for that signal, which is further reflected in the ability to synchronize with other cells. Therefore, c influences both the rate of intracellular synthesis of signaling molecules and the synchronization of the signaling processes between two cells, so the parameter c is taken to be a part of both $\mathrm{L}(\mathrm{X}_n)$ and $\mathrm{P}(\mathrm{X}_n)$. However, the relative ratio of these two influences depends on the current model setting. For example, if for both cells, X_n is strongly influenced by the intracellular concentration of signals that can provoke relatively smaller responses, then the form of equation will be

$$x_{n+1} = (1 - c)rx_n(1 - x_n) + cy_n^p\,, \tag{6a}$$

$$y_{n+1} = (1 - c)ry_n(1 - y_n) + cx_n^{1-p}\,. \tag{6b}$$

We now analyze our coupled system, given by (6a) - (6b). For $0 < x, y < 1$ and $0 < p < 1$, we have $0 < x < x^p < 1$. So, for small c, the dynamic of our investigated system is similar to the dynamic of the following systems obtained by minorization, as follows

$$\begin{aligned} x_{n+1} &= (1 - c)rx_n(1 - x_n), \\ y_{n+1} &= (1 - c)ry_n(1 - y_n), \end{aligned} \tag{7}$$

and

$$\begin{aligned} x_{n+1} &= (1 - c)rx_n(1 - x_n) + cy_n, \\ y_{n+1} &= (1 - c)ry_n(1 - y_n) + cx_n. \end{aligned} \tag{8}$$

If we apply a majorization, the considered system becomes

$$\begin{aligned} x_{n+1} &= (1 - c)rx_n(1 - x_n) + c, \\ y_{n+1} &= (1 - c)ry_n(1 - y_n) + c. \end{aligned} \tag{9}$$

For all of these systems, it is obvious that they do not depend on the parameter p. Because $f(x, y) = g(y, x)$, where f, g are the components of F in (8), their dynamics are symmetric to the diagonal Δ, $\Delta = \{(x, y) : y = x\}$, as was analyzed in [15]. This symmetry also exists for system (6a) - (6b), if $p = 0.5$.

Recalling the aforementioned conditions for p, x and x^p, we consider only systems (7) and (9) because the behavior of the system (6a) - (6b) comes from the properties of the mentioned ones.

It is seen that the systems (7) and (9) consist of uncoupled logistic maps on the interval $(0,1)$ (in (7)) or on the interval $(\delta, 1-\delta)$ (in (9)), where $\delta < 0$ is the smaller solution of the equation $x = (1-c)rx(1-x)+c$, i.e.

$$\delta = [(1-c)-1-\sqrt{[r(1-c)-1)]^2+4cr(1-c)}]/[2r(1-c)]. \tag{10}$$

From the property of the logistic equations, we get the following expression

$$\Gamma = [r(1-c)+4c-4\delta]/(1-2\delta), \tag{11}$$

where Γ in system (9) takes the role of $r(1-c)$ in (7). Comparing $r(1-c)$ with (11), we can conclude that the expression (11) is always greater, implying that bifurcations and chaos first appears for system (7), and then for (9). Finally, combining expressions (10) and (11) we get

$$\Gamma = \frac{r^2(1-c)^2+4cr(1-c)-2r(1-c)+2+2\sqrt{[r(1-c)-1]^2+4cr(1-c)}}{1+\sqrt{[r(1-c)-1]^2-4cr(1-c)}} \tag{12}$$

3. Analysis of the Coupled Maps Representing the Intercellular Exchange of Substances Using Methods of Nonlinear Dynamics

In order to further investigate the behavior of the coupled maps, we perform a numerical analysis of the coupled system (6) through its parameters c, r and p, using the largest Lyapunov exponent and cross sample entropy as measures of the chaotic behavior and border between synchronized and unsynchronized system states in intercellular exchange of substances.

3.1. Lyapunov Exponent of the Coupled Maps (6) for Constant Logistic Parameters

We calculate the Lyapunov exponent through an analysis of orbits. The orbit of the point X_0 is the sequence $X_0, F(X_0), \ldots, F^n(X_0), \ldots$ where $F^0(X_0) \equiv X_0$ and, for $n \geq 1$, $F^n(X_0) = F(F^{n-1}(X_0))$.

We say that the orbit is periodic with period k if k is the smallest natural number such that $F^k(X_0) = X_0$. If $k=1$, then the point X_0 is the fixed point. The periodic point X_0 with period k is an attraction point if the norm of the Jacobi matrix for the mapping $F^k(X) = (f_k(x,y)), (g_k(x,y))$ is less than one, i.e., $\| J^k(X_0) \| < 1$, where

$$\mathrm{J}^k(\mathrm{X}_0) = \begin{bmatrix} \frac{\partial f_k}{\partial x} & \frac{\partial f_k}{\partial y} \\ \frac{\partial g_k}{\partial x} & \frac{\partial g_k}{\partial y} \end{bmatrix}_{\mathrm{X}=\mathrm{X}_0} . \tag{13}$$

Here, we define $\| \mathrm{J}^k(\mathrm{X}_0) \|$ as max $\{|\lambda_1|, |\lambda_2|\}$, where λ_1 and λ_2 are the eigenvalues of the matrix. It is worth noting that for a periodic point with period k

$$\mathrm{J}^k(\mathrm{X}_k) = \mathrm{J}^k(\mathrm{X}_0) = \mathrm{J}(\mathrm{X}_{k-1}) \ldots \mathrm{J}(\mathrm{X}_1)\mathrm{J}(\mathrm{X}_0), \tag{14}$$

where

$$\mathrm{J}(\mathrm{X}) = \begin{bmatrix} (1-c)r(1-2x) & cpy^{p-1} \\ c(1-p)x^{-p} & (1-c)\,r(1-2y) \end{bmatrix}. \tag{15}$$

In particular, for the scalar equation $x_{n+1} = f(x_n)$, the norm is $|(f^k(x))'|_{x=x_0} = |f'(x_{n-1}) \ldots f'(x_1) f'(x_0)|$. The mapping for system (6a) - (6b) on the diagonal ($x_n = y_n$) is given by $d(x_n) = (1-c)rx_n(1-x_n) + cpx_n^p$. Then, we have $d'(x_n) = (1-c)r(1-2x_n) + cpx_n^{p-1}$. In order to characterize the asymptotic behavior of the orbits, we need to calculate the largest Lyapunov exponent, which is given for the initial point X_0 in the attracting region by

$$\lambda = \lim_{n \to \infty} (\ln \| \mathrm{J}^n(\mathrm{X}_0) \| / n). \tag{16}$$

with this exponent, we measure how rapidly two nearby orbits in an attracting region converge or diverge. In practice, using (8), we compute the approximate value of λ by substituting in (16) successive values from X_{n_0} to X_{n_1}, for n_0, n_1 large enough to eliminate transient behaviors and provide good approximation. If X_0 is part of a stable periodic orbit of period k, then $\| \mathrm{J}^k(\mathrm{X}_0) \| < 1$ and the exponent λ is negative, which characterizes the rate at which small perturbations from the fixed cycle decay, and we can call such a system a synchronized one.

We considered a two-cell system, where each cell is able to release and uptake the same substance. According to the assumption in model design, the dynamical behavior of the substance concentrations x_n and y_n depends on three factors: (i) its own concentration c within the radius of interaction in the surrounding environment, (ii) the parameter r and (iii) the affinity p for binding on cellular receptors. The first factor is determined by the underlying feedback mechanism of the intracellular regulations, while the second represents the level of the suitability of the environment to the communication between two cells [15]. The third factor depends on protein disorder [1, 2]. The variation of the Lyapunov exponent λ as a function of concentration c is depicted in Figure 2 for $p = 0.5$ and $r = 3.95$.

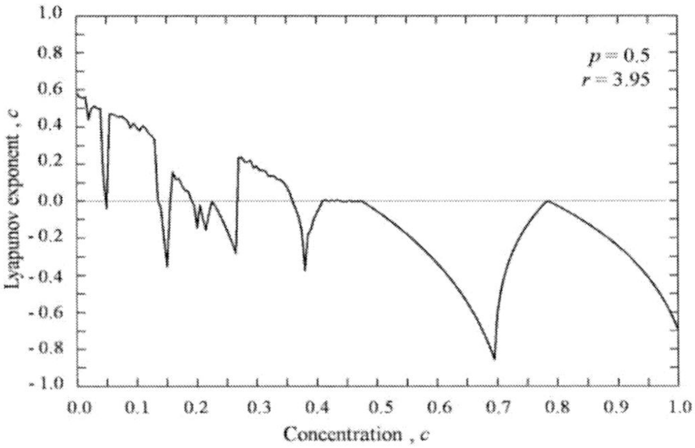

Figure 2. The Lyapunov exponent of the coupled maps, given as a function of the coupling parameter c ranging from 0.0 to 1.0, for affinity value p = 0.5. Each point on the above graph was obtained by iterating many times (2000 iterations) from the initial condition to eliminate transient behavior, and then averaging over another 500 iterations. The initial conditions were x = 0.3 and y = 0.5, with 200 *c* values.

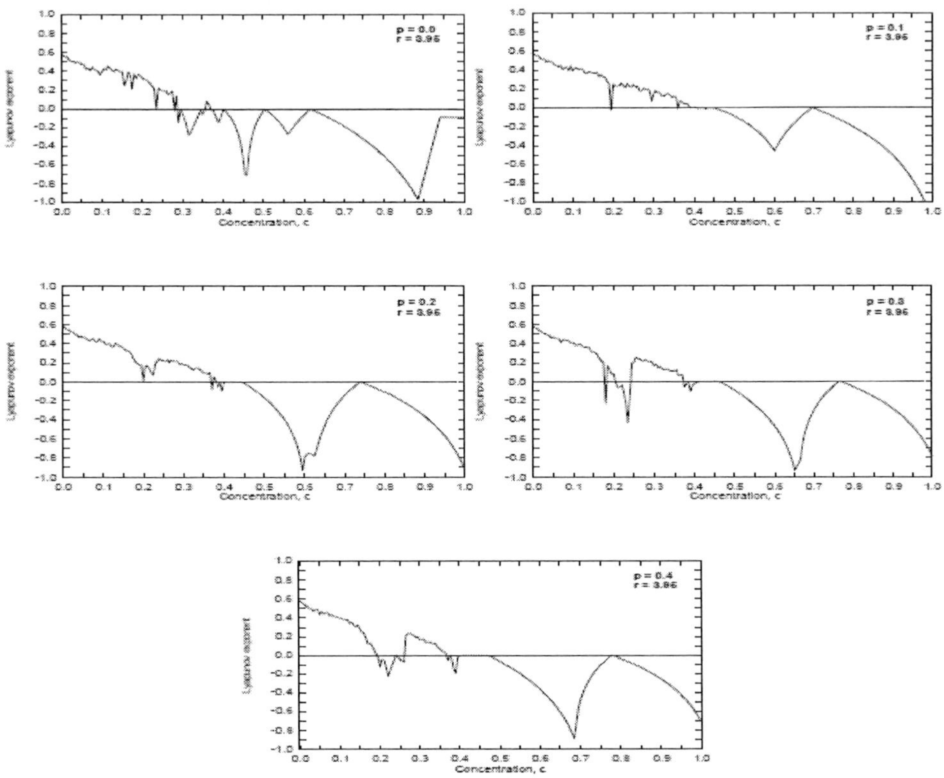

Figure 3. The Lyapunov exponent of the coupled maps, given as a function of the coupling parameter c ranging from 0.0 to 1.0, for different values of the affinity p. The same graphs can be obtained for values p = 0.6, 0.7, 0.8, 0.9 and 1.0 corresponding to those for p = 0.4, 0.3, 0.2, 0.1 and 0.0. Each point in the above graphs was obtained by iterating many times (2000 iterations) from the initial condition to eliminate transient behavior and then averaging over another 500 iterations. The initial condition was x = 0.3 and y = 0.5, with 200c values.

We see that when the value of c exceeds 0.4, complete synchronization (the Lyapunov exponent is less than zero) in intercellular exchange of substances is achieved. In contrast, for values of c smaller than 0.4, there exists a region of unsynchronized states in exchange with some windows where the exchange of substances between two cells is synchronized.

In this subsection, we further consider the behavior of coupling and estimate how a coupled map system can achieve synchronization in intercellular exchange of substances, depending on the parameter c (concentration), for a fixed value of r (in our case 3.95) and different values of the affinity p. For this purpose, we calculate the Lyapunov exponent of the coupled maps, given as a function of the coupling parameter c ranging from 0 to 1.0, for different values of the affinity p as is depicted in Figure 3, where the Lyapunov exponents are calculated for $p=$ 0.4, 0.3, 0.2, 0.1 and 0.0. If we look at all panels, it is seen that there is a border in values of concentration c (around 0.4) that splits the domain of concentration into two regions. The first one, located between 0 and 0.4, has unsynchronized states, including sporadic windows where synchronization is reached. In contrast to this, the second region (between 0.4 and 1.0) is a region where the process of exchange between two cells is fully synchronized. Because of the symmetry of the coupled system (6), the same results will be obtained for values $p=0.6$, 0.7, 0.8, 0.9 and 1.0 corresponding to those for $p=$ 0.4, 0.3, 0.2, 0.1 and 0.0.

3.2. The Lyapunov Exponent of the Coupled Maps (6) when All System Parameters Are Changeable

The effort towards formalizing a model of the process of exchanging substances between cell*s* is still in its infancy and much remains to be completed in order to build a mature theory. For a model to be useful, it must be able to predict characteristics and behaviors of the system. This means that the model has to be framed to explicitly describe constraints that bind the system. That effort is still ongoing. One of the hardest tasks in building those stairways is the issue related to stability and robustness. Exploring the difference between "stable" and "robust" touches on essentially every aspect of what we instinctively know and find interesting about robustness, not only in natural systems, but also in engineering and social systems.

Robustness is one of the fundamental characteristics of biological systems. Numerous reports have been published on how robustness is involved in various biological processes and on mechanisms that give rise to robustness in living systems (see, for example [8]). However, a mathematical foundation that provides a unique perspective on robustness has yet to be established. It is argued here that robustness is a measure of feature persistence in systems that compel us to focus on fluctuations, and often assemblages of perturbations, qualitatively different in nature from those addressed by stability theory. Moreover, to address feature persistence under these sorts of perturbations, we are naturally led to study issues including: the coupling of dynamics with organizational architecture, implicit assumptions of the environment, the role of a system's evolutionary history in determining its current state and thereby its future state, the sense in which robustness characterizes the fitness of the set of "strategic options" open to the system and the capability of the system to switch among multiple functionalities [16, 17].

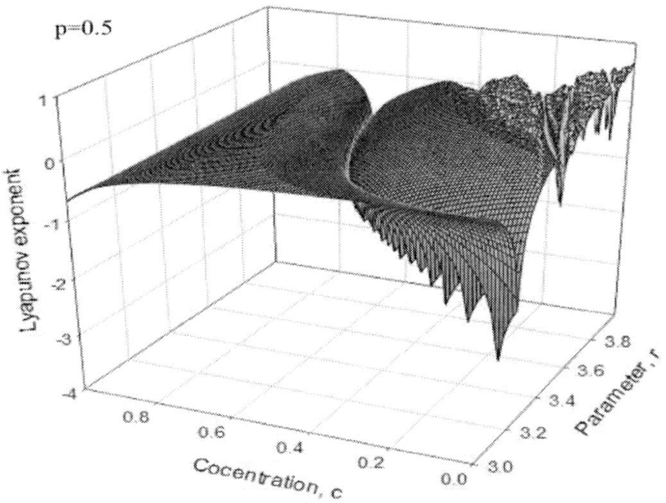

Figure 4. Lyapunov exponent of the coupled maps (6), given as a function of the (i) coupling parameter c, ranging from 0.0 to 1.0 and (ii) parameter r, ranging from 3 to 4, for value of the affinity p = 0.5. Each point in the above graphs was obtained by iterating many times (2000 iterations) from the initial condition to eliminate transient behavior and then averaging over another 500 iterations. Initial condition: x = 0.3, y = 0.5, with 200c values.

In this chapter, the following definition will be used: "robustness" is a property that allows a system to maintain its functions against internal and external perturbations. It is important to realize that robustness is concerned with maintaining *functions* of a *system* rather than *system states*, which distinguishes robustness from stability or homeostasis [8].

In addition to this issue, there is another interesting problem in nonlinear science. Although it is not the focus of this chapter, we will note it because it is always present in modeling of complex systems. This is the interplay between chaos and noise. This is important because complex systems are often under noise, and it is important to assess how the dynamics of a deterministic chaotic system is affected by noise. Some pioneering works on this problem comprise physics, biology and biophysics [18 - 21]. It is well known that chaos, in general, is robust under stochastic fluctuations, which can induce chaos for parameters just before a bifurcation to chaos. Namely, stochastic perturbations can induce chaotic dynamics where there is no naturally occurring chaos, far away from any bifurcation leading to chaos. This is possible due to the complex topology associated with two nearby unstable orbits [20]. Because the robustness, according to the aforementioned definition, is a property that allows a system to maintain its functions against internal and external perturbations, we investigate the synchronization as an indicator of robustness of the coupled maps (6) to perturbation of the c, r and p parameters. To do this, we calculate the Lyapunov exponent as a function of the (i) coupling parameter c, ranging from 0.0 to 1.0 and (ii) parameter r, ranging from 3 to 4, for value of the affinity $p = 0.5$. The changes in the Lyapunov exponent of the coupled maps (6), as a function of parameters c and r, is given in Figure 4. From this figure, it is seen that the model describing intercellular exchange of substances does not maintain its functions for lower concentrations c and higher values of the parameter r. Figure 5 shows the Lyapunov exponent of the coupled maps (6) for different values of the affinity $p = 0.4, 0.3, 0.2, 0.1$.

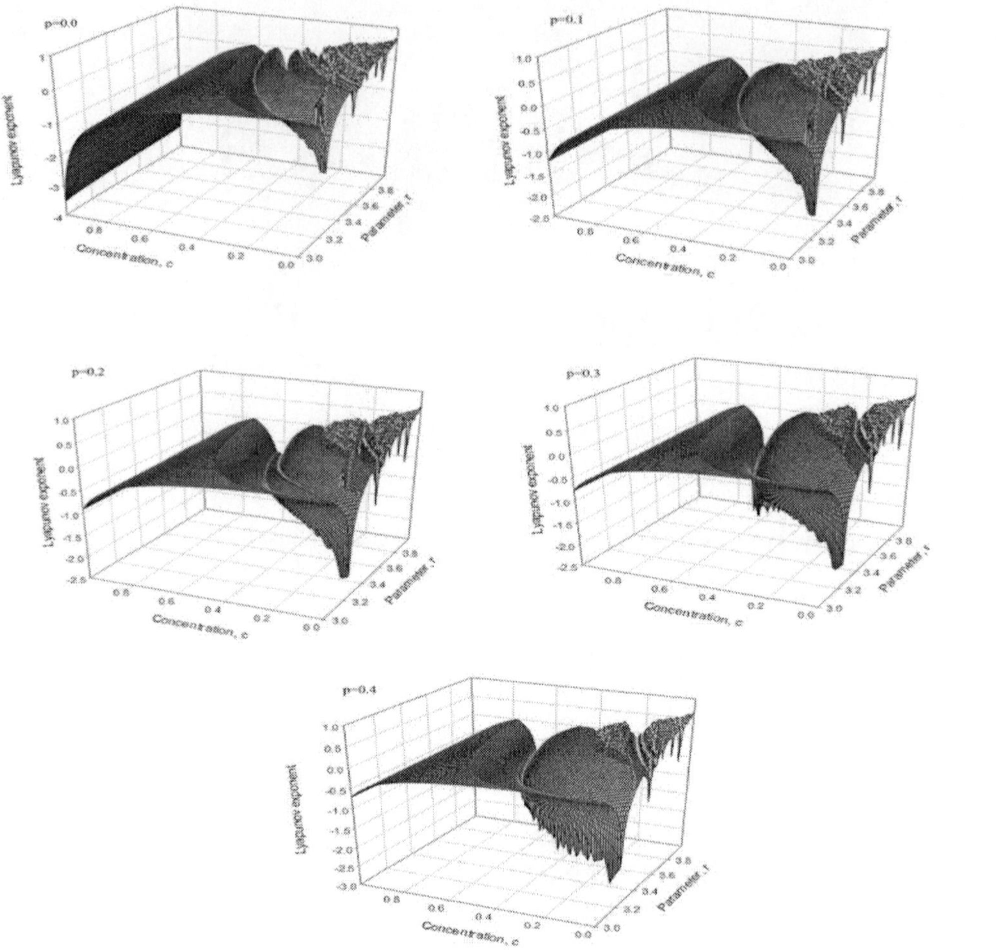

Figure 5. Lyapunov exponent of the coupled maps, given as a function of (i) coupling parameter c, ranging from 0.0 to 1.0 and (ii) parameter r, ranging from 3 to 4, for different values of the affinity p. The same graphs will be able to obtained for values p = 0.6, 0.7, 0.8, 0.9 and 1.0 corresponding to those for p = 0.4, 0.3, 0.2, 0.1 and 0.0. Each point in the above graphs was obtained by iterating many times (2000 iterations) from the initial condition to eliminate transient behavior and then averaging over another 500 iterations. The initial condition was x = 0.3, y = 0.5, with 200c values.

3.3. Entropy of the System of the Coupled Maps (6)

Estimation of the complexity of the system of coupled maps (6), through an analysis of concentration in cells (x_n, y_n) depending on intra- and intercellular parameters, is of great interest for modelling applications. In this chapter, we use the sample entropy (SampEn) as a measure of the complexity of the system considered. Sample entropy, a measure quantifying regularity and complexity, is believed to be an effective analysis method of diverse settings, which includes both deterministic chaotic and stochastic processes and is particularly operative in the analysis of physiological, sound, climate and environmental interfaces or cell signals which involve relatively small amount of data [22, 23]. Practically, we consider cross sample entropy $(Cross-SampEn)$, a measure of asynchrony, using a recently introduced technique for comparing two different time series to assess their degree of asynchrony or dissimilarity [17, 23, 24]. Let $u = [u(1), u(2), ...u(N)]$ and $v = [v(1), v(2), ...v(N)]$ fix input parameters m and ρ. $x(i) = [u(i), u(i+1), ...u(i+m-1)]$ and $y(j) = [v(j), v(j+1), ...v(j+m-1]$ are vector sequences, while N is the number of data points of the time series, $i, j = N-m+1$. For each $i \le N-m$ set $B_i^m(\rho)(v \| u) =$ (number of $j \le N-m$ such that $d[x_m(i), y_m(j)] \le \rho]$)$/(N-m)$, where j ranges from 1 to $N-m$.

Then,

$$B^m(\rho)(v \| u) = \sum_{i=1}^{N-m} B_i^m(\rho)(v \| u) / (N-m) \tag{17}$$

which is the average value of $B_i^m(v \| u)$. Similarly, we define A^m and A_i^m as $A_i^m(\rho)(v \| u) =$ (number of such $j \le N-m$ that $d[x_m(i), y_m(j)] \le \rho]$)$/(N-m)$.

$$A^m(\rho)(v \| u) = \sum_{i=1}^{N-m} A_i^m(\rho)(v \| u) / (N-m) \tag{18}$$

which is the average value of $A_i^m(v \| u)$. Finally, we have

$$Cross-SampEn(m, \rho, n) = -\ln\left\{ A^m(\rho)(v \| u) / B^m(\rho)(v \| u) \right\} \tag{19}$$

We applied $Cross-SampEn$ with $m = 5$ and $\rho = 0.05$ for x_n and y_n time series.

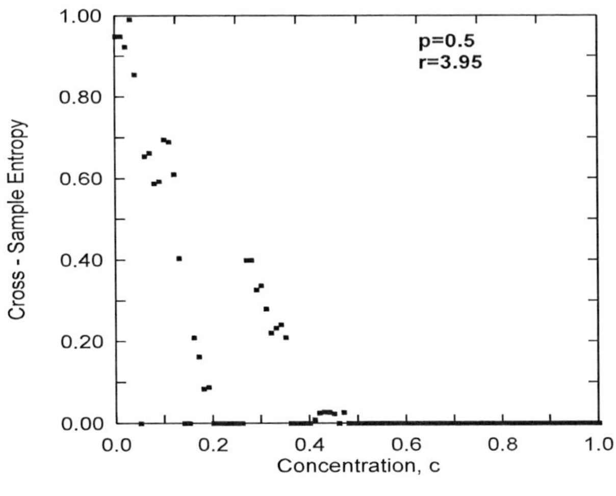

Figure 6. Cross sample entropy of the coupled maps, given as a function of the coupling parameter c, ranging from 0.0 to 1.0, for value of the affinity p = 0.5. The xn and yn time series in the above graphs was obtained by iterating many times (2000 iterations) from the initial condition to eliminate transient behavior and then averaging over another 2000 iterations. Initial condition: x = 0.3, y = 0.5, with 200c values.

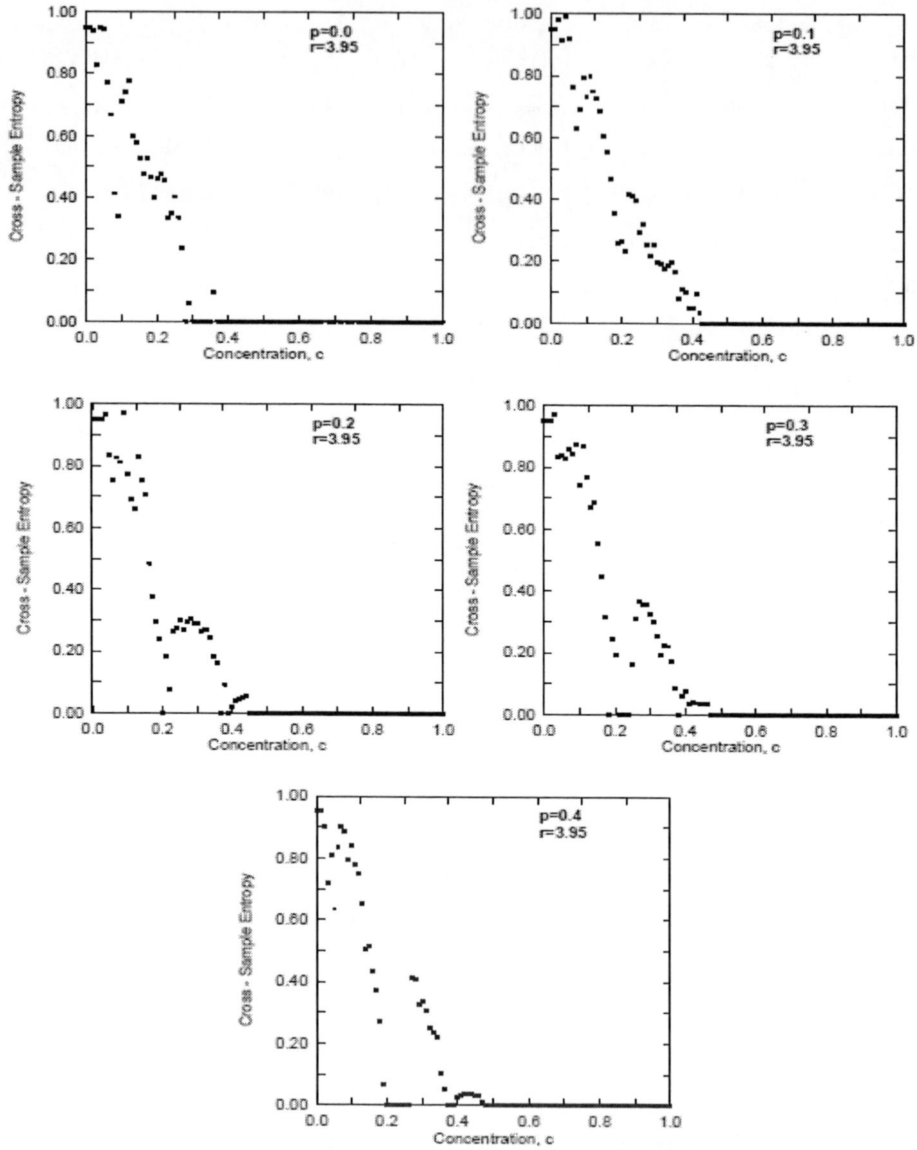

Figure 7. Cross sample entropy of the coupled maps, given as a function of the coupling parameter c (concentration), ranging from 0.0 to 1.0, for different values of the affinity p. The same graphs will be able to obtained for values p = 0.6, 0.7, 0.8, 0.9 and 1.0 corresponding to those for p = 0.4, 0.3, 0.2, 0.1 and 0.0. The xn and yn time series in the above graphs were obtained by iterating many times (2000 iterations) from the initial condition to eliminate transient behavior and then averaging over another 2000 iterations). Initial condition: x = 0.3, y = 0.5, with 200c values.

Figure 6 depicts the cross sample entropy of the coupled maps, given as a function of the coupling parameter *c,* ranging from 0.0 to 1.0, for a value of the affinity of $p = 0.5$ and $r = 3.95$. It is seen that there is high disorder in the system up to the concentration $c = 0.4$. After this value, there is a complete synchronization in the exchange of substances. We obtain similar behavior for different values of the affinity p (Figure 7). These data are in agreement with the analysis of the Lyapunov exponent performed in section 3.1, which indicates compatibility of the used measures.

4. EFFECT OF FLUCTUATIONS

To study the sensitivity of the system (6), which represents substance exchange, to fluctuations of environment parameters of different orders of magnitude, we used methods of nonlinear dynamics, i.e., maximal Lyapunov exponent and cross sample entropy. This system displays a wide range of behavior, as the parameters r and c are varied, including periodic, quasiperiodic, and chaotic motion. The variation of the Lyapunov exponent λ as a function of concentration c is depicted in Figure 8 for $p = 0.2$ and $r = 3.95$. The part of this curve in the elliptical area is chosen for analysis of the effect of fluctuations on synchronization of the system.

In this section, we will investigate the behavior of the coupled maps given by Equations. (6a) and (6b) in the presence of fluctuations of environmental parameters or other noise. As has been shown in the case of uncoupled nonlinear oscillators, the addition of parametric fluctuations has a pronounced effect on the dynamics of such systems [25].

In particular, the presence of noise introduces a gap in the bifurcation sequence of period-doubling systems, and renormalizes the threshold of the appearance of chaotic behavior in the model considered.

It is therefore of interest to investigate the effect of noise on the exchange of substances between two cells.

This interest is because cells are intrinsically noisy biochemical reactors: low reactant numbers can lead to significant statistical fluctuations in molecule numbers and reaction rates [26].

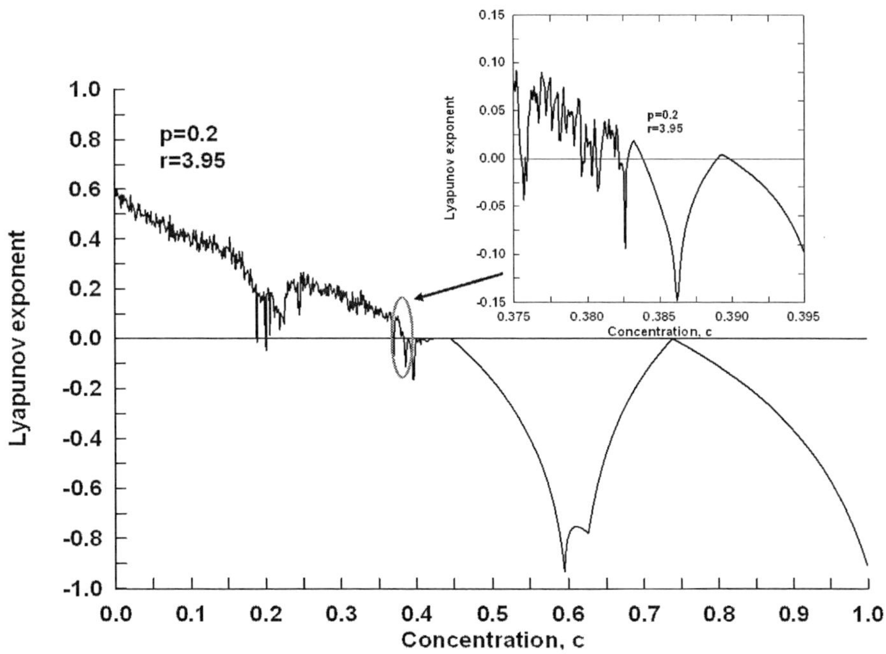

Figure 8. The Lyapunov exponent of coupled maps with no fluctuations [Eq. (6)] as a function of concentration c, ranging from 0 to 1. The ellipsis indicates the region used for analyzing the effect of fluctuations.

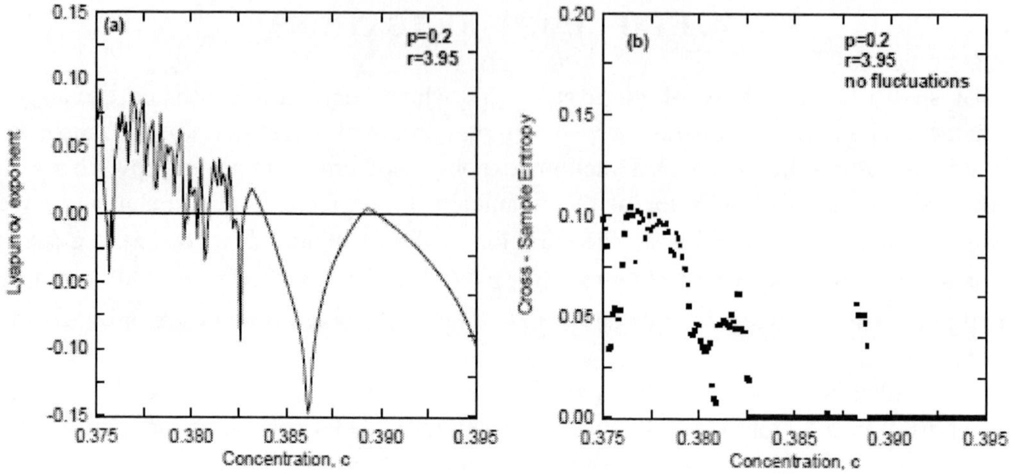

Figure 9. The Lyapunov exponent (a) and cross sample entropy (b) for coupled maps with no fluctuations (Eq. (6)) as a function of concentration c, ranging from 0.375 to 0.395.

The effect of fluctuations was modeled by adding uniformly distributed random numbers to the map of Eq. (6). Specifically, we considered the map

$$x_{n+1} = (1 - \mathrm{E}_n) R_n^{(1)} x_n (1 - x_n) + \mathrm{E}_n y_n^p \tag{20a}$$

$$y_{n+1} = (1 - \mathrm{E}_n) R_n^{(2)} y_n (1 - y_n) + \mathrm{E}_n x_n^{1-p} \tag{20b}$$

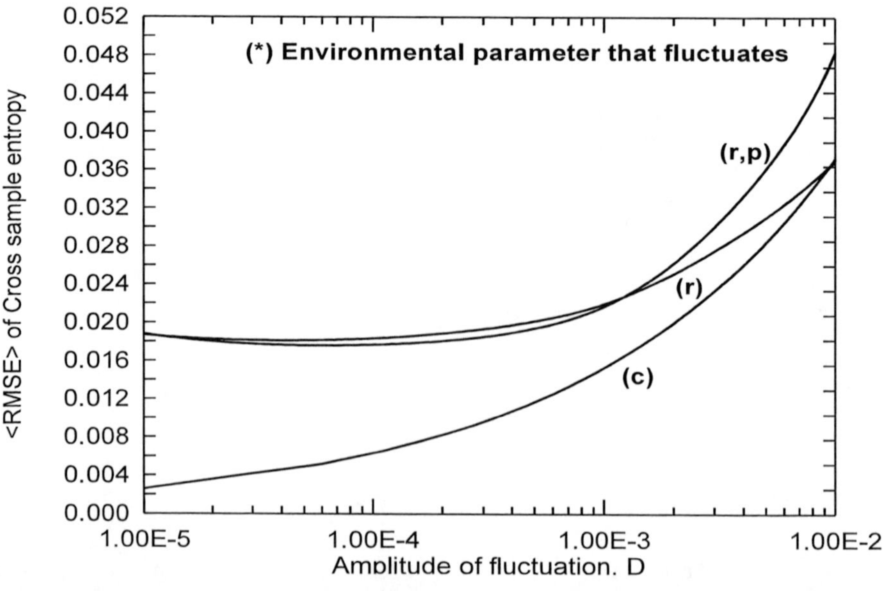

Figure 11. RMSE of the cross sample entropy for coupled maps [Eq. (20)] for fluctuation as a function of the amplitude of fluctuations ranging from 0.00001 to 0.01. The letters next to curves indicate fluctuations in: (c) concentration, (r) logistic parameter and both (r , c).

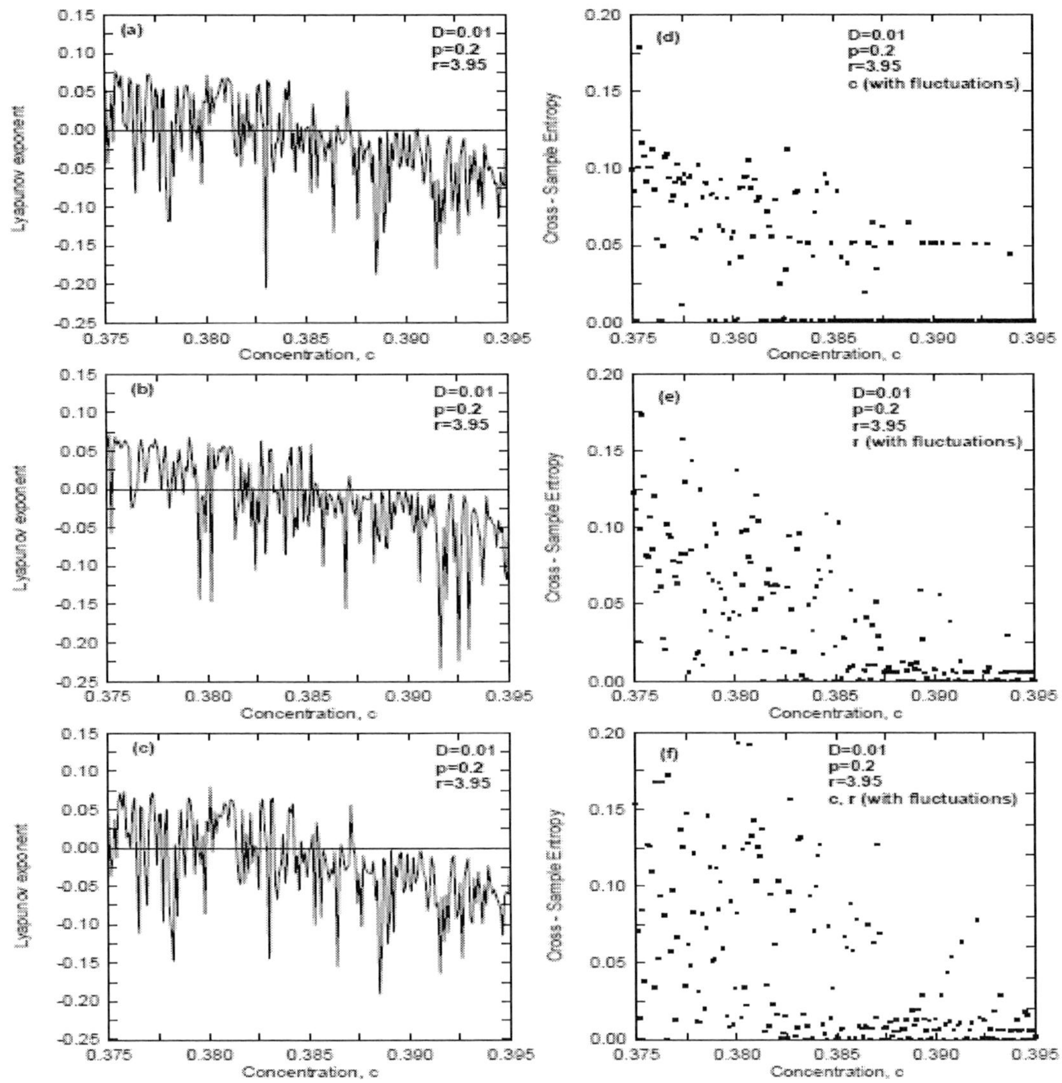

Figure 10. The Lyapunov exponent (a) - (c) and cross sample entropy (d) - (f) for coupled maps [Eq. (20)] for fluctuation with amplitude as a function of concentration c, ranging from 0.375 to 0.395.

where $R_n^{(i)} = r(1-\mathbf{D}\delta_n^{(i)})$, $i = 1, 2$, $\mathrm{E}_n = c(1-\mathbf{D}\delta_n)$, and δ_n and $\delta_n^{(i)}$ are random numbers uniformly distributed in the interval [- 1,1], and $\mathbf{D}$ is the amplitude of the noise. In numerical simulations, for the amplitude $\mathbf{D}$ we used the following values: 0.00001, 0.0001, 0.001 and 0.01. Figure 9a shows the Lyapunov exponent for coupled maps with no fluctuations [Eq. (6)] as a function of concentration c ranging from 0.375 to 0.395, while Figs. 10a - 10c depict this exponent for the largest amplitude of the fluctuation $\mathbf{D} = 0.01$ with the same interval range as c. Obviously, these fluctuations remarkably change the Lyapunov exponent if they occur in c and r. This disturbance is highly emphasized for the case when fluctuations occur in both parameters r and c.

We calculated $Cross-SampEn$ with parameters as in the previous section for x_n and y_n time series. Figures 10d - 10f show high disorder between them, particularly when fluctuations occur in the logistic parameter r and both r and c. Apparently, the disorder in entropy grows when the amplitude of the fluctuations increases. Additionally, we calculated the RMSE (root-mean-square-error) of the cross sample entropy with the state without fluctuations as the referent one. The cross sample entropy for coupled maps [Eq. (20)] for fluctuation as a function of the amplitude $\mathbf{D}$ of fluctuations ranging from 0.00001 to 0.01 is shown in Figure 11. Based on this figure, the highest increase in RMSE of the cross sample entropy is obtained when the fluctuations occurred in both environment parameters, r and c.

Conclusions

In this chapter, our focus was on modeling synchronization in intercellular exchange of substances. We gave a short overview of general mechanisms for exchange of substances between two cells, representing a cooperative communication process. We identified main parameters of the process and derived a system of two coupled logistic equations as an appropriate model of the given process. Then, we investigated synchronization of the model and its sensitivity to fluctuations of environmental parameters using methods of nonlinear dynamics, i.e. the largest Lyapunov exponent and cross sample entropy as measures. Results show that both measures are compatible and can be used interchangeably. Both show the existence of stability regions where noise in the form of fluctuations in concentration of signaling molecules in intercellular environment and fluctuations in affinity for uptake these molecules cannot interfere with the process of exchange. Since our model is inspired by the general scheme of intercellular communication, it naturally does not allow detailed modelling of some concrete, empirically variable intercellular communication process. Instead, it is designed to serve as a starting tool in general investigation of robustness in mutually stimulative populations, which can be readily extended to investigation of synchronization in larger networks of interacting entities [27, 28].

Acknowledgements

This chapter was realized as a part of the project "Studying climate change and its influence on the environment: impacts, adaptation and mitigation" (No. 43007), which is financed by the Ministry of Education and Science of the Republic of Serbia within the framework of integrated and interdisciplinary research over the period 2011 - 2014.

References

[1] A.K. Dunker, J.D. Lawson, C.J. Brown, R.M. Williams, P. Romero, J.S. Oh, C.J. Oldfield, A.M. Campen, C.M. Ratliff, K.W. Hipps, J. Ausio, M.S. Nissen, R. Reeves,

C. Kang, C.R. Kissinger, R.W. Bailey, M.D. Griswold, W. Chiu, E.C. Garner, Z. Obradovic, *J. Mol. Graph. Model.*, 19, 26-59 (2001).
[2] A.K. Dunker, C.J. Brown, J.D. Lawson, L.M. Iakoucheva, Z. Obradovic, *Biochemistry*, 41, 6573-6582 (2002).
[3] M.B. Elowitz, A.J. Levine, E.D. Siggia, P.S. Swain, *Science*, 297, 1183-1186 (2002).
[4] P.S. Swain, M.B. Elowitz, E.D. Siggia, *PNAS*, 99, 12795-12800 (2002).
[5] D. Longo, *J. Hasty, Mol. Syst. Biol.*, 2, 64 (2006).
[6] N. Barkai, B.-Z.Shilo, *Mol. Cell*, 28, 755-760 (2007).
[7] H. Kitano, *Nat. Rev. Genet.,* 5, 826-837 (2004).
[8] H. Kitano, *Mol. Syst. Biol.*, 3, 137 (2007).
[9] C.M. Waters, B.L. Bassler, *Annu. Rev. Cell Dev. Biol.*, 21, 319-346 (2005).
[10] P. Stoodley, K. Sauer, D.G. Davies, J.W. Costerton, *Annu. Rev. Microbiol.*, 56, 187-209 (2002).
[11] W.K. Purves, D. Sadava, G.H. Orians, C. Heller, Eds.; *Life – The Science of Biology;* 7th Edition, Sinauer Associates and Freeman, W.H.: Sunderland, CA, 2003; p. 1121.
[12] L. Deverney, *An Introduction to Chaotic Dynamical Systems*; Benjamin/Cummings Comp., Boston, US, 1986; p. 336.
[13] Y.-P. Gunji, M. Kamiura, *Physica*, D 198, 74-105 (2004).
[14] E. Ullner, A. Koseska, J. Kurths, E. Volkov, H. Kantz, J. Garcia-Ojalvo, *Phys. Rev. E*, 78, 031904 (2008).
[15] D.T. Mihailović, M. Budinčević, I. Balaz, M. Perišić, In *Advances in Environmental Modelling and Measurements*; D.T. Mihailović, B. Lalić Eds.; Nova Sciences Publishers: New York, US, 2010; pp. 89-100.
[16] E. Jen, *Complexity*, 8, 12-18 (2003).
[17] M.B. Kennel, R. Brown, H.D.I. Abarbanel, *Phys. Rev.,* A 45, 3403-3411 (1992).
[18] L. Billings, I.B. Schwartz, *J. Math. Biol.*, 44, 31-48 (2002).
[19] Z. Liu, W. Ma, *Phys. Lett.*, A 343, 300-305 (2005).
[20] W.M. Schaffer, B.E. Kendall, C.W. Tidd, L.F. Olsen, *J. Math. Appl. Med.*, 10, 227-247 (1993).
[21] B. Schwartz, D.S. Morgan, L. Billings, Y.-C. Lai, *CHAOS*, 14, 373-386 (2004).
[22] S.M. Pincus, *PNAS*, 88, 2297-2301 (1991).
[23] J.S. Richman, J.R. Moorman, *Am. J. Physiol-Heart*, C 278, H2039-H2049 (2000)
[24] D.E. Lake, J.S. Richman, M.P. Griffin, J.R. Moorman, *Am. J. Physiol.- Reg.*, I 283, R789-R797 (2002).
[25] T. Hogg, B.A. Huberman, *Phys. Rev. A,* 29, 275-281 (1984).
[26] M. Thattai, A. van Oudenaarden, *PNAS*, 98, 8614-8619 (2001).
[27] R.E. Amritkar, S. Jalan, *Physica*, A 321, 220-225 (2003).
[28] S. Jalan, R.E. Amritkar, C.-K. Hu, *Phys. Rev.*, E 72, 016211 (2005).

In: Essays on Fundamental and Applied Environmental Topics ISBN 978-1-61942-522-4
Editor: Dragutin Mihailovic

Chapter 2

FORMAL CONCEPT ANALYSIS AND CATEGORY THEORY IN MODELING INTERACTIONS OF LIVING SYSTEMS AND THEIR ENVIRONMENTS

S. Crvenković[1,*], D. T. Mihailović[2] and I. Balaz[3]
[1]Department of Mathematics and Informatics, Faculty of Sciences, University of Novi Sad, Novi Sad, Serbia
[2]Department for Field and Vegetable Crops, Faculty of Agriculture, University of Novi Sad, Novi Sad, Serbia
[3]Department of Physics, Faculty of Sciences, University of Novi Sad, Novi Sad, Serbia

ABSTRACT

All functional systems are able to receive inputs, perform some transformations on them, and produce proper outputs. When dealing with mechanical systems, identifying the input/output relation is usually unambiguous because it is purposely designed. However, in biological systems, the situation is more complicated. First, because of their complexity, many specific purposes can be identified, such as feeding, moving, or reproducing. Therefore, decisions regarding what signals or changes should be identified as inputs or outputs are governed by the purpose of investigation. In this case, the difference between mechanical and biological systems is only a matter of scale. Second, and fundamentally different, is the ability of biological systems to treat the same signal in different manners, depending on their current state or context. As a consequence, it is very difficult to straightforwardly identify functional inputs or outputs for any process within a biological system. Usually, this problem is avoided by assuming that the common form of processing within the defined context is the only one. In this way, a modeling approach is reduced to the standard form of sequential machines and can be performed efficiently and satisfy the power of prediction. However, we believe that this is accompanied by the cost of losing deeper insight into some of the specific and important features of biological system functions, such as flexibility and evolvability. In this chapter, after some theoretical considerations, we develop a formal representation of

* Corresponding author: E-mail address: sima@dmi.uns.ac.rs; Phone: +381 21 485 2859. Fax: +381 21 635 0458.

the manner in which organisms "observe" the environment using formal concept analysis (FCA) and category theory. FCA is a branch of applied lattice theory that defines a concept as a unit of two parts: extension and intension. The extension covers all objects belonging to the particular concept, and the intension comprises all attributes that are valid for all of those objects. Both attributes and objects are united by a triple (G, M, I), which is called a formal context, while the ordered set of all formal concepts of (G, M, I) forms a concept lattice that is always complete. By using FCA alone, we can clearly represent how the available set of attributes governs the process of concept formation. However, it is not powerful enough to examine the variation of "observations" because it deals only with isolated lattices. To analyze the algebraic properties of such changeable systems, we further consider concept lattices as objects in the category of lattices and morphisms that preserve lattice completeness. Based on this analysis, we elaborate on some of the functional consequences of the ability of biological systems to perform at different scales with the variable identification of environmental signals.

1. Introduction

In the past few decades, mathematical modeling of biological systems has become a quickly growing field. To understand the vast amount and complexity of data obtained by either biological or biochemical experiments, it is necessary to try to formalize them and construct mathematical models that can help us to predict the outcomes of some processes or elucidate the underlying mechanisms of some observed phenomena. A particularly interesting field for a great number of scientists is biomedical modeling, the subject of which can range from various cellular processes (e.g., cancer growth, angiogenesis, and tissue reparation) to the dynamics of interacting populations in epidemiology. However, almost all of these efforts are divided into two main branches: (i) modeling of particular isolated processes within an organism and (ii) population modeling. Both of these approaches share the same premise – the supposed objectivity of the observed data. When dealing with physical or even chemical systems, this can hold true, but in the realm of biological systems, it can be treated only as a very crude approximation that is valid for a limited number of modeling purposes. If we define our task within strictly pragmatic boundaries as a prediction yielding the approximation of some particular cellular processes or outcomes of some population dynamics, there will be no problems in applying currently used modeling tools. On the other hand, if we define our task using much broader terms, i.e., if we define it as the modeling of living systems in general, we should inevitably postulate completely different premises as the starting points of our journey. Thus, our primary goal is not to predict some existing cytosolic or morphogenic process but rather to capture the pattern of dynamics and organization of a cell (and organism) as a whole.

Therefore, one of the main premises that should be changed is the anticipated objectivity. Instead, we must introduce the notion of an observer who resides within the universe. This notion is not completely new in science – an umbrella term for this concept is endophysics [1] – but still requires attention devoted to its stronger consistency. However, we can clearly define some important qualities of such observer-based systems. First, the observer's environment is defined from the perspective of the system in which he or she is embedded. In other words, if we know that observation is not a simple act but a process of assimilation of external changes with internal operative schemes, then, in modeling we can make a clear distinction between the construction of the environment (by the observer) and the condition of that environment. The issue here is only construction, i.e., modalities of perceiving the environment by living systems. Second, it is a fundamental fact that each perceptive entity can observe or register only those changes in its vicinity that activate its functional operations. Thus, the externality of any such system is pre-structured in accordance with its functional purposes and objectives. The mere fact that something can be perceived means that it becomes connected with the needs and functions of the living system. Conversely, because each organism constructs its environment as a field of interests, only functionally relevant changes can appear in it. As an inherent consequence of such assimilation of reality, the external medium of living systems is transformed into a functionally treatable environment where external, arbitrary events can be conveyed into the internal scheme of reactions.

Third, such structuring (the interrelation of parts in an organized whole) of the environment is not directed only from organism to externality but also takes place within living systems, where each specific form of interactions constructs its own space. Regarding organized systems in general, Luhmann [2] defined this problem as a need for the replacement of the classical parts-whole paradigm with a paradigm of layered structuration, which could be defined as a system-environment distinction. Applying this new paradigm does not by itself imply any kinds of given hierarchies or primacies within a system, but it should be brought into the focus as a problem of local generation of perspectives, where each subsystem constructs its own environment and its own system-environment distinction.

Thus, to establish a more accurate estimation of the pattern of interactions of biological systems with their environment, as well as interactions among living systems, it is necessary to take into account the manner in which they "observe" the environment, separate it into different functional patches, and associate the patches with an internal functional schematism. The formal representation of such processes can be elegantly and efficiently performed by using formal concept analysis (FCA), which is a branch of the applied lattice theory. FCA was introduced by Wille [3] and defines a concept as a unit of two parts: extension and intension. The extension covers all objects belonging to a particular concept, and the intension comprises all attributes valid for all of those objects. Both attributes and objects are united by a triple (G, M, I), which is called a formal context if G and M are sets and $I \subseteq G \times M$ is a binary relation between G and M. Also, between these two closure systems (G and M), a dual isomorphism is established. The ordered set of all formal concepts of (G, M, I) forms the concept lattice of (G, M, I), which is always complete.

FCA has been academically and commercially applied over a wide range of domains, such as medicine, biology, psychology, musicology, archeology, law, civil and industrial engineering, library and information science, computer science, and mathematics. In these applications, the main achievements of concept lattices are due to the support of general tasks such as exploring, searching, recognizing, identifying, analyzing, investigating and deciding [4]. The mathematization of concepts may be understood as a first step in mathematizing the traditional philosophical logic, understood as a doctrine of the forms and functions of thinking based on concepts, judgments, and conclusions, which leads to contextual logic [5].

However, one of the most important characteristics of interaction with the environment in organisms is versatility. Therefore, an FCA-based formalism can be used within a much broader framework that allows for the comparison of substantially different structures. Category theory, as a branch of abstract algebra, deals with objects that are sufficiently abstract to allow for different interpretations within lower-level mathematical fields. In ecological modeling, it can be used as a tool for representing the global properties of complex systems. Thus, it can reveal fundamental similarities between structures.

In what follows, we present some possible applications of FCA and category theory in modeling the interaction of living systems and their environments. In Section 2, we consider concept lattices. This is accomplished through (i) an introduction, (ii) a description of the fundamental theorem on concept lattices (subsection 2.1), (iii) an example about the context of animals (subsection 2.2) and (iv) a demonstration of the use of FCA in constructing a subjective interface between biological systems and their environments. Section 3 is devoted to a short description of the categories in our stories. Following some general remarks (subsection 3.1), we provide the fundamentals of category theory in a condensed form and then detail an example of its application in ecological models following Rosen's idea [6] that category theory is the most general theory of modeling.

2. Formal Concept Analysis in Modeling the Interaction of Living Systems and their Environments

2.1. Concept Lattices

An order (or partial order) on a set P is a binary relation $\leq$ on P such that, for all $x, y, x \in P$,

(i) $x \leq x$,
(ii) $x \leq y$ and $y \leq x$ imply $x = y$,
(iii) $x \leq y$ and $y \leq z$ imply $x \leq z$.

A set P equipped with an order relation $\leq$ is said to be an *ordered set.*

Let P be an ordered set and let $S \subseteq P$. An element $x \in P$ is an upper bound of S if $s \leq x$ for all $s \in S$. A *lower bound* is dually defined. The set of all upper bounds of S is denoted by S^u and the set of lower bounds by S^l

$$S^u := \{x \in P \mid (\forall s \in S)\, s \leq x\}$$
$$S^l := \{x \in P \mid (\forall s \in S)\, s \geq x\}.$$

If S^u has a least element, x, then x is called the *least upper bound* of S. Equivalently, x is the least upper bound of S if

(i) x is an upper bound of S and
(ii) $x \leq y$ for all upper bounds y of S.

Dually, if S^l has a largest element x, then x is called the *greatest lower bound of S.* Least elements and greatest elements are unique, so least upper bounds and greatest lower bounds are unique when they exist. The least upper bound of S is also called the *supremum* of S and is denoted by $\sup S$; the greatest lower bound of S is also called the *infimum* of S and is denoted by $\inf S$.

We write $x \vee y$ in place of $\sup\{x, y\}$ when it exists and $x \wedge y$ in place of $\inf\{x, y\}$ when it exists. Similarly we write $\vee S$ instead of $\sup S$ and $\wedge S$ instead of $\inf S$ when these exist. Let P be a non-empty ordered set.

(i) If $x \vee y$ and $x \wedge y$ exist for all $x, y \in P$, then P is called a lattice.
(ii) If $\vee S$ and $\wedge S$ for all $S \subseteq P$, then P is called a complete lattice.

The theory of ordered sets and lattices provides a natural setting in which we can discuss and analyze hierarchies occurring within mathematics and in the "real" world.

A *concept* is considered to be determined by its extent and intent: the *extent* consists of all objects belonging to the concept, while the *intent* is the collection of all attributes shared by the objects. It is often difficult to list all the objects belonging to a concept and usually impossible to list all its attributes; therefore, it is natural to work with a specific *context* in which the objects and attributes are fixed. A *context* is a triple (G, M, I) in which G and M are sets and $I \subseteq G \times M$. The elements of G and M are called *objects* and *attributes*, respectively. $(g, m) \in I$ means "the object g has attribute m".

For $A \subseteq G$ and $B \subseteq M$, we define

$$A' = \{m \in M \mid (\forall g \in A)\, (g, m) \in I\},$$
$$B' = \{g \in G \mid (\forall m \in B)\, (g, m) \in I\}.$$

Therefore, A' is the set of attributes common to all objects in A, and B' is the set of objects possessing the attributes of B. The *concept* of the context (G, M, I) is a pair (A, B), in which $A \subseteq G$, $B \subseteq M$, $A' = B$ and $B' = A$. The *extent* of the concept (A, B) is A, while the *intent* is B. A subset $A \subseteq G$ is the extent of some concept if and only if $A'' = (A')' = A$, in which case the unique concept of which A is an extent is (A, A'). The corresponding statement applies to these subsets B of M, which are the intents of some concept.

The set of all concepts of the context (G, M, I) is denoted in the literature by $B\,(G, M, I)$. For concepts (A_1, B_1) and (A_2, B_2) in $B\,(G, M, I)$, we write $(A_1, B_1) \le (A_2, B_2)$ and state that (A_1, B_1) is a subconcept of (A_2, B_2) or that (A_2, B_2) is a superconcept of (A_1, B_1) if $A_1 \subseteq A_2$ (which is equivalent to $B_1 \supseteq B_2$). Assume that (G, M, I) is a context, and let $A, A_j \subseteq G$ and $B, B_j \subseteq M$, for $j \in J$. Then

(i)	$A \subseteq A''$,	(i)'	$B \subseteq B''$,
(ii)	$A_1 \subseteq A_2 \Rightarrow A_1' \supseteq A_2'$,	(ii)'	$B_1 \subseteq B_2 \Rightarrow B_1' \supseteq B_2'$,
(iii)	$A' \subseteq A'''$,	(iii)'	$B' \subseteq B'''$,

(iv) $\left(\bigcup_{j\in J} A_j\right)' = \bigcap_{j\in J} A_j'$ (iv)' $\left(\bigcup_{j\in J} B_j\right)' = \bigcap_{j\in J} B_j'$

Let Q be an ordered set, and $P \subseteq Q$. Then, P is *join-dense* in Q if for every element $s \in Q$ there is a subset A of P such that s is the supremum of A in Q, i.e., $s = V_Q A$. The dual of join-dense is *meet-dense*.

2.2. The Fundamental Theorem of Concept Lattices

Let (G, M, I) be a context. Then, $(B\,(G, M, I); \leq)$ is a complete lattice in which join and meet are given by

$$\bigvee_{j\in J}(A_j, B_j) = \left((\bigcup_{j\in J} A_j)'', \bigcap_{j\in J} B_j \right)$$

$$\bigwedge_{j\in J}(A_j, B_j) = \left(\bigcap_{j\in J} A_j, (\bigcup_{j\in J} B_j)'' \right)$$

Conversely, if L is a complete lattice, then L is isomorphic to $B\,(G, M, I)$ if and only if there are mappings $\gamma : G \to L$ and $\mu : M \to L$ such that $\gamma(G)$ is join-dense in L, $\mu(M)$ is meet-dense in L, and $(g, m) \in I$ is equivalent to $\gamma(g) \leq \mu(m)$ for each $g \in G$ and $m \in M$. In particular, L is isomorphic to $B\,(L, L, \leq)$ for every complete lattice L. The proofs of the previous assertion can be found in [7]. Therefore, the class of complete lattices coincides with the class of concept lattices.

If we are interested in algebraic laws satisfied by complete lattices, it is obvious that the class of all complete lattices could not be defined by a finite set of lattice identities. This is a consequence of the fact that inf and sup are basically infinitary operations. A lattice $\mathrm{L}(L, \wedge, \vee)$ is said to be *distributive* if the following law holds for L :

$$x \wedge (y \vee z) = (x \wedge y) \vee (x \wedge z).$$

An empirical experience seems to show that contexts arising from concrete problems in real life rarely happen to have distributive concept lattices. The following may be found in [8]. Define for any $A \subseteq G$ the *conditional incidence relation* $_A I$

$j\,_A Im$ iff every consequence of m that holds for all objects in A is valid for j.

Dually, for any set $B \subseteq M$ we define

$jI^B m$ iff every specialization of j possessing all attributes in B has property m.

We call the pair (A, B) *discriminating* if

$$I = {}_{A}I \cap I^{B}.$$

A concept lattice $(B\ (G, M, I); \leq)$ is distributive if and only if each concept of the context (G, M, I) or, equivalently, each pair (A, B) with $A \times B \subseteq I$ is discriminating.

The *arrow relations* of context (G, M, I) are defined as follows: for $h \in G$, $m \in M$ let

$$g \square\ m :\Leftrightarrow \begin{cases} (g, m) \notin I, & \text{and} \\ \text{if } g' \subseteq h' \text{ and } g' \neq h', \text{ then } (h, m) \in I \end{cases}$$

$$g \square\ m :\Leftrightarrow \begin{cases} (g, m) \notin I, & \text{and} \\ \text{if } m' \subseteq n' \text{ and } m' \neq n', \text{ then } (g, n) \in I \end{cases}$$

$$g \square\ m :\Leftrightarrow g \square\ m \text{ and } g \square\ m.$$

All the lattices in our examples are finite. We have the following: a finite lattice with standard context (G, M, I) is distributive if and only if

$$g \square\ m,\ g \square\ n \Rightarrow m = n.$$

A subrelation $J \subseteq I$ is *closed* if every concept of (G, M, I) is also a concept of (G, M, I). S is a complete sublattice of $(B\ (G, M, I); \leq)$ if and only if $S = (B\ (G, M, I); \leq)$ for some closed subrelation $J \subseteq I$ [7].

A class K of lattices is a *variety* if K is defined by a set of lattice identities. It is well known, from universal algebra, that a class K is closed under H (homomorphic images), S (sublattices) and P (direct product).

$$Var(K) = HSP(K).$$

A variety is finitely based if it can be defined by a finite set of identities. The following result is provided by McKenzie [9]: for any finite lattice L, the variety $Var(L)$ is finitely based.

Also, because we are dealing with finite lattices and every finite lattice is a complete lattice, it is good to know the following. The variety L of all lattices is generated by its finite numbers. Therefore, the class of finite concept lattices generates the whole variety of lattices. What is the meaning of this assertion? Think of any nontrivial lattice identity. There is a finite concept (G, M, I) such that $(B\ (G, M, I); \leq)$ does not satisfy this identity. The theory of lattices and techniques of universal algebra provide a powerful tool for the identification of the lattice identities of a given lattice. In the next subsection, we will illustrate the use of the FCA in the biological world.

2.3. The Context of Animals: An Example

The following example is adapted from [10]. Table 1 describes which of the mentioned attributes some animals have. This is indicated by crosses. An empty cell in Table 1 indicates that the corresponding animal does not have the corresponding attribute.

To explain the notion of a formal concept of a context, we look at the attributes of the FINCH and ask for all those animals (of this context) that have all of the attributes of the FINCH. Hence, we obtain the set A consisting of FINCH and EAGLE. This set A of objects is closely connected to the set B consisting of the attributes flying and bird: A is the set of all objects having all of the attributes of B, and B is the set of all attributes that are valid for all of the objects of A.

Table 1. Object versus attributes in the context of animals

Attribute Animals	preying	flying	bird	mammal
LION	X			X
FINCH		X	X	
EAGLE	X	X	X	
HARE				X
OSTRICH			X	

Between the concepts of a given context there is a natural hierarchical order, the "subconcept-superconcept" relation. For example, the preying, flying birds describe a subconcept of the concept of the flying birds. The extent of this subconcept consists only of the EAGLE, and the intent consists of the three attributes preying, flying and bird. The extent of the given superconcept consists of FINCH and EAGLE, only the intents flying and bird. In the following line diagram, we represent the conceptual hierarchy of all concepts of the context ANIMAL.

A line diagram consists of circles, lines and the names of all objects and all attributes of a given context (Figure 1). The circles represent the concepts, and the information of the context can be read from the line diagram by the following simple reading rule: an object g has an attribute m if and only if there is an upwards leading path from the circle named "g" to the circle named "m".

Hence, we recognize from the line diagram above that the FINCH has the attributes flying and bird. Using the reading rule, we can easily understand from the line diagram the extent and the intent of each concept by collecting all of the objects below each respective attribute above the circle of the given concept. Hence the object concept "FINCH" has the extent FINCH and EAGLE and the intent flying and bird. The extent of the top concept is (always) the set of all objects, while the intent of it does not contain any attribute (in this context). However, in other contexts, the intent of the top concept may not be empty, e.g., if we add to the given context the attribute "animal" with crosses in each row, then the top concept would be the attribute concept of "animal" and the intent of the top concept would contain only the attribute "animal".

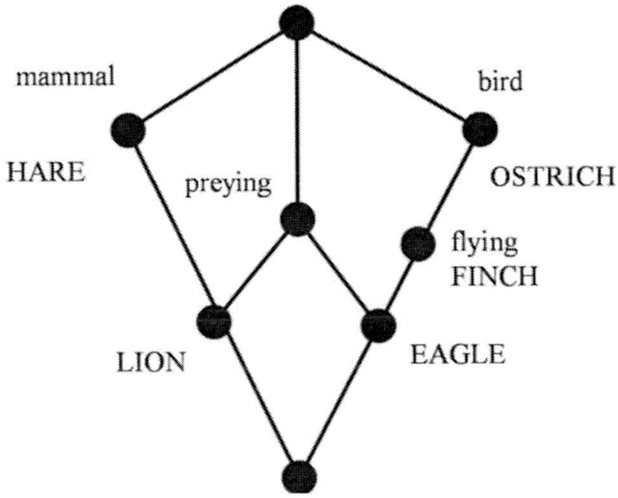

Figure 1. Conceptual hierarchy of all concepts of the context ANIMAL according to Table 1.

If we extend the set of objects, we must change the lattice of conceptual hierarchy. Namely, from the first diagram, it follows that flying animals are birds. The new object BEE is flying, but it is not a bird. Thus, we have to separate "flying" and "bird" to meet FINCH.

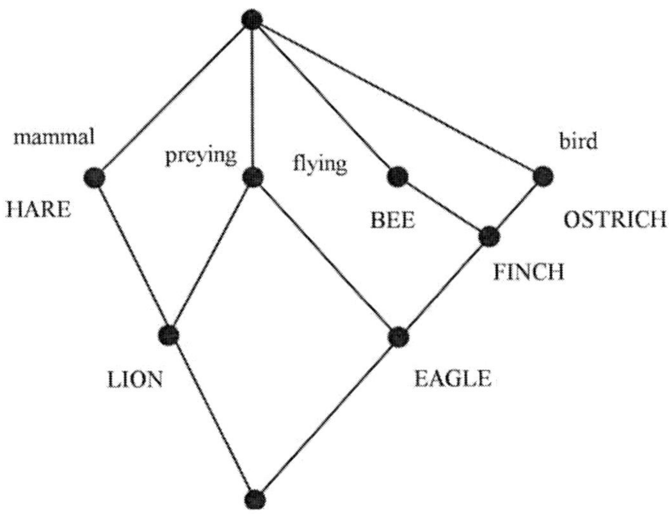

Figure 2. Modification of conceptual hierarchy to incorporate flying animals that are not birds.

According to the second diagram (Figure 2), "mammals" and "flying" imply "preying" and "bird", i.e., they meet at the bottom of the lattice, and "preying" and "bird" are above. Thus, we have to extend the lattice by adding a new vertex BAT. This is shown in the third diagram (Figure 3).

Extent and the intent of each concept by collecting all of the objects below each respective attribute above the circle of the given concept.

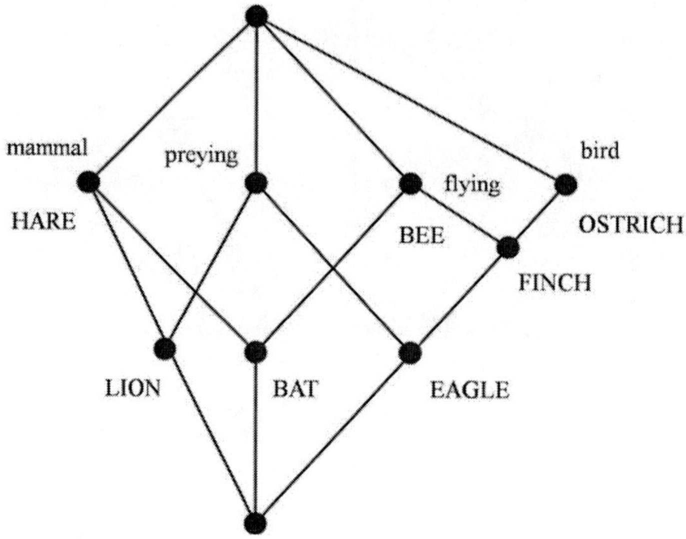

Figure 3. Further modification of conceptual hierarchy to separate flying birds from flying mammals.

Hence the object concept "FINCH" has the extent FINCH and EAGLE and the intent flying and bird. The extent of the top concept is (always) the set of all objects, while the intent of it does not contain any attribute (in this context). However, in other contexts, the intent of the top concept may not be empty, e.g., if we add to the given context the attribute "animal" with crosses in each row, then the top concept would be the attribute concept of "animal" and the intent of the top concept would contain only the attribute "animal". If we extend the set of objects, we must change the lattice of conceptual hierarchy. Namely, from the first diagram, it follows that flying animals are birds. The new object BEE is flying, but it is not a bird. Thus, we have to separate "flying" and "bird" to meet FINCH.

2.4. Use of FCA to Construct Subjective Interface between Biological Systems and their Environments

In this section, we present a simplified but illustrative example of an abstract living system's interaction with the environment. Our goal is to represent how subjective processing of the environment is achieved, what are the functional consequences and how this process influences the dynamics of the organization of living systems. This example is based on reasoning from [11].

Let us define set $O=\{o_1,o_2,o_3...o_n\}$ as a set of all environmental objects. To functionally treat them, living systems need to associate each external object with some attribute. Hence, we define a set of attributes $M=\{m_1,m_2,m_3..m_n\}$. Within FCA, objects and attributes are mutually defined, so we automatically obtain set $G\subseteq O:\{\forall g\in G\,|\,(g,m)\in I\}$, where I is a binary relation between M and G. In other words, every object (attribute) that does not form a binary relation with a corresponding attribute (object) simply does not exist from the perspective of a particular formal context. Thus, the concept of the "whole" environment changes its epistemological status and is reduced to an observable environment, i.e., to a set of treatable objects. The next step in the functional configuration of the environment is the separation of the homogenous observable environment into a set of concepts within a given context defined by (G,M,I). In biological (non-cognitive) systems, the attributes used in separating environmental factors from one another are scarce in number and can be divided into two main categories: physical influences (various types of radiation or temperature) and chemical influences (various types of molecules or ions). However, these attributes are further divided into a set of sub-attributes, which can be represented as a scale of values within a given attribute. In this way, a segment of the environment is encircled by a certain attribute and is further divided according to the given scale within that attribute, thus establishing a many-valued context (G,M,W,I) where G is a set of objects, M is a set of many-valued attributes, W is a set of attribute values and I is a ternary relation $I \subseteq G \times M \times W$.

Here, we will use the very simple model of a typical photosynthetic organism. As in all living cells, an interaction with the environment is performed through currently active receptors mainly located in a cell membrane. According to their configuration, they can be divided into photo-receptors and receptors for various types of external molecules. It is important to emphasize that the activity of molecular receptors is always connected with successive changes in the configuration of these receptors and the so-called active transport through membranes. On the other hand, water, along with dissolved ions, enters the cells spontaneously, either by diffusion or through aquaporins. Therefore, in our conceptual scheme we postulate the existence of two main attributes: radiation and molecule (Table 2). The attribute "molecule" is further divided into two sub-classes according to the size of the objects to which it can be applied. In the group of objects that are smaller than the supposed threshold there is water and the small ions dissolved in it, while all molecules larger than the threshold are in the second group (denoted by Larger than(1) and Smaller than(2), respectively). Attributes are further diversified by defining several types of radiation stimuli and several types of recognizable molecular structures. Finally, we introduce one more attribute, defined as a global regulator, which to preserve the simplicity of the model does not have any sub-attributes. Global regulators are ubiquitous in living systems and are activated in response to almost all kinds of external stimuli.

Although objects in this model correspond to physical objects, no specific description is assigned to them because they gain full "subjective" meaning only through the construction of concepts.

As has already been mentioned, each attribute defined here corresponds to the receptive ability of an organism to perform the following chain of actions: (i) recognition of some stimulus (to form concept), (ii) assimilation of stimulus and/or changing its own configuration, and (iii) activation (indirectly or directly) of some other molecules to process the received information.

Table 2. List of all attributes used in the model. In further text, the term Structure (X ...Q) will be denoted by the strings Str(X ... Q)

Attribute	Sub-attributes						
Molecule	Larger than(1)	Smaller than(2)	Structure X	Structure Y	Structure Z	Structure W	Structure Q
Radiation	700	450	UV				

Because this formal framework is unable to depict the sequential dynamics of some process, we will construct several different formal contexts and corresponding lattices, which will demonstrate how the internal network responds to external stimuli.

The first formal context that is defined shows the initial steps of interaction with the environment. According to the previously described process, the living system has a limited set of active receptors, and through them, it determines which kind of external objects can be visible and consequently can be functionally treated. In this example (Table 3), the living system can initially sense water and small ions (denoted as Mol. small. than[(1)]), three kinds of organic molecular structures (denoted as X, Y and Z), and three kinds of radiation sources (denoted as 700 nm, 450 nm, and UV).

Figure 4a shows that the complete environment is divided into two main concepts, which further diverge into several more specific concepts. Because each formed concept in this model is connected with a previously defined chain of events, Figure 4b shows how a simple feedback-governed metabolic chain would look like from an FCA perspective. The association of molecules with structures Z (Mol strZ) and corresponding receptors (concept <10>) activates the production of the next molecule (<12>), which in turn generates the product (<13>). Generation of the final product inhibits the production of strZ receptors, the intermediate product spontaneously degrades, and some other metabolic chain utilizes this final product (<13>). Because (<13>) is no longer available, the production of strZ receptors can start again.

In the second formal context, we introduced the activation of the supposed global regulator (Figure 5). As it can be clearly seen, the complete lattice is reconfigured, and the number of as well as the hierarchy between concepts drastically changes. It is very important to emphasize that the activation of global regulators is the inevitable consequence of the assimilation of almost any external stimulus.

More generally, an interaction with the environment through the construction of a subjective perspective for living organisms is connected with the temporality of the internal conceptual lattice. This means that finalizing the formation of one configuration simultaneously causes its degradation and reconfiguration.

Table 3. Formal context which represents initial steps in interaction with the environment

	Molecule	Radiation	Mol. larger than*	Mol. smaller than*	Mol. strX	Mol. strY	Mol. strZ	Rad. 700	Rad. 450	Rad. UV
Object 0		X						X		
Object 1		X							X	
Object 2		X								X
Object 3	X			X						
Object 4	X			X						
Object 5	X			X						
Object 6	X		X		X					
Object 7	X		X			X				
Object 8	X		X				X			

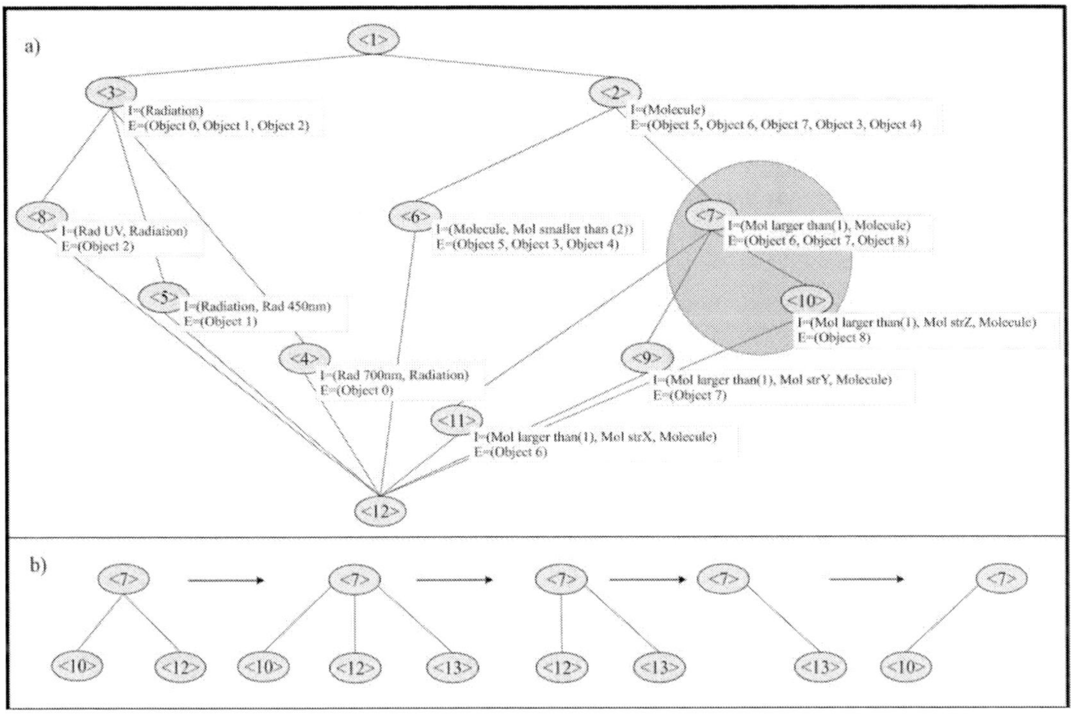

Figure 4. Lattice of the formal context given in Table 2 (lattice generation was performed using open-source software package "Galicia" [12]).

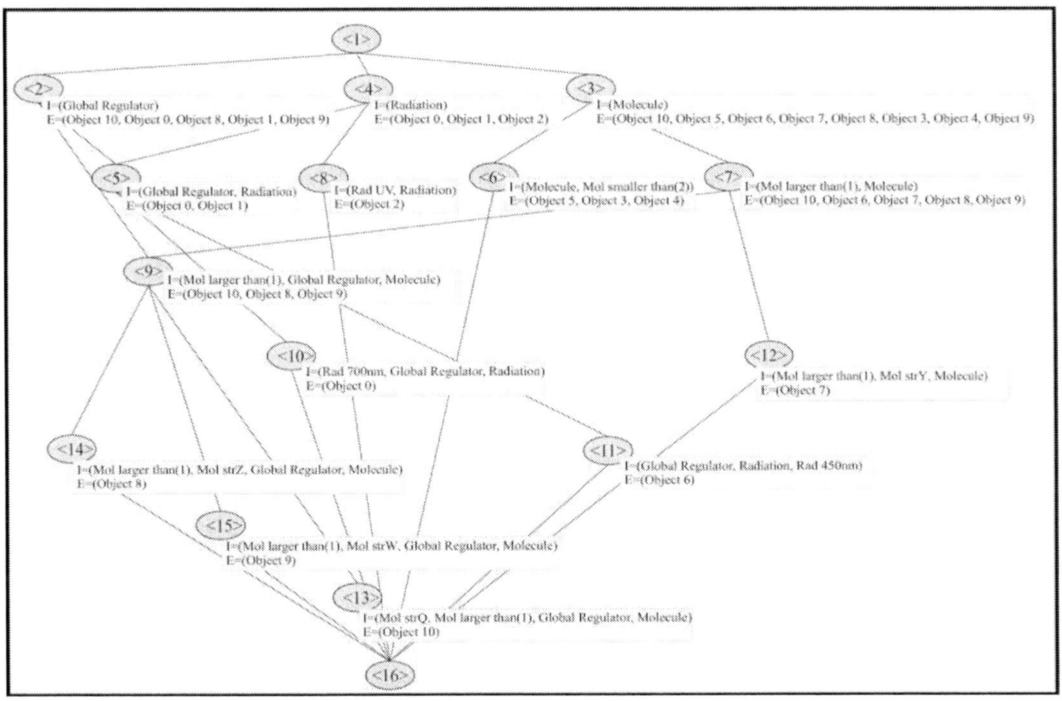

Figure 5. Lattice of the formal context for the same living system when global regulator is active.

It is obvious that this model is an over-simplified example of a photosynthetic organism, but it can still reveal some important characteristics of living systems. First, because they are able to form a subjective perspective, living systems operate only within a limited scope of an entire environment. Moreover, they shape the environment according to their ability to functionally process some external segments, thus forming the subjective environment. This subjective construction is highly variable because it is defined not as a reflection of the complete ability of a living system to sense its environment but only as a reflection of currently active receptors. Second, the formation of an internal conceptual lattice is temporalized by the very same process of its formation. When it is formed, its consequence is the activation of other molecules/receptors/regulators that each possess their own perspective, thus changing the current context and reconfiguring nearly the entire lattice.

3. Categories in our Stories

3.1. General Remarks

Following Mesarovic's *Mathematical Theory of General Systems* [13], if we observe interactions of agents with their surrounding environment, such a system can be defined as a set of interacting objects $S \subseteq O_1 \times O_2 \times O_3 \times ... \times O_n$. If we denote the population of agents under consideration as $P = \{p_1, p_2, p_3, ..., p_n\}$ and a set of external influences as $E = \{e_1, e_2, e_3, ..., e_n\}$ (to recall, these influences can be either other agents or extra-systemic influences), then the state of such formed systems at any particular moment in time can be defined as the Cartesian product $S \subseteq P \times E$. Because our system is not a static graph but a dynamical network of interactions where at each moment the hierarchical status of network elements can vary significantly, we have to define state of the population *P* as a mapping $\omega : e \to p$, $e \in E, p \in P$. Both e and *p* are defined as temporal sequences of events such that $E = \{e : T \to I\}$ and $P = \{e : T \to R\}$, where *T* is a set of time points *t*, *I* is a set of external stimuli on a particular agent such that at each time system receive stimulus *i*(*t*) and *R* is a set of responses, *r*(*t*). Furthermore, both *P* and *E* are formal systems. Therefore, the occurrence of *p* and occurrence of *e* at some particular time point *t* are governed not only by mapping ω but also by the internal rules of these systems, which are partially independent.

Thus, it is obvious that changes in an environment induce appropriate responses in agents through the model of coupled input/output pairs. In real systems, the reverse situation is also possible such that some external changes can be influenced by the activity of organisms, but for the sake of simplicity and because it is not directly related to the main topic of this presentation, we will not consider that problem. It is clear that a critical factor in building an evolvable model as described above is choosing the appropriate structure for the mapping $I \to R$. When dealing with models usually developed as prediction tools, it is sufficient to assume the attitude of analyzing a "black box". Therefore, we can propose a function that should summarize all available experimental data and obtain a set of more or less accurate predictions for various initial conditions. However, in such a case we will neglect the real meaning of the nature of mappings within *E* and *P*. Taking a slightly closer look at these relations and keeping in mind the notion of "operative assimilations", we can see that a somewhat hidden problem is that of how I is generated from the wholeness of external changes and what is the connection between generating I with a constitution of corresponding R. Although this connection can be efficiently represented using FCA [11], its evolvability demands a more advanced formal treatment to be fully comprehended.

First, we should more strictly define what can be considered an external environment. Let us define set $A=\{a_1,a_2,a_3,...,a_n\}$ as a set of all external changes. To functionally treat them, an agent needs to associate each observable external change with some attribute. Hence, we define a set of attributes $M=\{m_1,m_2,m_3,...,m_n\}$. From FCA, we could apply a rule that states that objects and attributes are mutually defined, so we automatically obtain the set of external influences $E\subseteq A:\{\forall e\in E|\,(g,m)\in O\}$, where O is a binary relation between M and E. In other words, every object (attribute) that does not form a binary relation with a corresponding attribute (object) simply does not exist from the perspective of a particular agent. In this way, the concept of "whole" environment changes its epistemological status and is reduced to an observable environment, i.e., to a set of treatable objects. Also, because E is a derivative of p, mapping $I\rightarrow R$ can be realized only as an indirect derivative of each p. Second, if we place additional attention on the mapping ω, we can see that from the perspective of both formal systems E and P, it is unentailed. It belongs to neither formalism and accordingly cannot be derived from anything within these formalisms. We have in hand two possible solutions to this problem. On the one hand, we can leave it as it is. Thus, the appropriate function for the focal problem can be applied, and the model will be valid until the field conditions change beyond the defined boundaries. If this happens, we have to create a new formal system by choosing another function and/or by modifying the organization of the whole system. On the other hand, we can incorporate ω into another formalization that compares formalisms [6]. This discipline is called category theory and was developed by McLane [14].

3.2. Fundamentals of Category Theory in a Condensed Form

A *category* is a quadruple $\mathrm{A}(O,hom,id,\circ)$, where

(1) O is a class of A -*objects*,
(2) for each pair (A,B) of A -objects, a set $\hom(A,B)$ is the set of A -morphisms, from A to B [$f\in\hom(A,B)$ is expressed as $f:A\rightarrow B$ or $A\xrightarrow{f}B$],
(3) for each A -object A, a morphism $A\xrightarrow{id}A$ is called the A -*identity* of A,
(4) a *composition law* associated with each A -morphism $A\xrightarrow{f}B$ and each A - morphism $B\xrightarrow{g}C$ an A -morphism $A\xrightarrow{g\circ f}C$, called the *composite* of f and g,

impose the following conditions:

(a) composition is associative, i.e., for morphism $A\xrightarrow{f}B$, $B\xrightarrow{g}C$ and $C\xrightarrow{h}D$, the equation $h\circ(g\circ f)=(h\circ g)\circ f$ holds,
(b) A -identities act as identities with respect to composition, i.e., for A -morphism $A\xrightarrow{f}B$, we have $id_B\circ f=f$ and $f\circ id_A=f$,

(c) the sets $\hom(A, B)$ are pairwise disjoint.

If A= $(O, hom, id, \circ)$ is a category, then

(1) the class O of A= objects is denoted by Ob(A).

(2) the class of all A= morphisms is denoted by Mor(A) (or Hom(A)) is defined to be the union of all the sets $\hom(A, B)$ in A=.

(3) if $A \xrightarrow{f} B$ is an A= morphism, we call A the domain of f [A=dom(f)] and call B the codomain of f [A=cod(f)].

(4) the composition $\circ$, is a partial binary operation on the class Mor(A). For a pair (f, g) of morphisms, $f \circ g$ is defined if and only if the domain of f and co-domain of g coincide.

If we have $h = g \circ f$, then sometimes it is denoted by $A \xrightarrow{f} B \xrightarrow{g} C$ or by saying that the following triangle commutes

$$\begin{array}{ccc} A & \xrightarrow{f} & B \\ h\square & & \downarrow g \\ & & \mathrm{C} \end{array}$$

Similarly, the statement that the square

$$\begin{array}{ccc} A & \xrightarrow{f} & B \\ h\downarrow & & \downarrow g \\ \mathrm{C} & \xrightarrow{k} & \mathrm{D} \end{array}$$

commutes means that $g \circ f = k \circ h$. A morphism $A \xrightarrow{f} B$ in a category is called an *isomorphism* provided that there exists a morphism $g : A \to B$ with $g \circ f = id_A$ and $f \circ g = id_B$. Objects A and B in category are said to be *isomorphic* provided that there is an isomorphism $f : A \to B$. If A= and B= are categories, then a *functor* from A= to B= is a function that assigns to each A=- object A a B=-object $F(A)$ and to each A=-morphism $A \xrightarrow{f} B$ a B=-morphism $F(A) \xrightarrow{F(f)} F(B)$ such that

(1) F preserves composition, i.e., $F(f \circ g) = F(f) \circ F(g)$ whenever $f \circ g$ is defined;

(2) F preserves identity morphisms, i.e., $F(id_A) = id_{F(A)}$ for each A=- object A.

A functor F : A$\Rightarrow$B is called an *isomorphism* provided that there is a functor G : B$\Rightarrow$A such that $G \circ F = id_{\mathrm{A}}$ and $F \circ G = id_{\mathrm{B}}$. The categories A= and B=are said to be isomorphic if there is an isomorphism F : A$\Rightarrow$B .

3.3. Category Theory in Ecological Modeling

The ecological world is a space that is rich with complex systems. Each system, as an open one, interacts in a coherent way, producing new structures and building cohesion and new structural boundaries. It undergoes emergence and self-organization. The philosophy of modeling approaches in ecology is also rich, but it requires longer steps because of use of the category theory originally proposed by Rosen [6] in modeling complex ecological systems. Within the last fifty years, this theory, as the most general (non-axiomatic and self-closing type theory avoiding Russell's paradox) of modeling theories, has made remarkable contributions along many fronts of ecology and related sciences – from micro to global scales. Its further efficient use in ecological models is closely connected with the development of mathematics, biology, physics, and computer science as well as philosophy.

In modeling procedures, ecologists are interested in the relationship between plants and the environment – a relationship that spans length scales from the molecular level (in the case of photosynthesis) to several kilometers and more (when considering energy flow in plant fields). Therefore, in modeling quantities of the plant environment, we must bear in mind that (i) spatial variability in the plant environment is highly related to the variability in soil moisture, incoming solar radiation and stand structure; (ii) the nonlinear response of many biological and physical processes in the plant environment to the modeled quantity means that simple calculations based on temporal averages can produce large errors; (iii) the knowledge of the variability of local environmental quantities in uniform plant stands is also required in sophisticated modeling of biophysical processes in the plant environment; and (iv) the aforementioned key ecological modeling points additionally reflect model reliability through computing procedures. Ecological modelers in numerically oriented studies base their calculations on mathematical models for the simulation and prediction of different processes, which are exclusively nonlinear in describing relevant ecological quantities [6]. A theoretical description of any ecological (natural) system includes at least two important aspects. First, one should construct a concrete mathematical model of both the admissible states of the system and the transitions between these states. Second, one should establish the rules of selecting among the many theoretically admissible states of the system only those states that are realized in nature under the given external conditions [15]. The ecological world is an ideal space for considering complex systems. Within complex systems, each system, as an open one, interacts in a coherent way, producing new structures and building cohesion and new structural boundaries.

Figure 6, which was adapted from Rosen ([6] see Figure 2.3.1 p. 74), represents the modeling relation. It schematically depicts a modeling relation when a natural system (N) and a formal system (F) are given. As mentioned above, two arrows represent the respective entailment structures: inference in formalism (F) and causality in a natural system (N). The two established dictionaries encode the phenomena of N into the propositions of F and decode the propositions of F back to the phenomena in N, respectively. As mentioned above, there are two paths in the diagram: (1) and (2) + (3) + (4). According to [6], the first (path (1)) represents the causal entailment within N (what an observer will see by simply sitting and watching what is happening). Arrow (2) encodes the phenomena in N into the propositions in F. In this route, we must use these propositions as hypotheses through which the inferential machinery of the formal system F may operate (denoted by the arrow (3)); it generates theorems in F, entailed precisely by the encoded hypotheses.

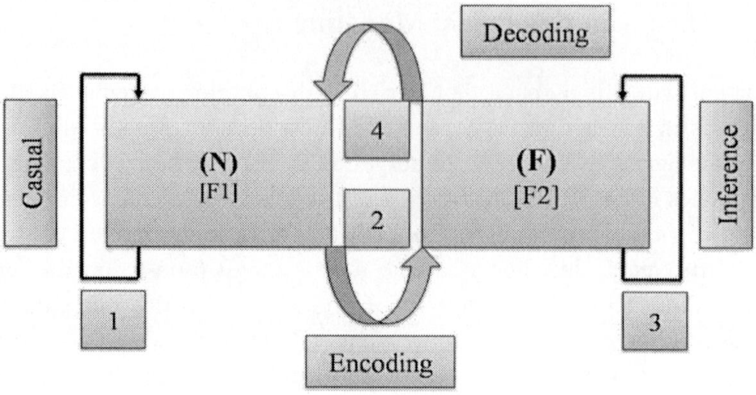

Figure 6. Schematic diagram representing both (i) the comparison of two formalisms F_1 and F_2 and (ii) modeling relation when we have given a natural system N and a formal system F [16]. Here, 1 represents causal entailment within the natural system (N); 2 represents encoding, where the observer's propositions about N are used as hypotheses in constructing formal system (F); 3 is the generation of theorems in F, which function as a model of N; and 4 is decoding, where the theorems of F are applied back to N in the form of predictions.

Finally, we have to decode these theorems back into the phenomena of N via the arrow (4). At this point, the theorems become *predictions* about N. Then, the formal system F is called a *model* of the natural system N if we always obtain the same answer regardless of whether we follow path (1) or path (2) + (3) + (4). This Rosen's diagram is a recognizable guide for modeling a community within complex systems. The process of modeling complex systems is a very comprehensive one.

A system is to be treated as a complex structure as per, for instance, Peter Checkland's definition: a system is a model of a whole entity; when applied to humans, the model is characterized fundamentally in terms of hierarchical structure, emergent properties, communication and control ([17], p. 318). The major components of complexity are openness and freeness, but the distinctive characteristic is "natural activity", such as self-organization, and is still of great importance as intra-activity but now joined by the phenomena of anticipation [18] and interactivity between systems found in global interoperability. The transition from connectivity to activity involves a type change and therefore requires a formal system with a built-in facility to cross between the levels. Thus, intraconnectivity between the components cannot give rise to interactivity between those components without some non-local integrity coming into play [19]. Non-locality is a principle that, among some others, is specific to a certain object area, such as the inorganic, living or human realm. This principle suggests that certain interaction between the elements of the system is treated as the transmission of information at infinite speed (see, for instance, [20]). Category theory, recommended by Robert Rosen as a modern tool for complex (and living) systems, is found to have a formal expressive power as a process beyond modeling for exploring the fundamental non-local concept of adjointness needed to understand advanced systems.

The arrow of category theory does not only have a formal meaning. According to [21, 22] it formalizes the principle of *constancy* (originally introduced by Heraclites and Parmenides) that is provided by a common source and target. Such an arrow refers to the situation in which a source and target are indistinguishable. In a defined system, the collection of entities can be identified as objects, while operations between them are defined by arrows.

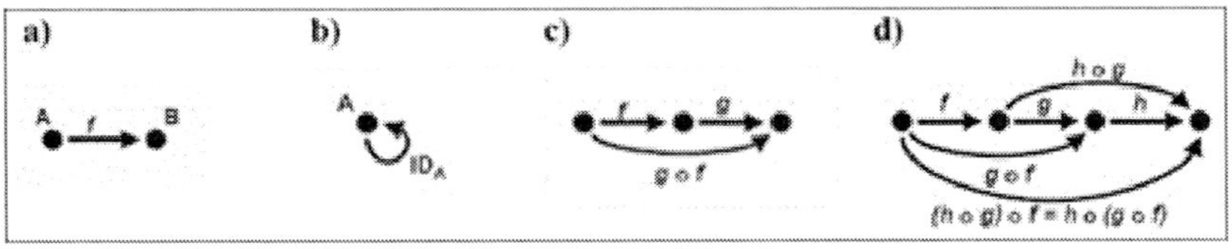

Figure 7. Schematic diagram representing the category theory essentials: (a) morphism (arrows), objects, domain and co-domain, (b) identity morphism, (c) composite morphism and (d) identity composition and associativity.

Figure 7 shows that there may be many possible arrows between objects. However, category theory holds that a unique limiting arrow may exist for all of these possible arrows that represent the resulting *intraconnectivity* of a local system. There is an order between the two entities established by the directions of arrows [19]. This means that the arrow limit between two entities is also a limit of all possible paths. Because of the existence of limits and all possible connectivity, this is classified by axiomatic categories as a Cartesian closed category.

Moving up one level, there is a grand limiting arrow for all of the aforementioned limits, existing as an identity functor characterizing the type and therefore the system as a category (Figure 8). A system as a category may then be drawn as a circular arrow, which is the identity functor that identifies the type of a system [19, 21]. The system is therefore one large arrow, i.e., a process in which the internal arrows are simply the components of one arrow. This then leads to *interconnectivity* between the systems. Also, the functor between two categories is conceptually the same as internal arrows between the arrows above. Within the framework of this theory, it is possible to repeat the abstraction to one level higher (a natural transformation) but not higher. This level is the level of *interactivity*.

It is important to note that the self-organization of a category theory-system (intra-activity) arises when the category-system pair is indistinguishable. Finally, category theory is a very useful tool when we meet difficult problems in some areas of mathematics, ecology, physics, computer sciences, biological nano-engineering and the self-organization of cell function in living systems [23], among many others. They can be translated into (easier) problems in other areas (e.g., by using functors, which map one category to another).

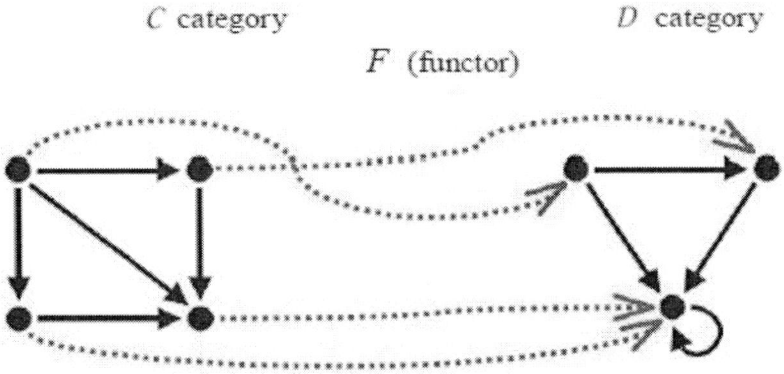

Figure 8. Schematic diagram representing the functor "action".

CONCLUSIONS

To be able to understand an enormous amount of data obtained by biological or biochemical experiments and the complexity associated with such information, it is necessary to try to formalize the data and construct mathematical models that can help us to predict the outcomes of some processes or elucidate the underlying mechanisms of some observed phenomena. In the construction of models of complex systems, one is usually concerned with the construction of a model of a concrete organism. Such a model depends on many data from which one expects to develop some general conclusions about the organism. Our approach in this chapter is to generalize the observed phenomena and represent it as a mathematical structure from which it is sometimes easy to obtain nontrivial conclusions. We use FCA to demonstrate how lattice theory can be used in modeling the subjectivity of the interface between an organism and its environment. We show that changes in both the environment and structure of perception can be represented as changes in the current context, which leads to the reconfiguration of lattices. From this example, it is also obvious that FCA is not powerful enough to examine the variability of "observations" because it deals only with isolated lattices. To analyze the algebraic properties of such changeable systems, we further consider concept lattices as objects in the category of lattices and morphisms that preserve lattice completeness. Based on such analysis, we elaborate some of the functional consequences of the ability of biological systems to perform the variable identification of environmental signals on different scales.

ACKNOWLEDGEMENTS

This chapter was realized as a part of the project "Studying climate change and its influence on the environment: impacts, adaptation and mitigation" (No. 43007), which is financed by the Ministry of Education and Science of the Republic of Serbia within the framework of integrated and interdisciplinary research over the period 2011 - 2014.

REFERENCES

[1] O.E. Rössler, *Endophysics: The World as an Interface*, World Scientific Pub Co., Singapore, New Jersey, London, Hong Kong, 1998; p. 204.

[2] N. Luhmann, *Social systems*, Stanford University Press: Stanford, US, 1996; p.684.

[3] R. Wille, *Ordered Sets*, 445 (1982).

[4] B. Ganter, R. Wille, *Formal Concept Analysis: Mathematical Foundations*, Springer-Verlag, 1997; p. 284.

[5] R. Wille, *Boolean Concept Logic*, ICCS, 317-331 (2000).

[6] R. Rosen, *Life itself, A Comprehensive Inquiry into the Nature, Origin, and Fabrication of Life;* Columbia University Press, US, 1991; p. 285.

[7] B.A. Davey, H.A. Priestley, *Introduction to Lattices and Order*, Cambridge Mathematical Textbooks (1990).

[8] M. Erne, *Algebra Univ.,* 30, 538-580 (1993).

[9] R.N. McKenzie, *Math. Scand.,* 27, 118 (1970).
[10] K.E. Wolff, In *SoftStat '93, Advances in Statistical Software 4*; Faulbaum, F. Ed.; Gustav Fisher Verlag, Stuttgart, DE, 1994; pp. 429-438.
[11] S. Crvenkovic, D.T. Mihailovic, I. Balaz, In *Environmental, Health and Humanity Issues in the Down Danubian Region, Multidisciplinary Approaches, Proceedings of the 9th International Symposium on Interdisciplinary Regional Research*, D. Mihailović, M. Vojinović Miloradov, Eds.; World Scientific: Singapore, SG, 2009; pp. 189-198.
[12] Galicia (2001) [http://www.iro.umontreal.ca/~galicia/].
[13] M. Mesarovic, Y. Takahara, *Mathematical Theory of General Systems*; Academic Press, US, 1972; p. 322.
[14] S. Mac Lane, *Categories for the Working Mathematician*, Springer-Verlag, 1971; p. 262.
[15] A.P. Levich and A.V. Solov'yov, *Cybernet. Syst.,* 30, 571-585 (1999).
[16] D.T. Mihailovic, I. Balaz, *Idojaras,* 111, 209-220 (2007).
[17] P. Checkland, *Sys. Res. and Behav. Sci.,* 17, S11-S58 (2000).
[18] G.J. Klir, *AIP Conf. Proc.*, 627, 37-48 (2002).
[19] N. Rossiter M. Heather, *7th International Conference of Enterprise Information Systems (ICEIS)*, Florida, USA, 92 (2005).
[20] J.S. Bell, *Physics,* 1, 195 (1964).
[21] E.G. Manes, M.A. Arbib, *Arrows, Structures and Functors, The Categorical Imperative* Academic Press, US, 1975; p. 185.
[22] D. Mihailovic, In *Environmental, Health and Humanity Issues in the Down Danubian Region, Multidisciplinary Approaches, Proceedings of the 9th International Symposium on Interdisciplinary Regional Research*, D. Mihailović, M. Vojinović Miloradov, Eds.; World Scientific: Singapore, SG, 2009; pp. 181-188.
[23] O. Wolkenhauer, J. Hofmeyr, *J. Theore. Biol.*, 246, 461-476 (2007).

In: Essays on Fundamental and Applied Environmental Topics ISBN 978-1-61942-522-4
Editor: Dragutin Mihailovic

Chapter 3

GENERATION OF INFORMATION IN BIOLOGICAL SYSTEMS: A SHORT EPISTEMOLOGICAL ESSAY

I. Balaž[1,*] and D. T. Mihailović[2]

[1]Department of Physics, Faculty of Sciences, University of Novi Sad, Novi Sad, Serbia

[2]Department for Field and Vegetable Crops, Faculty of Agriculture, University of Novi Sad, Novi Sad, Serbia

ABSTRACT

Modern information theory deals with the transfer of information between a source and a receiver. This theory applies to cases in which the information and the means of its encoding, decoding and transfer are already established. In man-made systems, successful functioning is a direct result of proper design regarding the source of information and the processing of received information. The same is true of natural systems, but only when their function is analyzed on short time scales. In such cases, due to evolutionary design by natural selection, the production of all informational components has already been encoded, and information processing is performed along a defined path. However, at longer time scales, it becomes evident that natural systems are able to autonomously change their patterns of information processing. For example, if during evolution an organism becomes capable of utilizing a new food source, this means several things: (i) the organism can receive a new kind of information, (ii) it has developed a new channel of information processing, and (iii) this new informational channel is incorporated into an existing network of functional processes. This example suggests that at least some of the principles of functioning of natural systems are fundamentally different from those of present-day man-made systems. In this chapter, we will investigate the informational aspect of the functioning of biological systems as one of these fundamental differences. Our main focus will be on an analysis of the prerequisites that enable biological systems to generate new information sources and information channels. Further, we will interpret these findings in terms of Shannon's

* Corresponding author: E-mail address: ibalaz@polj.uns.ac.rs; Phone: +381 21 485 3290. Fax: +381 21 635 0552.

information theory and show how current quantitative measures of information can be used within the new context.

1. Introduction

The functioning of any system can be analyzed at different levels. From the perspective of dynamical systems, we can analyze the development over time of chosen variables and their interdependencies. On a more general level, more static representation can be used to analyze the controlling mechanisms, hierarchies of transformation priorities and segmentation of the system into functional sub-units. On even more general level, the system can become an arbitrarily abstract algebraic structure. At any of these levels, the system is either explicitly or implicitly composed of a set of states. How the states change depends on the generation, transfer and interpretation of information between parts of the system. Information theory, as defined by Shannon [1], deals with the possible states of a system and the probabilities of their distribution. It was originally developed as a mathematical theory of communication (MTC) in the field of electrical engineering, for use in the study of communication limits. It treats two fundamental problems: the maximal level of data compression and the maximal rate of data transmission. The first problem is considered through the entropy, H, whereas the channel capacity, C, defines the second problem. Because of its generality, this mathematical theory of communication almost instantly found applications in a variety of other fields, and over time, became well known as information theory. In this context, information is treated in a purely technical sense as a measure of the probabilities of random variables, without any reference to possible functional traits or to ambiguities in interpreting the source of information. Therefore, the semantic aspect of information is completely omitted. Within the MTC, there are no notions of the importance or usefulness of information. At the same time, in the biological world, these aspects are crucial because selection and evolution are based on the comparison of the success of individuals. In other words, concepts of usefulness and importance are foundations for the evolutionary development of new functional traits. Despite this, the concept of information as defined by the MTC has acquired a prominent role in many parts of biology. It is not only applied in domains where the concept seems naturally applicable, as in studies of perception, communication or cognition, but it also pervades all biological disciplines. One routine application is in the description of the role of genes in metabolism and development as a transfer and expression of genetic information. In evolutionary biology, it has also been argued that an informational "domain" exists alongside the domain of matter and energy [2].

The questions of whether information theory should be applied so widely in biological disciplines and how the concept of information should be understood are topics of lively discussions [3]. Several alternative approaches to the transfer of the MTC into biological disciplines have been developed, among which biosemiotics has become the most prominent. However, our opinion is that these alternatives lack the clarity, rigor and applicability of the MTC. Therefore, in this chapter our focus is on the development of an extension of the classic MTC communication scheme so that it can encompass a changeable semantic aspect of information transfer. In Section 2, we give a short overview of the MTC and analyze its general settings regarding the syntax, semantics and pragmatics of information. In Section 3, we analyze the main postulates of biosemiotic approach as a possible alternative to the MTC.

A merging of these two approaches is presented in Section 4, and concluding remarks are given in Section 5.

2. The Mathematical Theory of Communication and the Semantics of Information

As stated in the introductory paragraph, the MTC deals with the communication process, and its main concern is the problem of reproducing at one point, either exactly or approximately, a message originating at another point (Figure 1). An information source is any system that can produce various states. In this context, information is a relative category because the states of the system alone do not have any significance. To characterize a change of states as a change of message, some other system must be able to perceive and register all or a subset of these states. Therefore, in the MTC, information is considered to be a quantity that reflects the ability of the receiver to distinguish between different states of the system. Although a description of states can have any imaginable form, in the most general case, states are represented as strings of symbols. From this basic consideration, in the framework of the MTC, entropy (*H*) is derived as one of the fundamental measures of information. Shannon [1] denoted the information entropy of a discrete variable *X*, with possible values $\{x_1, x_2, ..., x_n\}$, as

$$H(X) = -\sum_{x \in X} p(x) \log_2 p(x),, \tag{1}$$

where *p*(*x*) is the probability of *X* for some $x \in X$. In other words, values of *X* can be interpreted as possible states of the system observed by the receiver. It follows that entropy does not measure the absolute value of information at the information source, but rather the ability of the receiver to predict possible outcomes of the source. However, entropy can be applied to an information source as an absolute value within a fixed context. In this case, we can determine numerous useful quantities such as mutual information and the minimum channel capacity required to reliably transmit a signal. Mutual information can be defined as

$$I[X;Y] = \sum_{x \in X} \sum_{y \in Y} P(x,y) \log_2 \left(\frac{P(x,y)}{P(x)P(y)} \right).. \tag{2}$$

Informally, this is a measure of the mutual dependence of sources *X* and *Y*. If these two sources are completely independently distributed, mutual information disappears because $P(x,y)=P(x)P(y)$. If this is not the case, mutual information measures how much knowing one variable reduces our uncertainty about the other variable. A closely related concept is that of channel capacity (*C*), which represents the maximum mutual information between the input and the output of a channel. In practice, if the information rate from the source is less than *C*, then it is theoretically possible to achieve reliable transmission through the channel, whereas if the information rate is larger than *C*, then reliable transmission is not possible regardless of the amount of signal processing performed by the transmitter and the receiver.

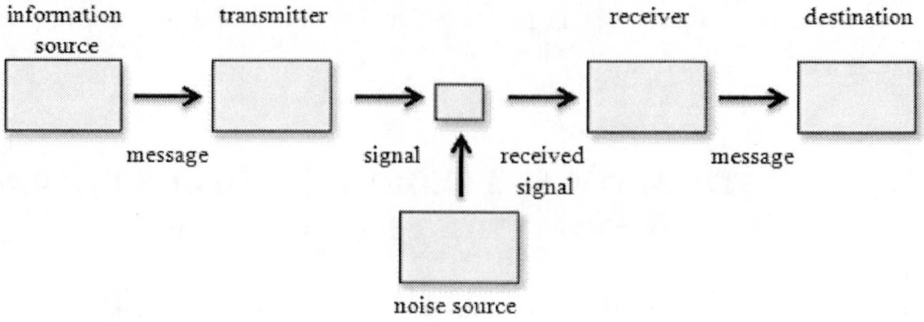

Figure 1. Schematic diagram of a general communication system.

Given these measures, an informational system can be optimally designed to be maximally efficient for a projected goal. However, we should keep in mind that entropy is a measure of how much information is not available when the receiver does not know the outcome of an information source. How can a receiver make predictions regarding this outcome? Ultimately, this depends on the semiotic context of communication. The structure of the source and the receiver determine the syntax of "proper" signals and thus the size of the information space of all messages that can be produced during communication. Rules that map each syntactic element of the signal to each syntactic element of a message determine the semantic aspect of communication. In natural systems, these rules are determined by the structure of the communication channel. The pragmatic aspect of communication is somewhat external to the MTC. In general, pragmatics is given by the conditional use of signals and the influence of these conditions on the meaning of the message. The MTC does not deal with meaning or any other functional notion, but only with the core of the communication process. Therefore, pragmatics is only partially included in the general scheme as a noise source. In summary, an implicit postulate of the MTC is that the information source and the destination are of the same nature regarding the syntax and the semantics of the information. Therefore, syntax and semantics are considered constants within this framework. With these two aspects fixed, the main focus of the MTC is on the efficiency of communication and thus on the core concept of information entropy along with information redundancy and channel capacity.

Although the MTC does not consider the meaning or the purpose of any particular information, the application of the MTC is strongly connected with these semantic concepts. For example, information redundancy refers to the amount of unnecessary data in the transmission of a message. In absolute terms, for purely abstract data, redundancy (D) is defined as

$$D = R - r \tag{3}$$

where $r = \lim_{n\to\infty}\frac{1}{n}H(x_1, x_2, .., x_n)$ is the entropy rate of a stochastic process that generates information and $R = \log|\mathbf{X}|$ is the absolute rate of a source where $|\mathbf{X}|$ is the cardinality of the message space. R represents the maximum possible rate of information that can be transmitted using a given alphabet. However, most information sources have some level of regularity in their generation of messages, which various models can represent. For example, in the English language, for any symbol of the alphabet, $H = -\log_2(26)$ or 4.7 bits, if we consider all symbols to be equiprobable and ignore spaces and punctuations [4]. In fact, the information content of each symbol is much smaller due to differences in the probabilities of occurrence of the symbols and the tendency of adjacent symbols and groups of symbols to form connections. Therefore, the average entropy rate of symbols in the modern English language is between 0.6 and 1.3 bits [4]. To calculate the actual value, we can apply various models that take into account letter frequencies, conditional probabilities of groups of 2 or more letters forming words, and other factors. All of these choices are strongly driven by the semiotic properties of the system in question and by our pragmatic needs. Consequently, the MTC is general enough to be applicable to any communication system, but at the same time, its explicit form relies on external semiotic assumptions. It can be used to analyze the informational aspect of any isolated communication system in which the source and the receiver are already mutually adjusted regarding the syntax and the semantics of the communication process. However, if we want to investigate the process of appearance and development of communication and how syntax and semantics appear, we must develop a more general strategy. This is particularly relevant in the domain of biological systems where function and evolution are tightly connected [5]. Numerous approaches have been developed to understand the informational aspect of the function of biological systems. The most straightforward of these simply translate concepts from the MTC to genetics and molecular biology, describing molecular processes in terms of information transfer and signaling. Following this approach, the causal role of genes is typically understood in terms of their informational role. However, such approaches are usually just convenient ways to describe certain processes in easily understood terms, without any in-depth analysis of the possible implications or applications. What are the underlying assumptions behind such information-related approaches? And what are the consequences of such assumptions? These questions are topics of serious debate [6, 7]. Another group of approaches is more focused on the whole semiotic context, and we will briefly review these in the next section.

3. Biosemiotics and Semiotic-Related Concepts

As noted above, the semiotic context of information can be much richer than is acknowledged by the MTC, but at the same time, this is difficult to represent formally. Probably the most important reason for this is the fact that a solid, widely accepted theoretic framework still does not exist. In any case, all of the main approaches to applying the semantic concept to information in biological systems use as a starting point the concept of biological function, in which the function of an entity derives from a history of natural selection. Only in this framework can we generate a purposive or normative description of the causal capacities of a functional entity based on the understanding that it has been developed and shaped through natural selection. Such a teleosemantic approach was initially approached by Sterelny et al. [8] and Maclaurin [9]. Since then, several lines of thought have been developed [10], some focused on the evolved functions of the genetic machinery as a whole and others on the natural selection of particular genetic elements. However, acknowledging that an entity has developed its function through evolution is not the same as applying the concept of semantic information to the changes it produces. One line of thought aimed at justifying the semantic aspect is based on the proposal that there is an intimate connection between evolutionary function and the semantic properties that have been selected to play a causal role in the developmental processes [8]. The argument is that because of these functional properties, genes represent the outcomes they are supposed to produce. According to this perspective, any non-genetic factors that have a similar developmental role, and have been selected to play that role, also have semantic properties. Thus, rich semantic properties are ascribed to genes and other structures. According to Shea [11] ordinary biological functions and functions that involve information use can be distinguished. Any object that has semantic properties plays a role that involves a kind of mediation between two "cooperating devices," a producer and a consumer. In the case of an indicative signal or representation, the representation is expected to affect the activities of the consumer in a way that will maintain the performance of the consumer's biological functions only under certain conditions. In the case of an imperative representation, the representation is expected to affect the consumer's activity by inducing the consumer to bring some state of affairs about. Therefore, genetic messages can be treated as having both indicative and imperative content.

A somewhat independent line of thought is focused not so much on the foundations but more on the description and identification of possible semantic structures in biological systems. This is called biosemiotics, and its central point is the concept of organic codes [12]. An organic code consists of structures that (i) represent the rules of correspondence between two independent worlds, (ii) give meaning to information structures and (iii) represent the collective rule that is partially independent of the individual properties of the constituent substructures. In living systems, the existence of organic codes is based on specific organic molecules, so-called "adaptors," that are able to perform two independent acts of transduction. In this context, the process of biochemical transformation can be identified as semiotic only if the following conditions are met [13]: (i) syntax is determined by the structural relationships between constituent elements, for example, the sequence of nucleotides, (ii) semantics is determined by the code that defines the mapping of each syntactic element of the signal onto the syntactic element of the message, for example, the binding sites on ribosomes for mRNA and the corresponding amino acids and (iii) pragmatics is determined by the conditions that influence the transfer of signals, for example, the set of biochemical parameters that trigger activation of gene transcription. Therefore, signs are semantically bound to the pragmatic conditions of communication, which leads to three types of semiosis in biological systems [14]. The first of these is the so-called *manufacturing semiosis*, the purpose of which is to produce objects that cannot come into existence in any other way. The second type of semiosis is derived from analysis of signal transduction. Here, the starting point is the fact that signaling within cells creates a correspondence between first and second messengers by means of receptor molecules that have all of the defining features of adaptors, but this is not a manufacturing semiosis because the second messengers already exist in the cell and are not manufactured by the transduction process. Signal transduction functions to create a specific signaling association between first and second messengers and therefore represents the *signaling semiosis*. Finally, the third type of semiosis, entitled *interpretive semiosis*, represents the process of interpretation that arose with the origin of animals. Although there are three types of semiosis, according to Barbieri [14], the concept of semiosis can be defined without any reference to interpretation because there is no necessary link between these two concepts. Semiosis can be defined exclusively in terms of coding, and in this form, it is directly applicable to the cell. A semiotic system is made of signs, meanings and coding rules, all produced by the same code maker. In the case of the cell, the translation apparatus is the codemaker, i.e., the seat of the genetic code, the transfer-RNAs are the adaptors that implement the rules of the code, and the sequences of codons and amino acids have the defining features of signs and meanings because they are codemaker-dependent entities. Thus, we can conclude that the cell is indeed a semiotic system but its semiosis is based on coding, not on interpretation [12].

At this point we might question the purpose of applying such a perspective. Are there any questions about the functioning of living systems that can be more easily addressed from the semiotic perspective rather than from the usual standpoint of absolute objectivity? A detailed answer would demand at least another chapter, but here we will point out some of the most important points:

1. It has been emphasized numerous times that the MTC is not suitable for complex biological systems [1, 3, 15]. Knowing that information transfer in biological systems is highly specific and context-dependent, the development of a formal framework would open possibilities for the development of artificial autonomous semiotic systems.
2. Because the evolution of living systems and the evolution of semiotic systems can be considered two aspects of the same process, the existence of a formal framework would enable the development of a broad range of evolutionary experiments with the aim of better understanding general evolutionary mechanisms as well as the necessary conditions for the emergence of semiotic systems.
3. If the agents that give rise to the emergence of life-like systems were considered to be semiotic systems, we would be able to better incorporate evolutionary features into abstract models because semantic rules can be defined as syntactic derivatives within the set of pragmatic boundary conditions.

In summary, the biosemiotic aspect of the functioning of living systems has recently received increasing attention. Although the theoretical treatment has not yet been fully developed, our standpoint is that more precise formal treatment should be developed synchronously, as this can offer numerous clarifications and additional insights. Because the MTC is widely accepted, with a well-established mathematical basis, our strategy here is to rely on the basic MTC scheme and to put it into the more general semiotic context. In the next section, we will outline this approach.

4. The Emergence of Semantics in Information Processing in Biological Systems

To talk about information processing in biological systems, we must first identify its material foundation. According to Floridi [16], if σ is an instance of information, then the following must be true: (a) σ consists of data, (b) the data in σ are well formed (clustered according to a given syntax) and (c) the well-formed data in σ are meaningful. Therefore, a single datum can be defined as a putative fact regarding a difference or a lack of uniformity within some context. With this in mind, the scheme of a communication system shown in Figure 1 can be interpreted in several different ways, depending on the focal point.

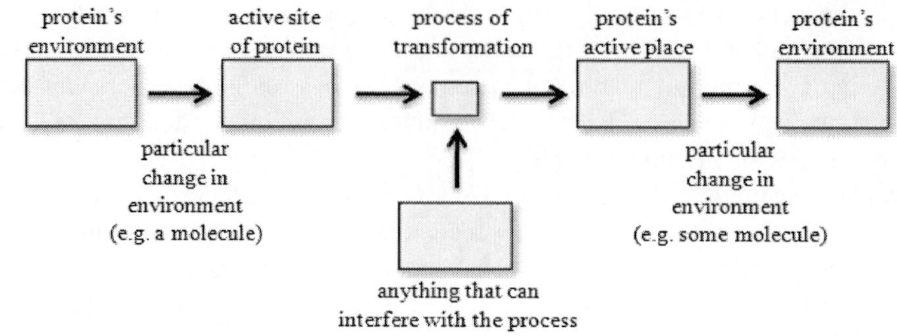

Figure 2. Schematic diagram of a general communication system, applied to a single protein.

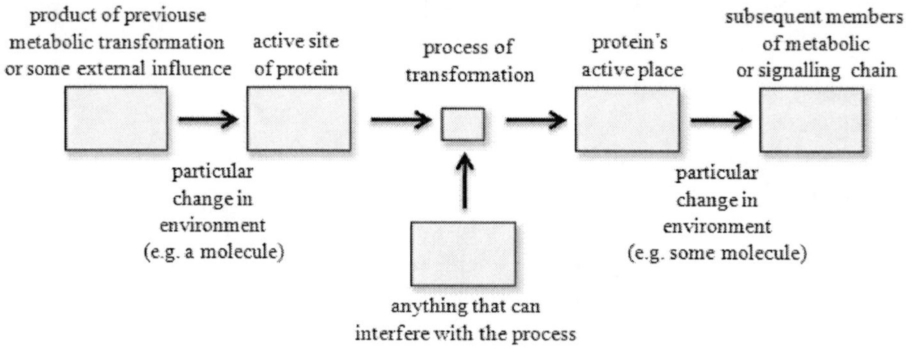

Figure 3. Schematic diagram of a general communication system, applied to a single protein as a part of metabolic network.

In the following passages, we will start our analysis with proteins, as elementary functional units within the cell, and continue into higher hierarchical levels.

At the most basic level, proteins can be regarded as state machines that are able to register certain external changes, to act according to them by performing transformations of certain molecules and to produce and release the products of transformation (Figure 2). Stated differently, if a substrate must have a certain structure to be recognized, and if the product of transformation has a structure defined by the nature of the protein, then the protein can be regarded as a code that associates one syntactic structure with another syntactic structure.

It has been observed that many proteins contain regions without well-defined structure, or intrinsically disordered regions (e.g., [17]). Intrinsic disorder has been associated with particular functions including cell regulation, signaling and protein, DNA, and ligand binding. Many proteins are intrinsically disordered in native form and fold only upon binding. This means that the same protein can bind to several different molecules and perform several different actions depending on context. Therefore, a large portion of cellular proteins cannot be described simply as functions acting on certain sets, but as sets of functions that choose their domains according to the current context. However, it is still not appropriate to talk about semantic information here because enzymes catalyze reactions but they do not instruct catalysis. In other words, at this level of description, we are still dealing merely with the execution of function. To provide a semantic context, a structure must exist that is able to mediate between two systems by conveying a representation of the changes in one system to another.

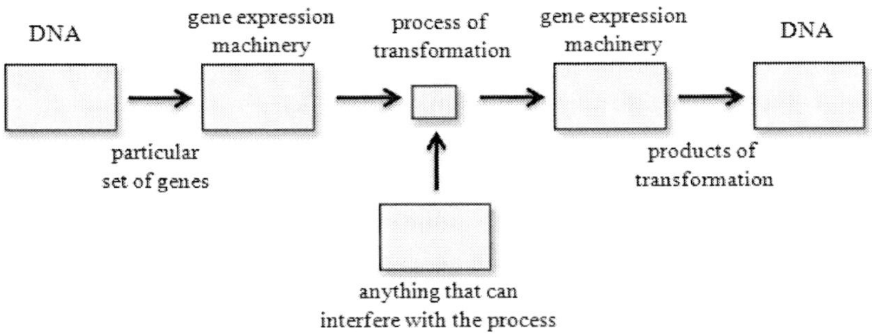

Figure 4. Schematic diagram of a general communication system, applied to a single metabolic pathway. Gene expression machinery encompasses transcription/translation mechanisms as well as regulation mechanisms.

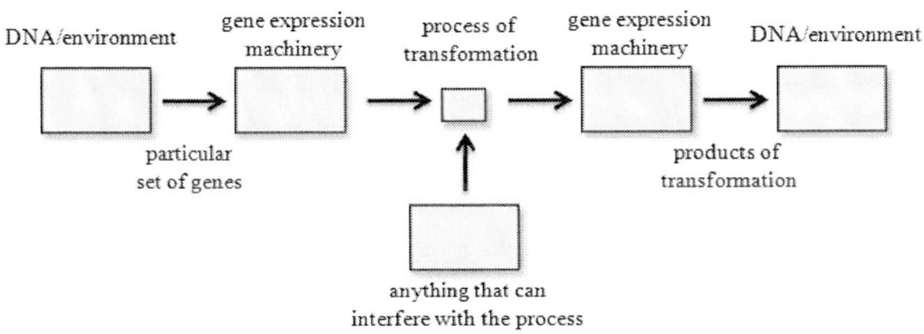

Figure 5. Schematic diagram of a general communication system, applied to the functioning of a whole cell.

Because the signal transmission scheme described above is a functional process that is only a part of a functional system, we can further generalize (Figure 3). Here, we take into account the fact that both the information source and the destination are part of the same pragmatic context for a given message. This simply means that metabolic reactions are always the products of or the precursors to some other metabolic or regulatory process.

If the above description is applied to each step of a hypothetical metabolic pathway, and then compressed, we will obtain a self-circular scheme (Figure 4).

Finally, we should incorporate the environment into our model (Figure 5). The environment is an intrinsic source of signals that can trigger action in an organism. Indeed, any genetic activation or suppression is at some point caused by changes in environment. It should be emphasized that the environment alone cannot be considered a source of information, but only as a source of signals. Information concerns the representation of signals from some source. Therefore, in the above consideration, we have presupposed that an information source is a unity of a data source (e.g., the structure of a physical object) together with a mechanism of assigning meaning to that data source. The environment, as a data source alone, consists of pre-epistemic data that have not yet been observed or interpreted by the system itself. To become data (from the perspective of the living system), environmental signals must be first observed (directly or indirectly) and then processed. During this process, observation filters out all data that are not well formed for a given system (e.g., the cell). This process inherently creates syntax, or, more precisely, syntax is a by-product of the act of functional observation. Processing is performed locally by receptors, enzymes and other molecular structures, but also globally through DNA regulation.

In summary, we can represent the process as *occurrence of differences* $\xrightarrow{observation}$ *data* $\xrightarrow{meaning}$ *information*. The question is, what structure gives meaning? A universal answer probably does not exist because the definition of "meaning" varies widely. In the case considered here, the situation is simplified because we are not dealing with cognitive systems. Thus, we will simply consider data to be meaningful if they can be incorporated into the functionality of the system. At this stage, the scheme presented above can be readily applied to the functioning of a complex man-made machine (Figure 6). Is there any difference between man-made machines and living systems regarding information processing? Both of them are organized systems that are able to observe and process data into information and to act accordingly utilizing received information in combination with their own sources of instructions.

Despite these resemblances, there are two important differences between the communication systems in living organisms and those of contemporary man-made machines. The first difference concerns the nature of the source of information. In living organisms, it is now widely accepted that gene expression cannot be interpreted in terms of simple extraction of stored information. Instead, various molecular mechanisms exist that perform interpretation and editing of genetic information during gene expression. This leads to a necessary revision of the long-standing dogma of "one gene, one protein, one datum." However, a newly emerged question is whether DNA can be regarded as a stored set of instructions, analogous to a computer program [18 - 20]. Formally speaking, for any specific DNA, we can observe all possible reading, editing and splicing processes. Therefore, we can collect a complete and finite list of the proteins that can be synthesized by a certain organism. In this sense, DNA can be considered to be a stored set of instructions. However, the focus here is on the control of processes and on the analysis of a sequence of activation of instructions. In algorithms, the sequence and positions of conditionals are fixed. In contrast, in organisms, this situation is rare, and instructions often compete with each other for execution, which has a profound effect on the development of functional processes. According to Luhmann [21], we can define a process as a mutually connected succession of events where the scope of possible selections is re-established at each stage. A specific characteristic of processes understood in this manner is their anticipatory structure.

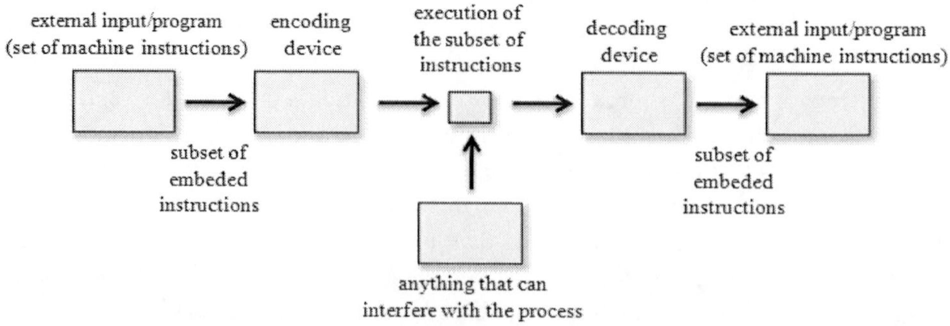

Figure 6. Schematic diagram of a general communication system, applied to the functioning of a man-made machine.

Such a succession of events is inherently an accumulation toward less and less probable states (less probable from the perspective of the beginning of the process), each of which is a necessary (structurally but not logically) consequence of previous stages. However, there is a very important difference between elementary processes (e.g., individual enzymatic transformations) and processes developed from these (e.g., metabolic pathways). The former can be identified as irreversible transformations towards a determined state (in an ideal case) or a group of very similar states. In living systems, this type of flow enclosed in a rigid structure is a basis for the further development of functionality. These irreversible sequences can be rearranged into higher-order structures (metabolic pathways), thus gradually relativizing processual indispensability with every superposed level of functionality constructed. In such a structure, individual events are no longer necessary for continuing the line of transformations (as is the case for elementary processes) but instead may become one of the possible realizations of functionality. As a consequence, the development of any process is uncertain and depends on numerous local conditions. Process flow is not predefined, and a system can achieve sufficiently high fluidity in shifting from one operative schematism to another without the need to construct specific regulative pathways for each situation.

The second difference between communication systems in living organisms and those in man-made machines concerns the unification of syntax and semantics. In formal languages, syntax and semantics are distinct from each other. Syntax deals with the rules of ordering of symbols, while semantics deals with the meaning of syntactically proper groups of symbols. The unification of syntax and semantics would lead to a self-referential paradox [22]. However, this paradox can be avoided if the self-reference is replaced by self-circularity, which can yield a dynamical circulation that connects two different logical levels. This approach is adopted in the notion of autopoiesis [23, 24]. The central concept in autopoietic theory is the assumption that an organized system determines its own components, which in turn assemble into organized system. At the biochemical level, a cell regulates the production of its molecular components, which perform various functional tasks such as enzymatic transformations, building of organelles, energy transfer, and others. Together, these functional processes constitute the cell. In this scheme, the organized system is well-defined and not susceptible to any substantial modification. Therefore, as Luisi [25, 26] pointed out, autopoiesis lacks an implementation of evolution and that life needs the coupling of an autopoietic system to its environment. He suggests that life needs the continual modification of a system itself, which results from the discrepancy between the system and its environment. Because the environment is inherently changeable, a system also has to be flexible enough to assimilate some of these changes. Moreover, it should be able to arbitrarily change both the semantic and the syntactic aspects of information. As explained above (Figure 3), the incorporation of enzymatic transformations into a biochemical network of functional processes adds a semantic layer to the structurally determined syntax. How do these aspects impact the generation of information in living systems? The construction and re-construction of material structures, changes in the observed identification of their elements, the assimilation of external changes into operative signals, their divergence into different possibilities of process development and convergence into local normative enclosedness and the continuous reorganization of operative horizons all constitute a dynamic system that is trapped in multi-level (spatial, temporal and organizational) reconstructive cycles. Owing to such organizational patterns, the existence and the maintenance of such structures are one and the same. They cannot at any moment passively exist, based on the mere fact that they have already formed. Because a living system is not a finished system, but rather a system under continuous self-construction, the maintenance of such structures is fully dependent on their functioning, which in turn continues only because of the ongoing formation of preservation mechanisms materialized by the temporalized re-constitution of elements. Therefore, the functioning of a system cannot be reduced merely to a set of internal transformations, but is inseparably composed of the elements of the system and their transformations. By constantly producing itself, namely by reflexive re-production, the local non-parametric state of the system (i.e., the context) emerges as a new regulatory force. Additionally, each deconstructed functional element (e.g., protein or subsystem) indicates not only the loss of one functional unit but also the loss of one object of interaction with the environment, along with the spatial and temporal structurations imposed by it, thus changing the balance of power between different subsystems. Therefore, when talking about living cells, we should not talk about the mere allocation of material structures but rather of the internal spatial and temporal structurations that undergo reconfigurations with each external change. In this sense, we can identify a permanent mutual dependence of dyadic modifications: the changes in 'objective' material composition (i.e., the uptake of nutrients or metabolic transformation) and the changes in the perception of such 'objective' changes each regulate the other. On the other hand, the mere fact that each living system is composed of perceptive elements, and consequently is characterized by parametric differentiation, leads to the constitution of well-known cybernetic regulatory models. There is, however, an important difference here relative to the usual perspective. Strictly speaking, in contrast to the classical model, which supposes the ultimate primacy of uniform and unambiguous parameters that penetrate the whole system and constitute its eternal organization, in organisms we can speak of cybernetic-like regulation only in short, separate segments. Unambiguous parameters are certainly achieved in living systems, and this should be clearly stressed, but this happens only in temporally short segments that are crystallized within, and in accordance with, a much broader situation. The entire intra-systemic environment constitutes one comprehensive controlling form, which is segmented rather than centralized and is not linked with (pre)defined signaling pathways. Internality, through temporalization, determines the dynamics of re-productive cycles and, through contextualization, determines the actual modes of identifying elements and is therefore able to continuously change the status of parameters. In this way, internality comprises or encircles all processes, including the cybernetic ones, which are then reduced from the eternal level to a transitory formation of non-ambiguity, in never-ending decomposition.

There may be a tendency to identify such systems as unstable. Without intending to open a discussion about the roots of such thinking or its correctness, only two factors will be emphasized here. First, by developing an operative schematism in dealing with the environment, a system actually stabilizes the environment. This is because the receptive reduction of the environment is not only a reduction in the sense of depletion but also an enhancement, in the sense that 'non-objective' relations can then be ascribed to elements in the environment, thus laying a basis for further development of their functional treatment. Additionally, the stabilization of the environment allows further construction of intra-systemic functional strategies. Second, with the temporalization of elements, an increase in local perturbations need not be followed by expansion throughout the whole system because it can be compensated by new self-constructions. It is true that such an organized system is not grounded in the steadiness of material structures or the non-ambiguity of rules, but these are not the only factors in systemic perseverance and functioning. Moreover, we believe that development of an operative schematism and temporalization of elements are the main organizational preconditions for achieving evolvability. Regarding the informational aspect, in order for a system to be autonomously evolvable, it should be open to the possibility that noise can eventually become information. This can happen only if both the codes and their inputs are changeable. In living systems because codes are "coded" in DNA that is susceptible to mutations, it follows that both the codes and the input to these codes are changeable. This forms a situation of mutually defined constraints between horizontal (within one organism or one generation) and vertical (across generations) information transfer: (1) the constraints of vertical information processing on horizontal information processing are represented by the scope of the possible actions determined by DNA and (2) the constraints of horizontal information processing on vertical information processing are represented by the scope of "acceptable" mutations. However, because horizontal information processing can be characterized as semiotic (i.e., it allows variability in systemic interpretations that are not errors), the constraints imposed on vertical information processing are more flexible than in typical man-made machine systems.

Conclusions

In this chapter, written as an epistemological essay, we have focused on the functioning of living systems from the standpoint of information generation and transfer. We gave a short overview of MTC, which is considered to be the backbone of current information theory. Because the MTC is clearly not applicable to the functioning of living systems, we also presented several alternative views that emphasize teleosemantic and biosemiotic approaches. The main concern of these theories is the meaning and the functional role of information, which are completely ignored in the MTC. We have attempted here to analyze the possibilities for adding a semiotic aspect to the MTC schematic, which provides a mathematically well-developed theoretical basis. We analyzed each step of the generalized metabolic pathway and found a high level of parallelism between the processes of biochemical transformations therein and the process of signal transduction as proposed by Shannon (1948). However, we also outlined several profound differences, the most important of which are (i) the competitive nature of the execution of rules in biological systems and (ii) the unification of syntax and semantics. Together, these properties enable the development of dynamic self-circulatory systems open to evolutionary change.

Acknowledgements

This chapter was realized as a part of the project "Studying climate change and its influence on the environment: impacts, adaptation and mitigation" (No. 43007), which is financed by the Ministry of Education and Science of the Republic of Serbia within the framework of integrated and interdisciplinary research over the period 2011 - 2014.

REFERENCES

[1] C.E. Shannon, *Bell Syst. Tech. J.*, 27, 379-423 and 623-656 (1948).
[2] R. Dawkins, *The Selfish Gene*; Oxford University Press: Oxford, US, 1976; p. 366.
[3] Hariri, B. Weber, J.III Olmsted, *J. Theor. Biol.,* 147, 235-254 (1990).
[4] C.E. Shannon, *Bell Syst. Tech. J.*, 30, 50-64 (1951).
[5] J. Maynard-Smith, *Philos. Sci.*, 67, 177-194 (2000).
[6] P. Griffiths, R. Gray, *J. Philos.*, 91, 277-304 (1994).
[7] W. Harms, *Information and Meaning in Evolutionary Processes*; Cambridge University Press: Cambridge, US, 2004; p. 280.
[8] K. Sterelny, K. Smith, M. Dickison, *Biol. Philos.,* 11, 377-403 (1996).
[9] J. Maclaurin, *Biol. Philos.,* 13, 37-59 (1998).
[10] P. Godfrey-Smith, In *Where Biology Meets Psychology: Philosophical Essays;* V. Hardcastle, Ed.; MIT Press: Cambridge, MA, US, 1999; pp. 305-331.
[11] N. Shea, *Biol. Philos.,* 22, 313-331 (2007).
[12] M. Barbieri, *Naturwissenschaften*, 95, 577-599 (2008).
[13] S. Artmann, In *A Companion to the Philosophy of Biology;* S. Sarkar, A. Plutynski, Eds., Blackwell Publishing Ltd: Oxford, GB, 2008; pp. 22-39.
[14] M. Barbieri, *Biosemiotics*, 2, 19-30 (2009).
[15] C. Adami, *Phys. Life. Rev.* 1, 3-22 (2004).
[16] L. Floridi, In *The Blackwell Guide to Philosophy of Computing and Information;* L. Floridi, Ed., Blackwell Publishing Ltd: Oxford, GB, 2004; pp. 40-61.
[17] C. Haynes, C.J. Oldfield, F. Ji, N. Klitgord, M.E. Cusick, P. Radivojac, V.N. Uversky, M. Vidal, L.M. Iakoucheva, *PLoS Comput. Biol.* 4; 2(8):e100 (2006).
[18] G. Marcus, *The Birth of the Mind: How a Tiny Number of Genes Creates the Complexities of Human Thought*; Basic Books, New York, US, 2004; p. 288.
[19] L. Moss, In *East Lansing: Philosophy of Science Association;* D. Hull, M. Forbes and K. Okruhlik, Eds., PSA, 1992; Volume 1, pp. 335-348.
[20] K. Sterelny, *Philos. Sci.*, 67, 195-201 (2000).
[21] N. Luhmann, *Social Systems*. Stanford University Press: Stanford, US, 1996; p. 684.
[22] Y.-P. Gunji, T. Shirakawa, T. Niizato, T. Haruna, I. Balaz, *Progr. Theoret. Phys. Supp.,* 173, 26-37 (2008).
[23] H. Maturana, *Int. J. Man.-Mach. Stud.,* 7, 313-332 (1975).
[24] F. Varela, *The Principles of Biological Autonomy*; North Holland: New York, US, 1979; p. 328.
[25] P.L. Luisi, *The Emergence of Life, from Chemical Origins to Synthetic Biology*; Cambridge University Press: Cambridge, US, 2006; p. 332.
[26] P.L. Luisi, *Chem. Biodive*r., 4, 603-621 (2007).

In: Essays on Fundamental and Applied Environmental Topics ISBN 978-1-61942-522-4
Editor: Dragutin Mihailovic

Chapter 4

MONTE CARLO RAY-TRACING METHOD FOR USE IN THE CALCULATION OF THE ALBEDO FOR DIFFERENT GEOMETRIES

D. Kapor[1,*], D. T. Mihailović[2], A. Ćirišan[3] and A. Firanj[2]

[1]Department of Physics, Faculty of Sciences, University of Novi Sad, Novi Sad, Serbia
[2]Department for Field and Vegetable Crops, Faculty of Agriculture, University of Novi Sad, Novi Sad, Serbia
[3]Institute for Atmospheric and Climate Sciences, ETH Zurich, Zurich, Switzerland

ABSTRACT

Environmental modeling at various scales demands the parameterization of physical processes over heterogeneous environmental interfaces. A previously introduced novel approach to calculating the aggregated albedo of a heterogeneous grid-cell with regions of different heights is analyzed using different geometries ("trilateral prism", "slope" and "canyon") in a grid-cell commonly encountered in environmental models. The loss coefficient, which describes the relative flux lost due to absorption on the vertical lateral boundaries of patches lying on the higher levels, can sometimes be derived as a rather complex analytical expression. In most situations, the analytical form of the loss coefficient is not accessible, and the Monte Carlo ray tracing (MCRT) method has been determined to be the most reliable approach for calculating the aggregated albedo. As the dimensions of the problems increase, this method becomes highly efficient and less time-consuming than deterministic methods. The albedo was calculated over a grid-cell consisting of the aforementioned geometries on a horizontal surface, leading to some extremely interesting results. Finally, the differences in albedo obtained by conventional techniques and by the MCRT method were emphasized by using the Land Air

* Corresponding author: E-mail address: dvk@df.uns.ac.rs; Phone: +381 21 485 2809. Fax: +381 21 459 367.

Parameterization Scheme (LAPS) to calculate the effective surface temperatures of the grid-cells of all considered geometries.

1. Introduction

Grid-based environmental models often require the determination of the albedo over a heterogeneous environmental interface [1, 2]. The structure of the grid-cell on the environmental interface can consist of patches covered with surfaces of different origin. Our subject here will be the environmental interface between the atmosphere and land, which consists of patches of plant communities, bare soil, rocky landscapes or liquid parts, thus creating a very heterogeneous picture within the grid-cell. In atmospheric models, the most common approach to calculating the albedo is to take a simple average over the grid-cell. However, a physics-based analysis indicates that there is significant deviation of the albedo above a heterogeneous surface, seriously affecting the calculated values of quantities that describe surface biophysical processes such as land surface energy budgets, canopy photosynthesis and transpiration, snow melt, urban surface physics, and others [3 - 6]. This recognition stresses the importance of understanding the general behavior and limitations of the approaches used to aggregate the albedo over a heterogeneous grid-cell in current land surface models. In the papers by Mihailović et al. [7] and Kapor et al. [8], the assumptions that underlie aggregation over a very heterogeneous surface on which various surfaces are present at different heights were considered theoretically, which led to the proposal of a method (in the further text, indicated as the PM - proposed method) for deriving a general expression for the aggregated albedo. The suggested expression for the albedo was compared to the standard approach (SA) to a two-patch grid-cell with a simple geometric and height distribution.

In this review, we continue the work initiated in the paper [7], in which we emphasized that the PM method will be applied to more realistic geometries. Now, we deal with the numerical calculation of the albedo in different geometries commonly encountered in grid-cell architectures in environmental models of different scale. This method was recently included in the LAPS (Land Air Parameterization Scheme) radiation module [9], particularly for the purposes of modeling over steep mountains in numerical weather prediction models. More precisely, a method for accounting for the effects of different height levels and surface types present in a given grid-cell is considered in this chapter for the "trilateral prism", "slope" and "canyon" geometries (Figure 1). Urban environments will not be the subject of this review because such calculations usually demand the study of a complex of objects rather than a single object, as in the geometries we shall study here [10 - 12].

Another objective of this review is to demonstrate the advantage of calculating the albedo in environmental models of more complicated solid geometries by the Monte Carlo ray-tracing (MCRT) technique over computing the albedo either classically or analytically. To meet this goal, we emphasize (i) the agreement of the result obtained using the MCRT method with very complicated analytical expressions where they exist and (ii) the difference in CPU times required for the MCRT calculation and for standard numerical integration (SNI). Here, we discuss the aggregated albedo values obtained by the MCRT method and their impact on the maximum effective surface temperature of the grid-cell calculated by the LAPS land surface scheme.

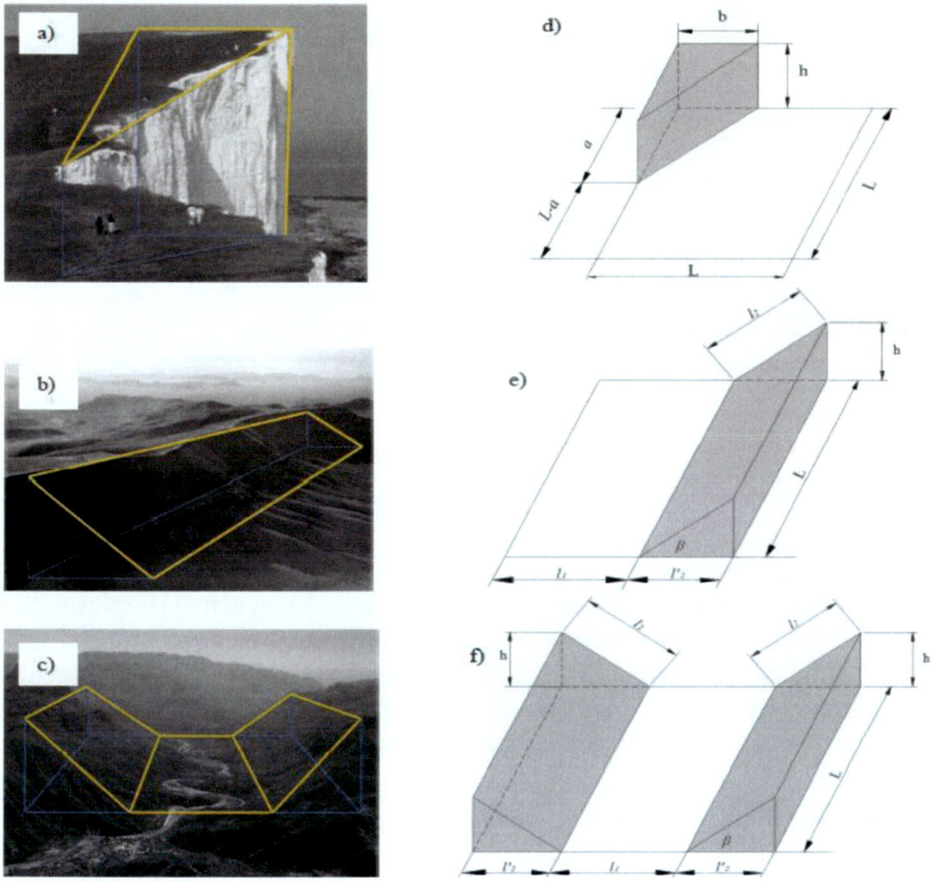

Figure 1. Schematic representation of natural and artificial solid surfaces [(a) - (c)] and their approximate geometries in the parameterization of the albedo over the heterogeneous surfaces [(d) "trilateral prism", (e) "slope" and (f) "canyon" geometry].

Section 2 briefly summarizes the theoretical approach to aggregating the albedo over heterogeneous surfaces (subsection 2a) and the basic features of the MCRT method (subsection 2b). Section 3 deals with the numerical calculation of the loss coefficients of the aforementioned geometries using the MCRT method. In Section 4, we depict and discuss the influence of the calculated albedo on the effective temperatures of grid-cells of various geometries covered by different natural covers surrounding a solid component. In Section 5, we provide some concluding remarks.

2. Short Review of the Aggregation of the Albedo over the Heterogeneous Surfaces

2.1. Theoretical Approach

In this subsection, we briefly summarize the primary theoretical features of the method suggested in [7]. The basic idea of our approach relies on the observation that part of the radiation reflected from the lower surface will be absorbed by the lateral sides of surfaces lying at a higher level (Figure 2) and thus does not contribute to the albedo. The amount (ratio) of the reflected energy lost is obtained by calculating the solid angle within which these lateral sides are seen from each point of the lower surface.

We assume here that the radiation reflected from a given surface is diffuse and homogeneous, neglecting multiple scattering effects and the dependence of the albedo on the zenithal angle. This assumption is based on the further assumption that the reflecting surface is sufficiently rough that the reflected radiation is isotropic and the amount of energy reflected within any solid angle is proportional only to the solid angle itself. Thus, we can neglect the orientation of the radiation falling on that surface. To define the flux lost due to this absorption, we introduce the loss coefficient k_l, which is determined by the radiant energy flux ratio. Following Liou [13], the basic expression to calculate the radiant energy flux is (dE/dt)

$$(dE/dt) = I\ dS \cos\theta\ d\Omega \tag{1}$$

where is the total radiation intensity obtained by integrating a monochromatic intensity distribution over the entire spectral range, is the infinitesimal element of the surface from which radiation is reflected, gives the direction of the radiation stream, and is the solid angle element to which the infinitesimal amount of radiant flux emitted from the infinitesimal surface element is confined. The amount of flux emitted from the lower horizontal surface into the upper space is $(dE/dt)_h = IS_1\pi$, where is the area from which the radiation is reflected. The total energy radiated from the lower horizontal surface towards the lateral surface is and is derived using the following expression:

$$(dE/dt)_l = I \iint_S dxdy \int_{\Phi_l(\vec{r})}^{\Phi_u(\vec{r})} d\Phi \int_{\theta_l(\vec{r},\Phi)}^{\theta_u(\vec{r},\Phi)} d\theta \cos\theta \sin\theta \tag{2}$$

where the boundaries of the integration for a given point are determined over the local azimuthal and zenithal angles in terms of the coordinates (Figs. 1 and 2 in Kapor et al. [8], or in more detail, Figs. 3, 4, 6 and 7 in Mihailović et al. [7]). The subscripts *u* and *l* denote the lower and upper bounds of integration, respectively. The ratio of Eqs. (1) and (2) give the *loss coefficient* k_l ($0 < k_l \le 1$).

$$k_l = (dE\ /\ dt)_l\ /\ (dE\ /\ dt)_h\ . \tag{3}$$

If we assume that the grid-cell of the area is divided into two components of areas and with corresponding albedos and α_2, respectively, then the average albedo over the grid-cell following the SA method commonly used in environmental models is

$$\bar{\alpha}_c = \alpha_1 \sigma_1 + \alpha_2 \sigma_2 \tag{4}$$

where is the fractional cover, which is calculated as the ratio of a patch's area to the total grid-cell area $S(\sigma_i = S_i/S, i = 1,2)$. Introducing the *loss coefficient*, the average albedo should be calculated as

$$\bar{\alpha}_n = (1 - k_l)\,\alpha_1 \sigma_1 + \alpha_2\,\sigma_2\,. \tag{5}$$

The definition of the *loss coefficient* is conceptually analogous to the idea of the sky-view factor introduced by Oke [14]. A detailed overview of the recently proposed evaluation of the *loss coefficient* needed to calculate the average albedo is available for different geometries in Mihailović et al. [7] and Kapor et al. [8].

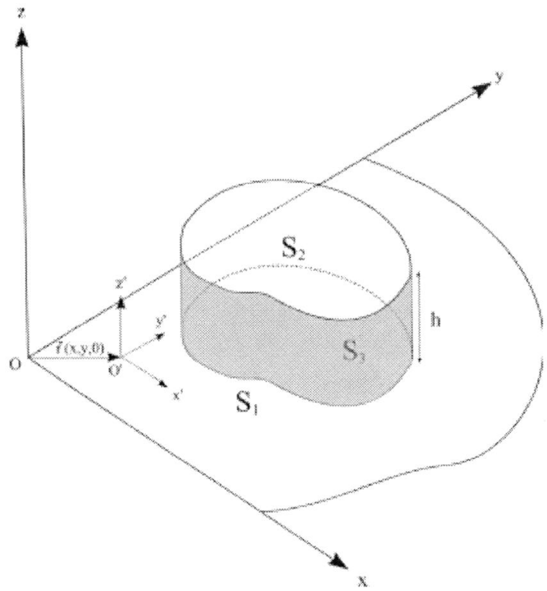

Figure 2. Schematic representation of an arbitrary grid-cell geometry consisting of two surfaces located at heights 0 and h. The notation follows the text [7, 8].

2.2. The MCRT Method Calculation of the Grid-Cell Albedo

Our previous studies indicated that in the case of very complicated grid-cell geometries that prevent analytically solving for the *loss coefficient*, the MCRT method is reliable and is the most efficient method for calculating the aggregated albedo. Note that the simple Monte Carlo method is less efficient [15] than the MCRT method, which has previously been shown to reproduce analytical results to high precision. Let us first explain the general idea of the calculation procedure.

The main idea of the MCRT method is to follow the path of an appropriately chosen ray of light after it has undergone diffuse, isotropic single scattering from the lower surface of the grid-cell. The purpose of this observation is to determine whether the ray may be absorbed by the vertical boundary, i.e., the lateral side of the surface lying at a higher level.

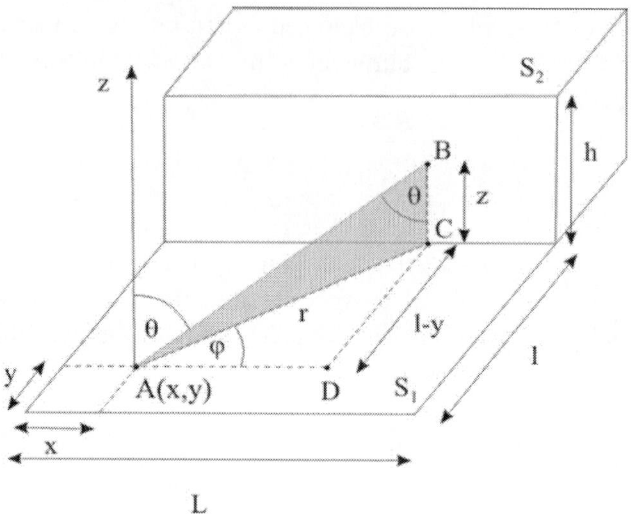

Figure 3. Schematic diagram of the procedure for applying the MCRT method to calculate the aggregated albedo.

Averaging the behavior observed in this way over a large number ($N=10^6$) of light paths, we can determine the value of the *loss coefficient* from radiative flux loss within a given grid-cell geometry. The details of a particular Monte Carlo procedure strongly depend on the geometry of the grid-cell, so we will illustrate the particular procedure related to the simplest grid-cell geometry shown in Figure 3.

The point $A(x,y)$, which belongs to the lower surface and represents the point of intersection between this surface and the incoming beam, is randomly sampled by generating two random numbers and that are uniformly distributed within the interval (0, 1). The area of the lower surface is $S_1 = L \times l$. Therefore, we write that and that $y = r_2 \times l$. In agreement with our basic assumption – diffusive and single ray scattering – we choose a random direction in the upper half-space (θ, φ) [with $\theta \in (0, \pi/2); \varphi \in (0, 2\pi)$] to trace the scattered beam. Using uniform random numbers and from the range (0, 1) allows us to choose this random direction as and $\theta = r_4 \times \pi / 2$. The reflected beam is treated as a straight line, and the vertical area is a plane. The intersection of the line and the plane can be derived using general expressions of analytical geometry [16], but the following is based on computer-oriented exposition by Bourke [17]. The intersection of the line and the plane occurs when

$$N(PP + t(PP1 - PP)) - (N \cdot P0) = 0 \tag{6}$$

where the points and are two known points on the line, $P0$ is the known point on the plane, is the plane normal and is the line coefficient. The final expression for the line coefficient, represented in the coordinate system, is

$$t = \frac{Nx(P0x - PPx) + Ny(P0y - PPy) + Nz(P0z - PPz)}{Nx\cos\varphi\sin\theta + Ny\sin\varphi\sin\theta + Nz\cos\theta} \tag{7}$$

where subtraction of and is performed in terms of azimuthal and zenithal angles. The coordinates of the intersection point are the following

$$\begin{aligned} x &= PPx + t\cos\varphi\sin\theta \\ y &= PPy + t\sin\varphi\sin\theta \\ z &= PPz + t\cos\theta. \end{aligned} \tag{8}$$

Now, if the point lies within the borders of the vertical area, then the diffusively scattered beam will be absorbed (in the single scattering approximation), and this case is then counted as positive for absorption. This procedure is repeated times, and the *loss coefficient* is estimated as

$$k_l = N_a \left(\textit{number of cases which were positive for absorption}\right) / \\ N \left(\textit{number of conducted numerical experiments}\right). \tag{9}$$

3. Calculation of the Loss Coefficient

We calculated the *loss coefficients* and aggregated albedos of the heterogeneous grid-cells of the geometries shown in Figs. 1d - 1f using (i) the MCRT method and (ii) the standard numerical integration (SNI) method. All of the experiments were performed for different values of the reduced heights and lengths.

3.1. The "Trilateral Prism" and "Slope" Geometries

In this subsection, we first analyze the geometry with an elevated corner area having a triangular base, called the "trilateral prism" geometry (Figure 1d). As the second example, the "slope" geometry is introduced, where reflection from the lower, horizontal surface occurs along with reflected light that first reaches the sloped surface (Figure 1e), reflects from it, and then falls onto the horizontal area. This process should be also taken into account in the *loss coefficient* calculation, as it will also affect (decrease) the albedo value.

Let us describe the geometries in more detail. In the top left corner of a square grid-cell is an elevated surface at a height with respect to the rest of the grid (Figure 1d). This area has a triangular shape with dimensions $a \times b$, where is a fixed arm of exactly half the length of the cell edge while the arm takes different values depending on the angle of the triangle.

In the case of the sloped surface, the horizontal axis is divided into two parts. The first part corresponds to the edge of the horizontal surface and is called l_1, while is the base of the sloped surface (Figure 1e). The sloped surface forms an angle with the horizontal plane and reaches a height from the horizontal surface. To calculate the *loss coefficient* calculation, it is necessary to compute two contributions to the coefficient, and the final value for is the sum of the two. The first contribution relates to the light reflected from the bottom horizontal surface onto the sloped surface, and the other contribution describes the reflection of light from the sloped surface to the horizontal surface.

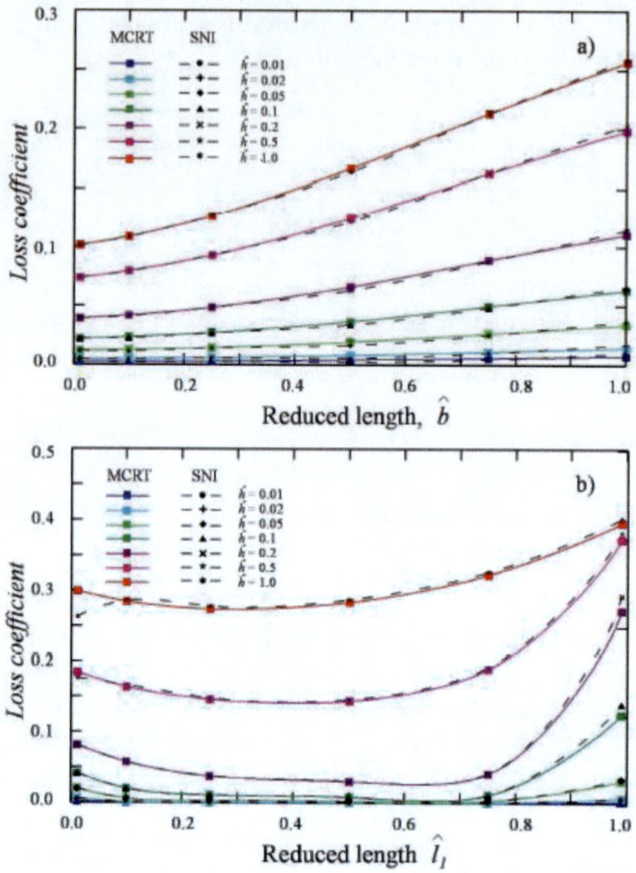

Figure 4. Dependence of the *loss coefficient* on the reduced lengths and $\hat{l}_1$, for different values of the reduced height $\hat{h}$, obtained by the standard numerical integration (SNI) approach and by the MCRT method: (a) the "trilateral prism" geometry and (b) the "slope" geometry.

The analytical solution in these complex geometries was too complicated to be determined, so a direct simulation was performed using the MCRT method to determine the *loss coefficient* needed to calculate the average albedo. Figures 4a - 4b present the *loss coefficient* calculated by the MCRT method for different values of the reduced lengths and (0.0, 0.1, 0.25, 0.5, 0.75, 0.99) and different heights of the surface. The reduced dimensionless quantities $\hat{b} = b / L$ (in the "trilateral prism" geometry), (in the "slope" geometry) and were introduced to simplify the *loss coefficient calculation*, where is the area of the grid-cell. To confirm the accuracy of the MCRT method, the SNI procedure is used to obtain the *loss coefficient* of both geometries. As can be seen, the two methods agree quite well.

3.2. The "Canyon" Geometry

The geometry considered in this subsection is one of the most complicated geometries that we have analyzed in our studies. As previously mentioned, two mountain slopes with a valley between them form the shape of our canyon (Figure 1f). The grid-cell has a rectangular base covering an area of $L \times (L + l_2^{'})$. The lateral dimension of the valley is l_1, and its width is L. The base lengths of both slanted surfaces are $l_2^{'} = L - l_1$. These surfaces are sloped at some angle and have heights of h. This geometry is interesting because incident radiation can be reflected from the horizontal surface and also from the sloped surfaces, which will further reduce the albedo values.

Here, we again used the MCRT and the SNI to calculate the loss coefficient. The simulations were performed for several values of the reduced length (0.01, 0.10, 0.25, 0.50, 0.75 and 0.99) and of the visible horizontal surface $\hat{l}_1 = l_1 / L$. Also, the value of the reduced height takes the following values: 0.01, 0.02, 0.05, 0.1, 0.2, 0.5 and 1.0. The value of the *loss coefficient* varies with the reduced length and the reduced height as shown in Figure 5. As we can see, when the sloped surfaces have small heights and the distance between them is small, the loss coefficient does not increase significantly. Nevertheless, as the height increases, remarkable values of are obtained, particularly as the value of the reduced length approaches unity.

A comparison of the aggregated albedo of the "canyon" geometry with those of the other considered geometries presents the most interesting results. To see this, we calculated the albedo with both methods (the SA and PM) with grid-cells (valley and sloped surfaces of the canyon) covered completely with grass or rock and with a combination of the two.

We chose the following combinations: (a) the entire area covered with rock (R), (b) the sloped surfaces of the canyon covered with grass and the valley covered with rock (R-G) and (c) the whole area covered with grass (G). These different scenarios are shown in Figure 6. In the simulations we used (i) albedo values of 0.10 and 0.30 for the grass and rock, respectively, taken from [4], (ii) reduced heights of 0.01, 0.02, 0.05, 0.1, 0.2, 0.5 and 1.0 and (iii) reduced lengths of 0.01, 0.1, 0.25, 0.50, 0.75 and 0.99.

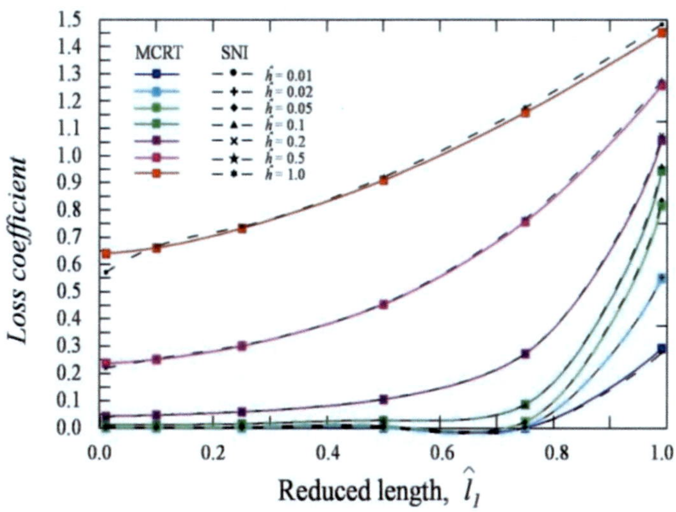

Figure 5. Dependence of the *loss coefficient* on the reduced length for different values of the reduced height obtained by the standard numerical integration (SNI) approach and by the MCRT method for the "canyon" geometry.

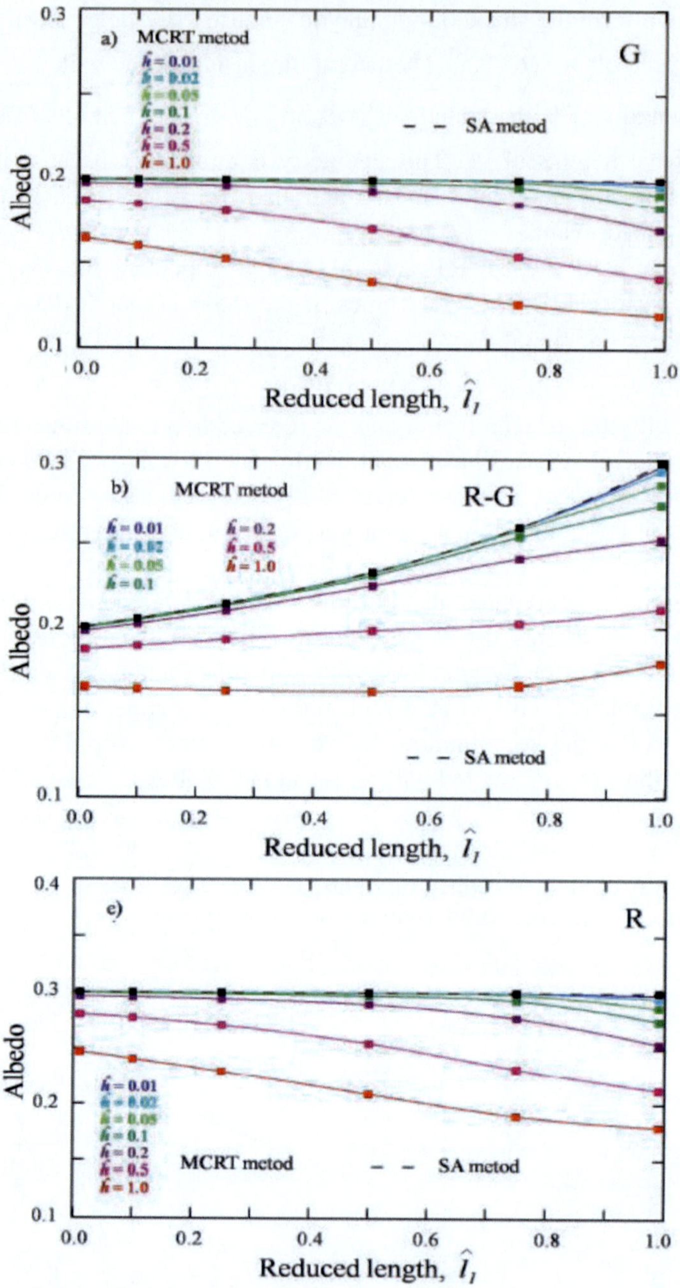

Figure 6. Dependence of the aggregated albedo on the reduced length for different values of the reduced height and different underlying surfaces obtained by the PM (MCRT) and SA albedo calculation methods for the "canyon" geometry. (a) The entire area covered with rock (R), (b) the sloped surfaces of the canyon covered with grass and the valley covered with rock (R-G), and (c) the entire area covered with grass (G).

Figure 6 clearly shows that the values obtained by the PM method (using the MCRT) are less than those obtained by the SA method, in particular for the higher values of the reduced length $\hat{l}_1$.

4. Influence of the Albedo Calculation on the Effective Temperatures of Heterogeneous Grid-Cells Consisting of Different Natural Covers

The differences the value of the aggregated albedo calculated by the SA and the MCRT method may produce differences in the effective surface temperature calculated over the grid-cell. These differences will be illustrated by performing a set of experiments based on the SA method and a set based on the PM method (using the MCRT method) to calculate the albedo. In these experiments, we computed the effective surface temperature over the grid-cell using the land surface scheme LAPS, which is designed to be run either as a standalone model or as part of an atmospheric model [9]. The numerical tests over grid-cells with different geometries were performed with forcing meteorological data for July 17, 1999 in Philadelphia, PA.

The corresponding prognostic equation, the so-called "force-restore" equation, for calculating the effective surface temperature of the heterogeneous grid-cell has the form

$$C_{eff} \frac{\partial T_{eff}}{\partial t} = R_{net} - \bar{H}_{agr} - \overline{\lambda E}_{agr} - \bar{G}_{agr} \tag{12}$$

where is the effective heat capacity per unit area, is the time, is the net radiation, is the sensible heat flux, is the latent heat flux and is the soil heat flux. Let us note that these are all actually flux densities, but according to the commonly used terminology we call them "fluxes". The net radiation is calculated as the sum of the incoming and outgoing radiation, including the following terms

$$R_{net} = (1 - \alpha_{agr}) R_{short} + R_{long} - \varepsilon \sigma_B T_{eff}^4 \, . \tag{13}$$

where is the shortwave radiation, R_{long} is the atmospheric counter-radiation, is the aggregated albedo, is the emissivity (0.95 in this study) and is the Stefan-Boltzmann constant. To calculate from Eq. (12), we applied a flux aggregation method over the grid-cell. This procedure obtains an average flux over the grid-cell from the particular fluxes calculated by the LAPS scheme [9] over the patches constituting the grid-cell.

In these calculations, we also aggregated the other parameters following the procedure of parameter aggregation by Mihailović et al. [18] and Lazic et al. [19]. The aggregated sensible and latent heat fluxes $\bar{H}_{agr}$ and for two surfaces are given by the expressions

$$\bar{H}_{agr} = H_1 \sigma_1 + H_2 \sigma_2 \tag{14}$$

and

$$\overline{\lambda E}_{agr} = \lambda E_1 \sigma_1 + \lambda E_2 \sigma_2 \tag{15}$$

where and are the sensible heat fluxes over surfaces 1 and 2, respectively, and and λE_2 are the corresponding latent heat fluxes. and are the fractional covers of the surfaces constituting the grid-cell. The soil heat fluxes and are estimated from the net radiation depending on the type of the underlying surface (fR_{net}, f =0.05, 0.10 and 0.25 for the grass, the forest, and the concrete and rock, respectively). The aggregated values of the soil heat flux were calculated as

$$\bar{G}_{agr} = (G_1\sigma_1 + G_2\sigma_2)R_{net} . \tag{16}$$

For the sake of simplicity, the experiments were performed with two and three (for the "canyon" geometry) surfaces of different albedos. The purpose of the experiments was to establish the daily changes of the effective surface temperature over the heterogeneous grid-cell caused by the difference in the albedos calculated by the PM method and the SA method. The effective surface temperature simulations were performed using a 600 s time step. The albedo was calculated by the PM method from Eq. (5) after the loss coefficient was obtained by the MCRT method. To obtain the albedo by the SA method, we used Eq. (4). In both cases, the obtained value of the albedo is introduced into Eq. (13) and then LAPS was run for the different geometries.

We must repeat that we assume isotropic, diffuse reflection so that the reflected light is isotropic and does not depend on the zenithal angle of the incoming light. In a more detailed study, one should also consider the daily variation of the zenithal angle when one follows the daily temperature changes.

4.1. The "Trilateral Prism" and "Slope" Geometries

In this subsection, we first deal with the "trilateral prism" geometry (Figure 1d) with the previously described dimensions. We considered the prism to have a triangular shape with an area of and a reduced height $\hat{h}$. The reduced value was fixed, while the reduced length $\hat{b} = b / L$ took values of 0.25, 0.5, 0.75 and 1. The fractional coverages of the elevated surface and the rest of the grid-cell were and $\sigma_1 = 1 - \hat{a} \times \hat{b}/2$, respectively. The simulations were performed similarly to those in Subsection 3.1, still using the G surface type, but also using the surface type R instead of C.

We analyzed the differences in the daily course of the effective surface temperature obtained by the two methods for grid-cells covered with surface types G and R on the surface next to the "trilateral prism" made of concrete. We used the following fractions of and σ_2 : 93.75 % (σ_1) - 6.25 % (σ_2), 87.5 % (σ_1) - 12.5 % (σ_2), 81.25 % (σ_1) - 18.75 % (σ_2) and 75.0 % (σ_1) - 25.0 % (σ_2). The LAPS model was run for the simulations with G-R and R-R and used the same albedo values used in the previous subsection. The reduced heights ($\hat{h}$) of the trilateral prism patch were 0.01, 0.02, 0.05, 0.1, 0.2, 0.5 and 1.0. In the G-R simulations the grass was 0.5 m tall.

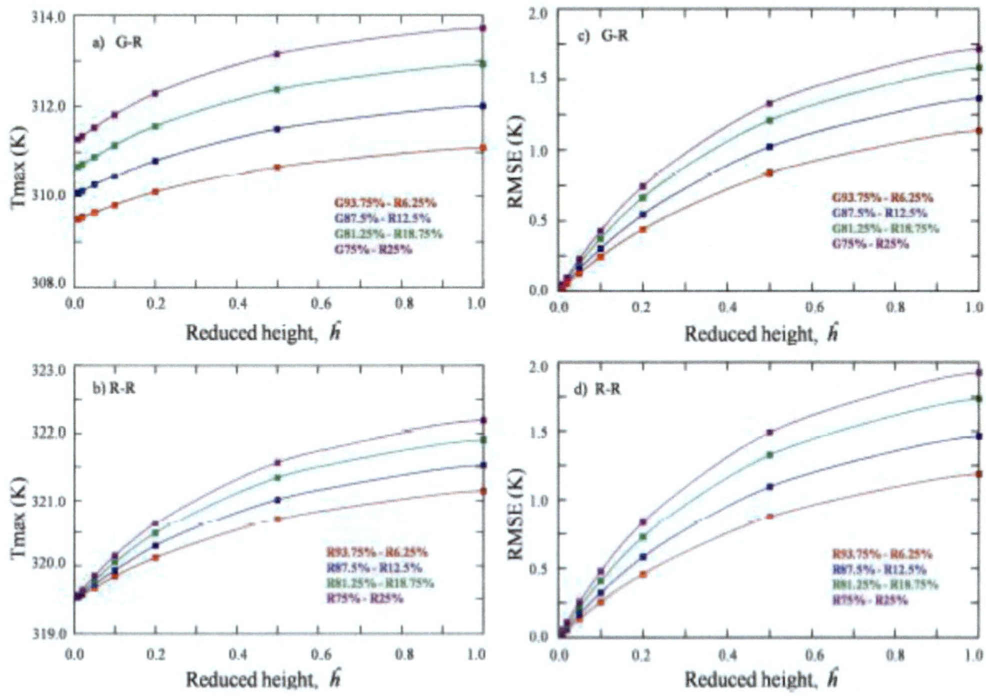

Figure 7. The maximum effective surface temperatures obtained by the PM (performed by the MCRT) (left panels, a - b) and the RMSE values (right panel, c - d) versus the reduced height for different values of the fractional coverages in the “trilateral prism” geometry. The RMSE is calculated on the basis of differences between the PM and SA albedos over all points in the simulated daily courses.

Figures 7a - 7b show the differences in the maximum effective surface temperature obtained by the PM method for different values of the reduced height and fractional cover in the “trilateral prism” geometry. When the cell is covered with grass and rock or when the entire grid-cell is covered with rock, the increase in the maximum temperature ranges from about 1.5 K to 2.5 K (Figs. 7a and 7b). The root mean square error (RMSE) (Figs. 7c - 7d) of the differences between the PM and SA albedo calculations over all points in the simulated daily courses was computed, and the case in which the entire grid-cell area was covered by the rock presented the largest differences. As the area of the surface lying at the higher level is larger, it increases the RMSE value in the observed case.

What happens when the sloped surface is introduced? In the “sloped” geometry, the horizontal axis is divided into two parts. The first part corresponds to the visible horizontal surface marked with l_1, while is the base of the sloped surface of height (Figure 1e). The simulations were performed with the same surface types as the “trilateral prism” simulations. The albedos were calculated for reduced lengths of 0.25, 0.5, 0.75 and 0.99 and reduced heights of 0.01, 0.02, 0.05, 0.1, 0.2, 0.5 and 1.0. The fractional coverages were calculated as and $\sigma_2 = 1 - \hat{l}_1$. The analyses of (i) the RMSEs of the daily effective surface temperature course differences obtained by the two methods and (ii) the maximum effective temperatures obtained by the PM method were performed for grid-cells covered with each surface type, i.e., G and R filling the surface next to the rocky sloped surface and with the following and values: 75.0 % (σ_1) - 25.0 % (σ_2), 50.0 % (σ_1) - 50.0 % (σ_2), 25.0 % (σ_1) - 25.0 % (σ_2)

and 99,0 % (σ_1) - 1,0 % (σ_2). The LAPS model was run for the G-R and R-R simulations with the same albedo values used in the previous subsection. In the G-R simulations, the grass was 0.5 m tall. However, in these simulations, the only values for the reduced height were 0.2, 0.5 and 1.0.

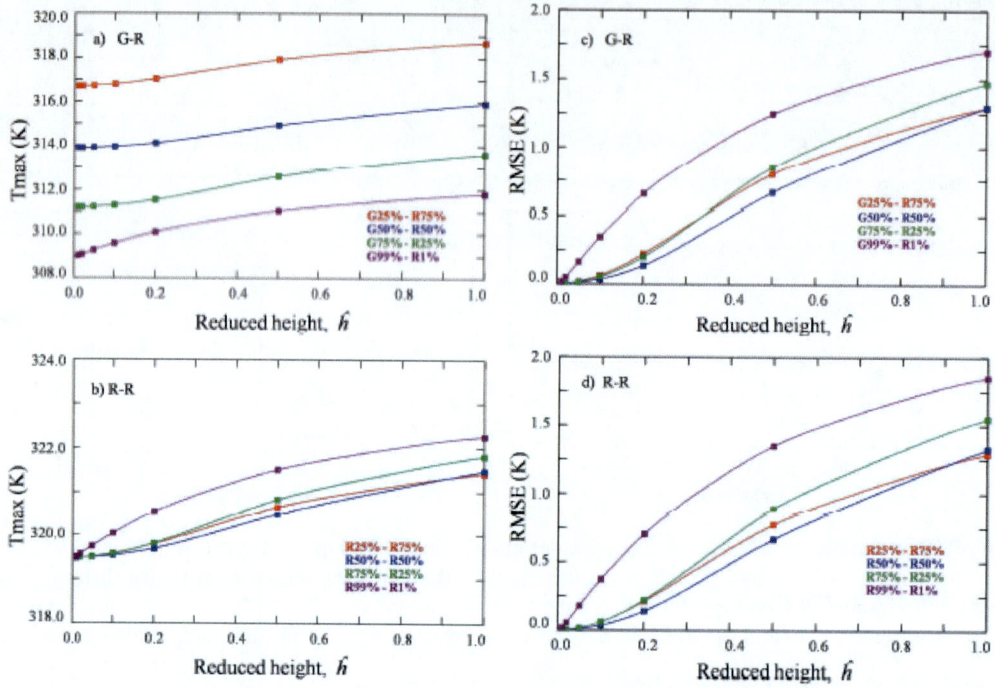

Figure 8. The maximum effective surface temperatures obtained by the PM (performed by the MCRT) (left panels, a - b) and RMSE values (right panel, c - d) versus the reduced height for different fractional coverages in the "slope" geometry. The RMSE was calculated from the differences between the PM and SA albedo calculation methods over all points in the simulated daily courses.

In the G-R and R-R simulations, the maximum effective temperature increases range from 1.5 K till 2.5 K (Figs. 8a and 8b). The RMSE is drastically larger in the case with 99 % horizontal surface coverage (Figure 8c - 8d). This result is understandable because in that simulation, reflection and absorption are observed from the largest surface and therefore produce the largest decrease in the albedo compared with the other simulations. In both cases, the geometries in which the horizontal and sloped surfaces cover the same areas of the cell produced the minimum reduction in albedo and the minimum temperature differences between the two methods of calculating the albedo.

4.2. The "Canyon" Geometry

In this subsection, we analyze the "canyon" geometry. As already mentioned, two mountain slopes with a valley between form our canyon (Figure 1f). The grid-cell has a rectangular base covering an area of $L \times (L + l_2')$.

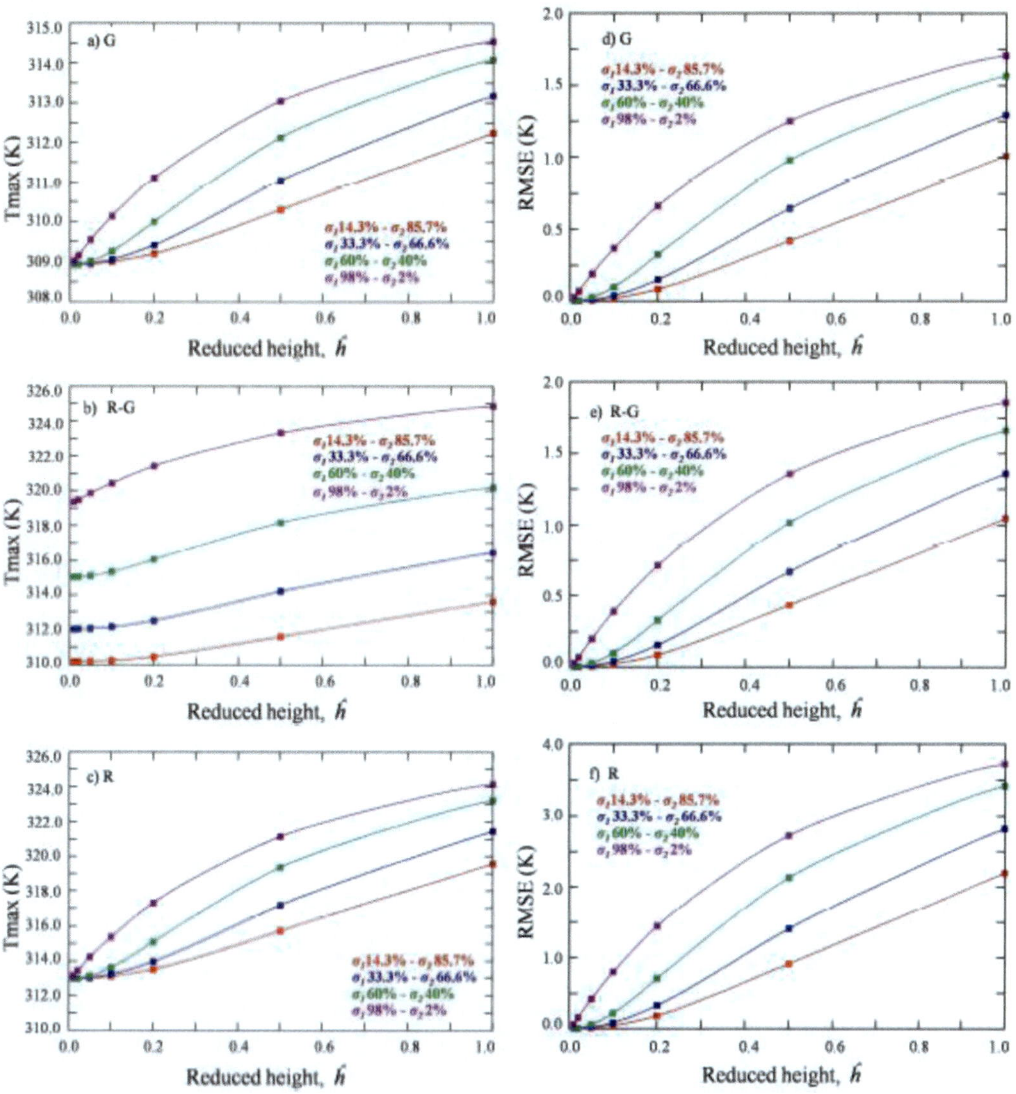

Figure 9. The maximum effective surface temperatures obtained by the PM method (performed by the MCRT) (left panels, a - c) and the RMSE values (right panel, d - f) versus the reduced height for different fractional coverages (σ_1 and σ_1) in the "canyon" geometry. The RMSE is calculated from the differences in the daily temperature courses of the PM and SA albedo calculation methods. Notation: (a) the entire area covered with grass (G), (b) the sloped surfaces of the canyon covered with grass and the valley covered with rock (R-G) and (c) the entire area covered with rock (R).

The horizontal lateral dimension of the valley is l_1, its width is L, and both slanted surfaces are of height h. The simulations were performed for several values of the reduced length (0.25, 0.50, 0.75 and 0.99) of the visible horizontal surface; the reduced base length of the slanted surface is $\hat{l}_2^{'} = 1 - \hat{l}_1$. Also, the simulations varied the reduced height with the following values: 0.01, 0.02, 0.05, 0.1, 0.2, 0.5 and 1.0. The fractional coverage of the horizontal area was $\sigma_1 = \hat{l}_1 / (1 + \hat{l}_2^{'})$, while the surface had an area of $\sigma_2 = 2\hat{l}_2^{'} / (1 + \hat{l}_2^{'})$. In the simulations, we considered the following situations: (i) the entire canyon area were covered with grass (G), (ii) the valley between the sloped surfaces were covered with rock, while the sloped surfaces were covered by grass (R-G) and (iii) the entire area was covered with rock (R). The fractional coverages σ_1 and σ_2 took the following values: 14.3 % (σ_1) - 85.7 % (σ_2), 33.3 % (σ_1) - 66.7 % (σ_2), 60.0 % (σ_1) - 40.0 % (σ_2) and 98.0 % (σ_1) - 2.0 % (σ_2).

We considered the differences between the effective surface temperatures in a canyon obtained by the PM and SA methods for different values of the reduced height and the fractional cover of each patch. Figures 9a - 9f present (i) the maximum temperature (obtained by the PM method) and (ii) the RMSE of the effective surface temperatures.

The maximum effective surface temperature obtained from the albedo calculated by the PM method increases as the reduced height and the fractional cover of the horizontal surface increase. Next, the entire area is covered with grass or rock, the maximum temperature increases by 3.5 K to 5.5 K for the largest reduced heights (Figs. 9a - 9c on the left panels). The largest difference in the maximum temperature values, approximately 10 K, is observed in the R case. In the R-G simulations, the deviation in the maximum temperature values is even larger as the fractional coverage of the patches varies, from about 4 K to 6 K for the largest values of the reduced height $\hat{h}$. To more precisely determine the difference between the temperature profiles obtained by the PM method and the SA method, the RMSE was also calculated. Very similar RMSE (2 K - 4 K) values were found for the R-G simulations (Figure 9e), the G simulations and the R simulations.

CONCLUSIONS

Our attention was focused on methods to capture energy transfer processes through and over heterogeneous underlying surfaces. The MCRT method was used to study the effect of different height levels and surface types in grid-cells of the following geometries: "trilateral prism", "slope" and "canyon".

The aggregated albedo calculated with this method was significantly reduced from that calculated through the conventional techniques, indicating the importance of geometrical effects in effective albedo calculations. The differences in the calculated albedos evidently led to a redistribution of the surface energy and remarkable differences in the effective surface temperature of the grid-cell. The proposed method is very useful for environmental models that include calculation of the surface fluxes, in particular when the grid-cell architecture varies.

ACKNOWLEDGEMENTS

This chapter was realized as a part of the project "Studying climate change and its influence on the environment: impacts, adaptation and mitigation" (No. 43007), which is financed by the Ministry of Education and Science of the Republic of Serbia within the framework of integrated and interdisciplinary research over the period 2011 - 2014.

REFERENCES

[1] D.T. Mihailović, I. Balaž, *Idojaras*, 111(2-3), 209-220 (2007).

[2] B. Cushman-Roisin, C. Gualtieri, D.T. Mihailović, In *Fluid Mechanics of Environmental Interfaces;* C. Gualtieri, D.T. Mihailović, Eds.; TaylorandFrancis Ltd: London, GB, 2008; pp. 1-16.

[3] Z. Hu, S. Islam, L. Jiang, *Bound.-Layer Meteorol.*, 93, 313-336 (1999).

[4] M.Z. Jacobson, *Fundamentals of Atmospheric Modelling;* Cambridge University Press: The Edinburgh Building, Cambridge, GB, 1999; p. 645.

[5] D. Kapor, D.T. Mihailović, T. Tošić, S.T. Rao, C. Hogrefe, *Proceedings of iEMSs*, Lugano, CH, 2, 448-453 (2002).

[6] P. Schwerdtfeger, *Environ. Model. Softw.*, 17, 51-60 (2002).

[7] D.T. Mihailović, D. Kapor, C. Hogrefe, J. Lazić, T. Tošić, *Environ. Fluid Mech.*, 4, 57-77 (2004).

[8] D.V. Kapor, A.M. Ćirisan, D.T. Mihailović, In *Advances in Environmental Fluid Mechanics;* D.T. Mihailović, C. Gualtieri, Eds.; World Scientific Publishing Co Pte Ltd: New Jersey, US, 2010; pp. 145-165.

[9] D.T. Mihailović, J. Lazić, J. Leśny, J. Olejnik, B. Lalić, D.V. Kapor, A. Cirisan, *Theor. Appl. Climatol.*, 100, 299-323 (2010).

[10] E.S. Krayenhoff, A.V. Voogt, *Bound.-Layer Meteorol.*, 123, 433-461 (2007).

[11] Martilli, *Bound-Layer Meteorol.*, 130, 301-306 (2009).

[12] Rasheed, D. Robinson, A. Clappier, C. Narayan, D. Lakehei, *Int. J. Climatol.,* 31, 289-301 (2011).

[13] K.N. Liou, *An Introduction to Atmospheric Radiation;* 2nd Edition, Academic Press, Inc.: New York, US, 2002; p. 583.

[14] T.R. Oke, *Boundary Layer Climates;* 2nd Edition, Methuen, MA, 1987; p. 435.

[15] M.C. Sanchez, *MSc Thesis*, Mechanical Engeneering Department, Virginia Polytechnic Institute and State University: Blackburg, VA, 1998.

[16] W.H. McCrea, *Analytical Geometry of Three Dimensions;* 2nd Edition, Oliver and Boyd Ltd: Edinburgh and London, GB, 1960; p. 160.

[17] P. Bourke (1999). Intersection of a Plane and a Line [http://local.wasp.uwa.edu.au/~pbourke/geometry/planeline/].

[18] D.T. Mihailović, S.T. Rao, C. Hogrefe, R. Clark, *Enivron. Fluid Mech.*, 2, 315-337 (2002).

[19] J. Lazić, D.T. Mihailović, B. Lalić, I. Arsenić, C. Hogrefe, *Proceedings of iEMSs*, Lugano, CH, 2, 448-453 (2002).

In: Essays on Fundamental and Applied Environmental Topics ISBN 978-1-61942-522-4
Editor: Dragutin Mihailovic

Chapter 5

LIMITATIONS AND UNCERTAINTIES IN THE LOGARITHMIC WIND PROFILE ABOVE VERY ROUGH SURFACES

***B. Lalić*[1,*], *D. T. Mihailović*[1] *and D. Kapor*[2]**
[1]Department for Field and Vegetable Crops, Faculty of Agriculture, University of Novi Sad, Novi Sad, Serbia
[2]Department of Physics, Faculty of Sciences, University of Novi Sad, Novi Sad, Serbia

ABSTRACT

This chapter discusses the limited applicability of the logarithmic wind profile above very rough forest and urban surfaces. The impact of roughness element size on the deviation of the vertical wind profile from logarithmic was analyzed using micrometeorological measurements. This chapter also addresses the consequences of (i) different numerical values of von Karman's constant and (ii) arbitrary determinations of wind profile parameters, i.e., roughness length and zero plane displacement height, for theoretical and practical questions in boundary layer meteorology. Results obtained for the usual range of wind profile parameters indicate that the relative deviation of wind profile values from the chosen referent profile above very rough surfaces can vary from 8 to 18 % when $k = 0.40$ and reaches a maximum of 49.1 % when $k = 0.34$.

1. INTRODUCTION

Wind velocity profiles result from complex atmospheric forcings caused by solar radiation as well as ground heating and roughness. The governing equations and derived expressions describing how wind velocity changes with height have been the subject of many debates during the last 50 years. In these discussions, neither the logarithmic nor the power law has proven definitively superior [1]. However, some authors believe the logarithmic law

* Corresponding author: E-mail address: branka@*polj.uns.ac.rs;* Phone: +381 21 485 3204. Fax: +381 21 6350 552.

is better suited to the surface layer while the power law is better suited to the rest of the atmospheric boundary layer [2]. Despite the debates, the logarithmic wind profile has become one of the keystones of micrometeorology. As pointed out by Tennekes [3], "*... a great deal of folklore on its accuracy and applicability has become entrenched ... since it was first derived*". Many decades of field experiments (for example, HAPEX–MOBILHY [4, 5], Kansas [6], and ABRACOS [7]) and the significant amount of scientific energy invested in the improvement of boundary layer theory and modeling allows us to make fairly specific comments about the limits and uncertainties that follow the application of the logarithmic wind profile.

The logarithmic wind profile is commonly used to calculate the surface stress, τ and the friction velocity, u_*. This means that if the wind velocity is known at the referent level within the surface layer, when u_* is calculated from the logarithmic wind profile, one can calculate geostrophic wind in the whole atmospheric boundary layer [3]. In numerical weather prediction (NWP) models, u_* and the roughness length, z_0, are the most important variables, related to momentum and energy transfer between Earth's surface and atmosphere, which can be assimilated from a soil-vegetation-atmosphere transfer (SVAT) scheme as a part of the model.

However, the detailed analysis of wind profile measurements reveals that there exists a significant deviation between observed and calculated wind profiles over natural surfaces. For smooth surfaces, the difference is caused by different values of von Karman's constant, k, which is well documented in [8 - 10] and [11], among others. Above very rough surfaces (forests and urban areas), however, the difference is related to (i) the applicability of the logarithmic wind profile in the transition layer [12 - 18] and (ii) the variation of k in the rest of the surface layer [10, 19].

Bearing in mind both the importance and possible drawbacks of the logarithmic wind profile, the main purpose of this chapter is to focus on the applicability of the logarithmic wind profile and uncertainties in the determination of wind profile parameters and k. Section 2 is devoted to experimental evidence of the limitation of the applicability of the logarithmic wind profile above very rough surfaces and, relatedly, constant-flux approximation in the surface layer. In Section 3 we calculate the relative deviation of normalized wind velocity $(u(z)/u_*)$ for different values of the logarithmic wind profile parameters and for a range of k-constant values encountered in the literature. This deviation was analyzed for both smooth and very rough surfaces to assess the potential impact of uncertainties in the logarithmic wind profile parameters and k on the calculated wind profile within the surface layer. Concluding remarks are given in Section 4.

2. Limitations of the Logarithmic Wind Profile

In a stationary, horizontally homogeneous, adiabatic and barotropic atmospheric boundary layer, the equations of motion can be written in the form [9, 20]

$$-f(v-v_g)=\frac{d}{dz}(-\overline{uw}) \tag{1}$$

$$+f(u-u_g)=\frac{d}{dz}(-\overline{vw}) \tag{2}$$

where u and v are the components of the wind in the atmospheric boundary layer, z is the vertical coordinate, u_g and v_g are the components of the geostrophic wind, f is the Coriolis parameter and $\overline{uw}$ and $\overline{vw}$ are the corresponding components of the Reynolds stresses. In order to characterize the importance of Coriolis acceleration on atmospheric flows, the non-dimensional Rossby number, R_0, which includes the magnitude of the geostrophic wind, V_g, is often used ($R_0 = V_g/(fz_0)$) [20]. A small Rossby number signifies a system that is strongly affected by Coriolis force, while a large Rossby number signifies a system in which inertial and centrifugal forces dominate.

According to asymptotic similarity theory [20], in the limit as the Rossby number R_0 tends to infinity ($R_0 \to \infty$) Eqs. (1) and (2) admit two kinds of solutions. The first one, for finite zf/u_* and $z/z_0 \to \infty$ (well outside the surface layer), can be written in the form

$$\frac{u-u_g}{u_*}=F_x\left(\frac{zf}{u_*}\right) \tag{3}$$

$$\frac{v-v_g}{u_*}=F_y\left(\frac{zf}{u_*}\right). \tag{4}$$

The second solution, when z/z_0 remains finite while $zf/u_* \to 0$ (within the surface layer), has the form

$$\frac{u}{u_*}=F_s\left(\frac{z}{z_0}\right) \tag{5}$$

$$\frac{v}{u_*}=0. \tag{6}$$

According to continuity conditions and for large R_0 there is a region where both equations and all their derivatives are valid. According to Blackadar and Tennekes [20], that layer is called the matched layer or inertial sublayer, where $z/z_0 \to \infty$ and $zf/u_* \to 0$. Within the inertial sublayer, the continuity conditions of Eqs. (3) - (6), following Tennekes [3], can be satisfied only if $u(z)$ is written in the form

$$\frac{u(z)-u_g}{u_*}=\frac{1}{k}\ln\left(\frac{zf}{u_*}\right)+\frac{B}{k} \tag{7}$$

$$\frac{u(z)}{u_*} = \frac{1}{k} \ln \frac{z}{z_0} \tag{8}$$

where B is an empirically determined constant.

In vegetated or urban areas, it is suitable to set the z coordinate to the zero plane displacement height, d, just below the canopy/building average height, H. Then Eq. (8) becomes the well-known logarithmic wind profile above rough surfaces, i.e.,

$$\frac{u(z)}{u_*} = \frac{1}{k} \ln \frac{z-d}{z_0}. \tag{9}$$

Even if these expressions are treated only as a first approximation, in the asymptotic case when $R_0 \to \infty$, as Monin et al. [21] note, Eq. (8) is valid only if z/z_0 is large enough. But what is "large enough"? We try to answer this question using wind profiles measured above very rough urban and forest surfaces [12, 13, 22 - 24].

Another interesting feature that follows from Eqs. (1) - (4) is related to the constant-stress layer assumption. Replacing Eq. (3) - (6) into Eqs. (1) - (2), the following relation for the Reynolds stress is obtained:

$$\frac{d}{dz}\left(-\overline{uw}\right) = \frac{v_g}{f}. \tag{10}$$

This relation leads us to conclude that the constant-stress layer assumption ($\overline{uw}$= const.) is only one possible solution of this equation in cases when $v_g = 0$. This conclusion is supported by many of the aforementioned experiments. Below, we analyze in detail the measurements of the Reynolds stress above very rough urban surfaces described by Mulhearn [13] and Kastner-Klein and Rotach [24].

Turbulence above cities. Due to difficulties associated with the measurement of turbulent fluxes and wind profiles above urban areas, wind tunnel experiments are commonly used to improve the understanding of the flow above very rough urban surfaces. In one of these experiments, using a wind tunnel in the Commonwealth Scientific and Industrial Research Organisation (CSIRO) (Australia), it was noted that close to the rough surface, there is a region in which the mean velocity and shear stress vary horizontally as well as vertically [13]. The depth of this region for the wind velocity variation is of the same order of magnitude as the distance between roughness elements, while variations in shear stress observed were over doubled depth. In the neutral atmospheric boundary layer wind tunnel at the University of Karlsruhe (Germany), Kastner-Klein and Rotach [24] noticed that the influence of building pattern irregularities on the mean flow and shear stress distributions range up to $2.5\,H$ (Figure 1). Additionally, above $3.5\,H$ the impact of building distribution decreases but the variation of Reynolds stress with height is still significant (Figure 1). Obviously, above very rough urban surfaces, $3.5\,H$ is not far enough from the surface to allow applicability of the constant-stress layer assumption. Consequently, the logarithmic wind profile at this height is still not valid; i.e., z/z_0 is not large enough.

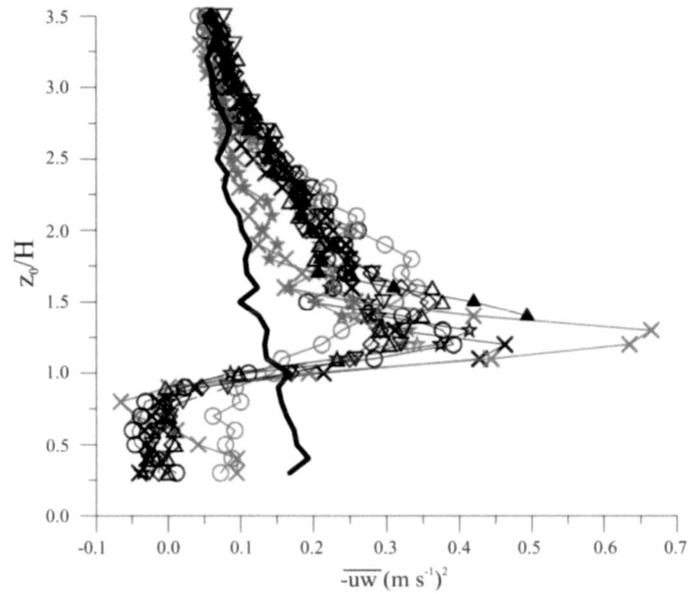

Figure 1. Mean shear stress profiles measured at the 11 sampling locations above and within an urban roughness layer [24].

Turbulence above forests. Many experiments have attempted to measure the micrometeorological characteristics of the forest canopy and the atmosphere above it [7, 12, 22, 23, 25], among others). To assess the domain of applicability of the logarithmic wind profile above very rough forest surfaces [26], we analyzed five wind profiles measured above pine (three profiles), maple (1 profile) and oak (1 profile) forests under neutral conditions (Table 1). For the purpose of further analysis, Eq. (9) will be written in the form

$$u(z) = \frac{u_*}{k} \ln \frac{z-d}{C_2} + \frac{u_*}{k} \ln \frac{C_2}{z_0}. \quad (11)$$

where C_2 is a constant, typically 1 m. If the logarithmic wind profile accurately reproduces the variation of wind velocity with height above the forest canopy, then measured values of wind velocity should be distributed along a straight line in a $\{\ln[(z-d)/C_2], u(z)\}$ coordinate system, where u_*/k is the slope and $u_*/k \ln(C_2/z_0)$ is the intercept.

Table 1. List of forest canopies where measurements were conducted. *LAI* denotes the leaf area index

Forest type	*H* (m)	*LAI* (m^2 m^{-2})	Literature
Pine	21.3	2.3	[12]
Maple	18.0	5.0	[22]
Oak	23.0	4.9	[23]

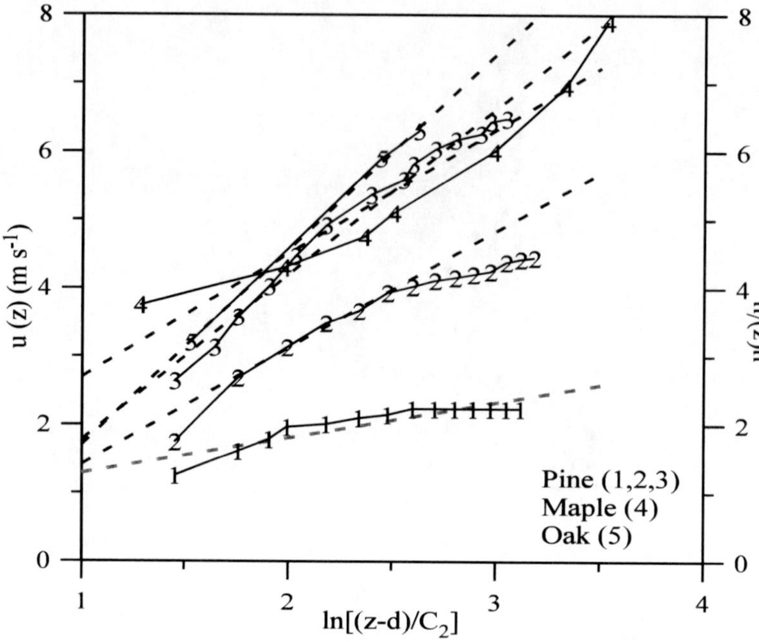

Figure 2. Observed and normalized wind profiles above different forest canopies (numbers) and corresponding best linear fits (dashed line).

To assess deviation of the observed wind profiles from logarithmic, we used the wind profiles measured above: (1) pine forest in a $\{\ln[(z-d)/C_2], u(z)\}$ coordinate system, and (2) maple and oak forest canopies in a $\{\ln[(z-d)/C_2], u(z)/u_*\}$ system. From these data we calculated the best linear fit for all forest canopies. The observed wind profiles and corresponding best fits are depicted in Figure 2, which shows that the linear relationship between and $u(z)$, i.e. is retained only when $\ln[(z-d)/C_2] > 2$, i.e., for heights above (e = 2.71828). Therefore, in the sense of asymptotic similarity theory, at heights above $C_2 e^2 + d$, the ratio is large enough for the logarithmic wind profile to be applied.

3. Sources of Uncertainty in Determination of the Logarithmic Wind Profile

Von Karman's constant. k was introduced more than 70 years ago to scale the logarithmic wind profile within the atmospheric surface layer. It has been assumed to be constant under neutral conditions [27]. Considerable time and effort have been spent on theoretical reasoning and numerous field and laboratory experiments to determine the value of k when the logarithmic shape is used to represent mean wind profiles in the surface layer. Table 2 indicates the range of empirically determined values of k. According to information reported by Frenzen and Vogel [11], the sources of these differences can be summarized as follows:

(i) Instrumental error: Kondo and Sato [19] note that after correcting for cup anemometer overspeeding the value (k = 0.42) obtained by Pruitt et al. [28] becomes 0.39, while the value (k = 0.41) reported by Hicks [29] becomes 0.40.
(ii) Violation of measurement conditions: in determining k, values of the wind velocity measured at heights $z < 100\,z_0$ should be omitted because the definition of k assumes that the stress at the surface is generated only by skin friction. Therefore, calculations based on wind velocity values measured very close to a rough surface tend to overestimate k [45].
(iii) Stability effect: measurements performed in weak stable stratifications can lead to incorrect estimates of k, because the flow may become decoupled from the surface.

Table 2 lists a spectrum of values for determined by a variety of calculations. However, Hinze [46], considering evidence from different experiments, notes that this constant varies according to environmental conditions. He used measurements of flow in roughened pipes [47] to calculate k over a wide range of Reynolds numbers (Re), finding that decreases asymptotically to 0.36 as increases to 10^6. Tennekes [8] fitted his second-order pipe-flow theory to the calculated values and found a minimum closer to $k = 0.34$. Current experimental estimates of range between 0.34 and 0.41. The effects of this uncertainty are not negligible in its use in environmental models. One reason is that many atmospheric boundary layer relations used, for example, in NWP models, involve the square of k. Another is its impact on the Monin-Obukhov stability function, φ_m. According to Frenzen and Vogel [48], the changes in are twice as large as those in over the range where Monin-Obukhov similarity is applicable. A recent review of values is given by Andreas et al. [49]. Their calculations, based on two large atmospheric data sets, suggest that k lies in the range 0.387 ± 0.003 for a broad range of (2 ≤≤ 100).

Table 2. Values of k obtained in field and laboratory conditions

Value	Surface/conditions	Reference/Source
0.40 - 0.41	Measured in the laboratory	[30]
0.40	~	[31]
0.408 ± 0.004	~	[32]
0.46	Measured in the field	[33]
0.37	~	[34]
0.35	~	[10]
0.434 ± 0.013	~	[35]
0.41	~	[36]
0.41	~	[37]
0.38	~	[38]
0.385 ± 0.021	~	[39]
0.39 ± 0.027	~	[19]
0.40 ± 0.01	~	[40 - 42]
0.375	~	[43]
0.364 ± 0.015	~	[44]

Table 3. Roughness length values for different surface types [50]

Surface	z_0-parameter (m)
Sea, sand, snow	0.0002
Concrete, desert, flat tides	0.0002 - 0.0005
Flat snow field	0.0001 - 0.0007
Rough ice field	0.001 - 0.012
Fallow ground	0.001 - 0.004
Short grass	0.008 - 0.03
Long grass, heather	0.02 - 0.06
Low crops	0.04 - 0.09
High crops	0.12 - 0.18
Pine forest	0.8 - 1.6
Town	0.7 - 1.5

Despite the experimental evidence, however, the most common value of k used in NWP models is 0.40. Therefore, sensitivity tests of the logarithmic wind profile to variation of k are important. We discuss these tests in Section 3.

Roughness length. Petersen et al. [51] defined z_0 as the aerodynamic characteristic representing the collective effect of the terrain surface and its roughness elements that leads to an overall retardation of the wind near the ground. The literature contains a variety of parameterizations for z_0 with varying degrees of complexity [27, 52, 53].

The most common parameterizations are (i) the constant value approach [50], (ii) an approach based on the dependence of z_0 on characteristics of the roughness elements (H, *LAI*) [27, 53], (iii) the drag partitioning approach, based on a derivation of z_0 from drag characteristics of the rough surface [52] and (iv) aggregation, based on a derivation of z_0 over a heterogeneous surface [54, 55] and commonly used in urban modeling. The first two parameterizations are often used in NWP models, and we consider them in detail. The first approach is based on predefined values of z_0 for each type of rough surface, as listed in Table 3.

The second approach is based on the assumption that a normalized value of z_0 (z_0 / H) is a constant or depends on the distribution of roughness elements (*LAI*, for example). The relationship z_0 / H can be determined experimentally for different rough surfaces using measured wind profiles.

To illustrate this, we calculated the relationship z_0 / H for the very rough forest surfaces given in Table 1 using measured wind profiles (Figure 2) and the best linear fit (Table 4). One simple but often used parameterization of z_0, incorporated in the Simplified Biosphere model (SiB) to follow growth of vegetation [56, 53], has the form

$$\frac{z_0}{H} = 1 - b_z \exp(-h_z LAI) \quad \text{where } b_z \text{ and } h_z \text{ are empirical constants.} \tag{12}$$

Table 4. The slope and intercept of the best linear fit and calculated normalized roughness length z_0 / H [26]

Profile	u_* / k	$u_* / k \ln(C_2 / z_0)$	z_0 / H
Pine 1	0.519	0.769	0.01
Pine 2	1.400	0.200	0.04
Pine 3	2.426	- 0.646	0.06

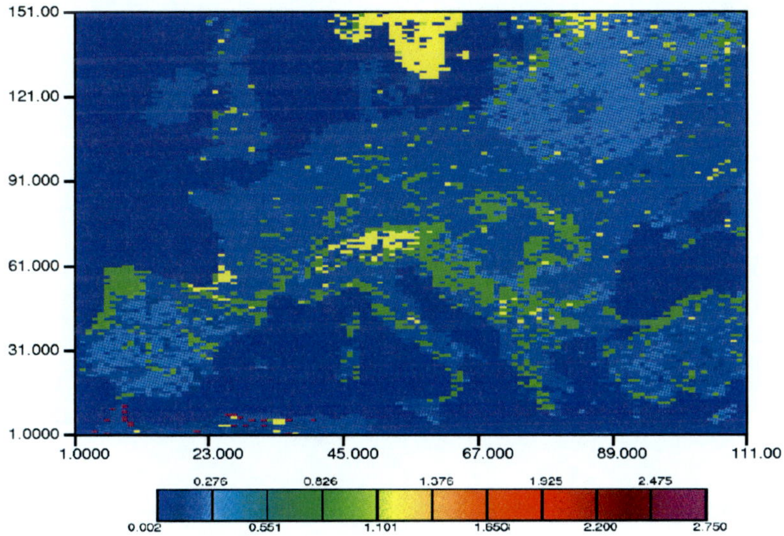

Figure 3. Roughness length distribution in WRF NMM model [57, 58].

Many relations in the surface layer module of NWP models make use of z_0 (Figure 3). The improper parameterization of z_0 appears to be a source of serious errors, for example, in determining stress from wind profiles and parameterization of resistances in surface flux calculations. Mihailovic and Ruml [59] considered the impact of the variation of z_0 in cases of smooth (bare soil) and rough (maize canopy) surfaces. Their results indicate significant impact of z_0 variation on the root-mean-square-error (RMSE) of latent heat flux (66 - 83 W m^{-2}) and ground temperature (3.7 - 2.8 °C) in the smooth case (Figure 4, upper panels), where RMSE was calculated on the basis of measured and modeled values. For vegetated surfaces, variation of maize roughness length, z_c, has no significant impact on the RMSE of the latent heat and foliage temperature.

Displacement height. d denotes the location above the ground where turbulent exchanges begin to occur [60]. A frequently used value for d in environmental models for prevailing natural surfaces is $2/3\,H$ [27]. For particularly dense distributions of roughness elements (e.g., some forest canopies) $d = H$ is often used. As d is subtracted from a function argument and z_0 appears in the denominator, the impact of this parameter on uncertainty in the logarithmic wind profile will be analyzed in the next part of this chapter, together with the effect of variation in z_0, through sensitivity tests.

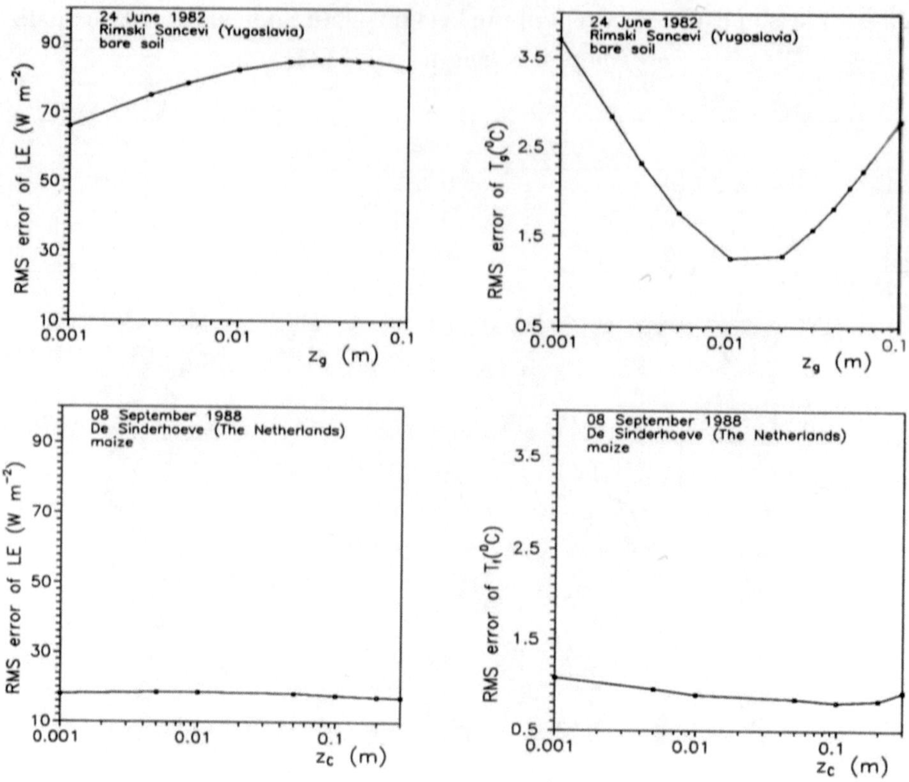

Figure 4. RMSE of sensitivity test outputs as a function of bare soil roughness length, z_g , and maize roughness length, z_c . Latent heat flux outputs, LE , are displayed on the left panels; RMSE of the temperature outputs for ground, T_g , and foliage, T_f , are located on the right panels [59].

4. Sensitivity Tests

To identify quantitatively the sensitivity of the logarithmic wind profile to variation of wind profile parameters and *k*, for smooth and very rough surfaces, the relative deviation of the normalized wind profile ($u(z)/u_*$), δ_{u/u_*} was calculated using the relation

$$\delta_{u/u_*} = \frac{\left| \dfrac{u(z)}{u_*} - \dfrac{u_{ref}(z)}{u_*} \right|}{\dfrac{u_{ref}(z)}{u_*}} 100\ \% \quad (13)$$

where u_{ref} denotes the referent value of the logarithmic wind profile calculated for common values of the profile parameters and *k*. The eight initial estimates of *k* in the range between 0.34 and 0.41 [11] are used for both smooth and very rough surfaces.

For the smooth surface, z_0 was varied in the range between 0.001 m and 0.01 m [50]. The referent profile was calculated using 0.40 and 0.001 m for *k* and z_0, respectively. Figure 5 shows that in the case of the smooth surface, varying z_0 has the greatest impact on δ_{u/u_*}. Between 1 m and 200 m from the ground, the influence of z_0 produces a decrease of δ_{u/u_*} of 15 %. For k = 0.40 and z_0 = 0.01 m, δ_{u/u_*} varies from 34.95 % (1 m above the ground) to 20.84 % (200 m above the ground).

Above very rough surfaces the value of z_0, as the height at which viscosity is replaced by turbulence, is determined by the height of the roughness elements. Thus, to estimate the impact of z_0 on the variation of δ_{u/u_*}, the normalized value (z_0/H) was taken into account as a variable in the range from 0.04 to 0.08 [53]. Accordingly, normalized values of d (d/H) are set, for each calculation of the profile, between 0.6 [27] and 1.0 [60]. The referent profile was calculated by letting k = 0.40, z_0/H = 0.06 and d/H = 0.8. Figure 6 shows that the most pronounced deviation (49.1 %) was obtained for values of parameters smaller than reference: k = 0.34, d/H = 0.60 and z_0/H = 0.04.

Table 5 shows that relatively small deviations in z_0 (± 0.02 compared to the reference value), with k and d held at referent values, can produce δ_{u/u_*} from 16.50 to 11.71 %. Finally, δ_{u/u_*} produced by varying k from 0.34 to 0.37 (with z_0 and d at referent values) is in the range from 17.65 to 8.11 %. The results in Table 5 refer only to δ_{u/u_*} produced by the variation of one parameter at a time.

Table 5. Maximum relative deviation δ_{u/u_*} (%) of the logarithmic wind profile above the very rough surface for selected values of d/H, z_0/H, and k

k	d/H	z_0/H	δ_{u/u_*} (%)
0.40	0.8	0.04	16.50
0.40	0.8	0.08	11.71
0.34	0.8	0.06	17.65
0.37	0.8	0.06	8.11

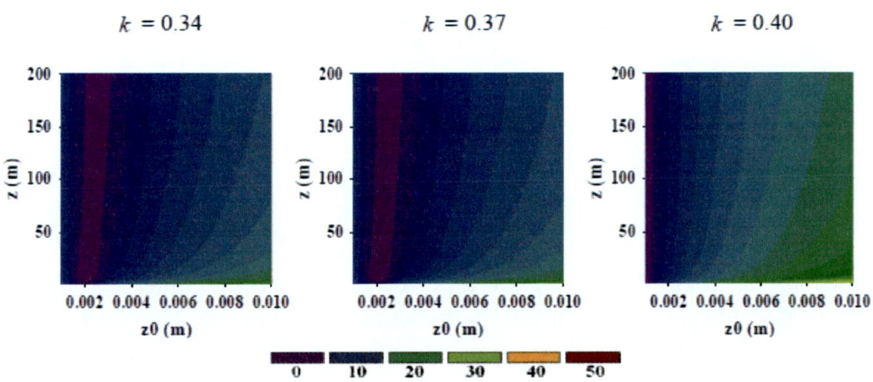

Figure 5. Relative deviation δ_{u/u_*} (%) of the logarithmic wind profile above the smooth surface.

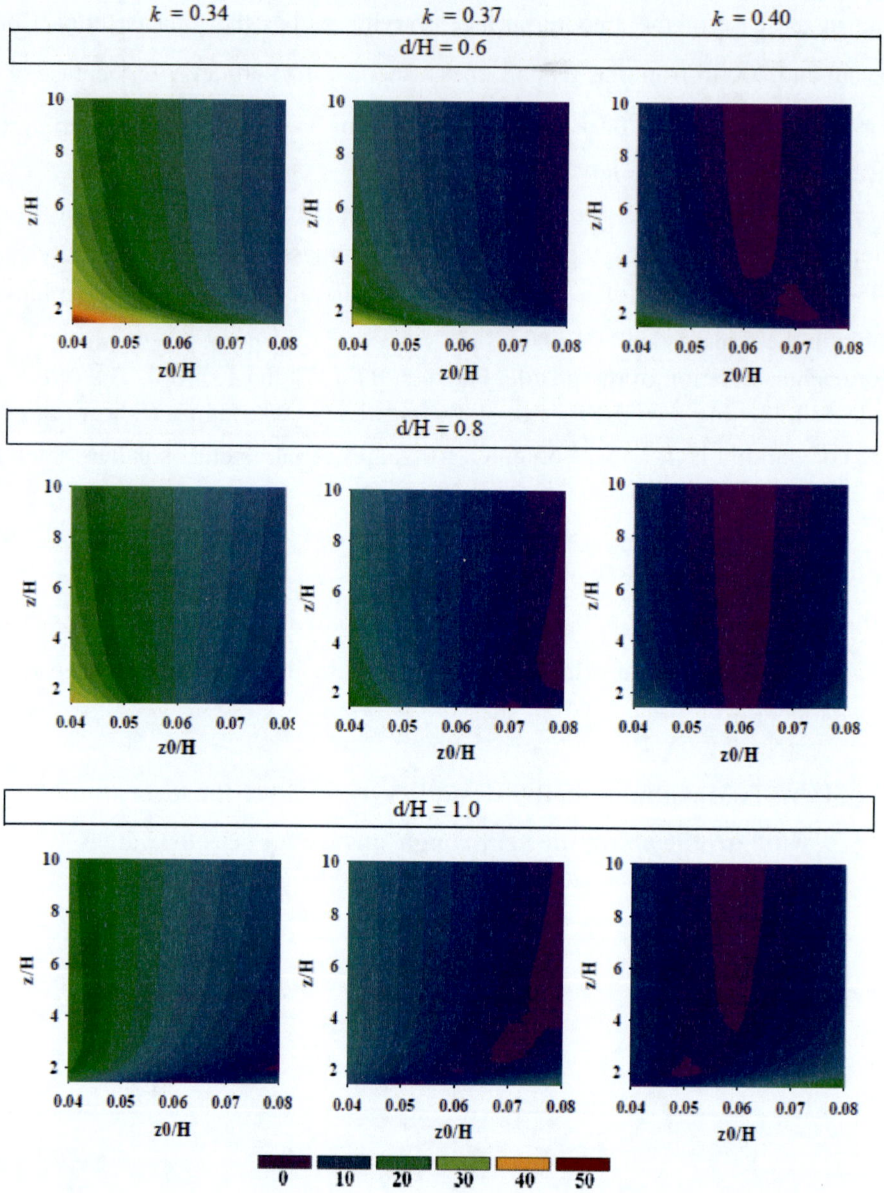

Figure 6. Relative deviation δ_{u/u_*} (%) of the logarithmic wind profile above a very rough surface.

Conclusions

The applicability of the logarithmic wind profile is an important question in surface layer meteorology. Many authors agree that within the surface layer but far enough from roughness elements ($z >> z_0$), the logarithmic wind profile reproduces wind velocity variation with height correctly. The main question is, "how far is enough?". The results of many experiments suggest that the answer depends mainly on the structure, density and dimensions of the rough underlying surface. For example, in the case of a very rough urban surface, at heights above 3.5 H the logarithmic wind profile and Reynolds's stress for momentum still differ significantly from observations. In contrast, for some rough forest surfaces, heights approximately equal to 2 H are "far enough" from roughness elements.

Many experiments have shown that the well-known shear-flow scaling factor k is not a universal constant. Additionally, for many practical applications the values of z_0 and d are often arbitrary. Based on the literature reviewed here and our numerical tests we conclude that even relatively small variations of profile parameters and k produce significant changes in the surface layer wind profile.

The consequences of improper application of the logarithmic wind profile and the aforementioned uncertainties in calculating u_* from the wind profile are evident. Following the results obtained, computed values of u_* may differ from its "true" value by 8 - 18 %, even if $k = 0.40$. In the worst case, this deviation may reach 49.1 %. Because in NWP models, k is pre-defined, while z_0 and d are either constant or simply empirically related, these limitations and uncertainties can cause errors in calculating many quantities in these models.

ACKNOWLEDGEMENTS

This chapter was produced as a part of the project "Studying climate change and its influence on the environment: impacts, adaptation and mitigation" (43007) financed by the Ministry of Education and Science of the Republic of Serbia within the framework of integrated and interdisciplinary for research for the period 2011 - 2014.

REFERENCES

[1] M.H. Buschmann, M. Gad-el-hak, *AIAA.*, 41(4), 565-572 (2003).
[2] T. Ishihara, K. Hibi, S. Oikawa, *J. Wind Eng. Ind. Aerod.*, 83, 95-107 (1999).
[3] H. Tennekes, *J. Atmos. Sci.*, 30, 234-238 (1973).
[4] J.F. Mahfouf, *Bound.-Layer Meteor.*, 53, 201-222 (1990).
[5] J.C. Andre, J.P. Goutorbe, A. Perrier, *Bull. Amer. Meteor. Soc.*, 67, 138-144 (1986).
[6] Y. Izumi, *Kansas 1968 Field Program Data Report*; Air Force Geophysics Laboratory, Bedford Massachusetts, USA, 1971; p. 79.
[7] O.M.R. Cabral, A.L.C. McWilliam, J.M. Roberts, *Amazonian Rainforest and Climate*; John Wiley: Chichester, GB, 1996; p. 638.
[8] H. Tennekes, *AIAAJ*, 6, 1735-1740 (1968).
[9] H. Tennekes, J.L. Lumley, *A First Course in Turbulence*; MIT Press: Cambridge, USA, 1972; p. 300.
[10] J.A. Businger, J.C. Wyngaard, Y. Izumi, E.F. Bradley, *J. Atmos. Sci.*, 28, 181-189 (1971).
[11] P. Frenzen, C.A. Vogel, *Bound.-Layer Meteor.*, 72, 371-392 (1994).
[12] W.L. Fons, *J. Forest.*, 38, 481-487 (1940).
[13] P.J. Mulhearn, *6th Australasian Hydraulics and Fluids Mechanics Conference*, 269-272 (1977).

[14] J.R. Garratt, *Quart. J. Roy. Meteor. Soc.,* 104, 199-211 (1978).
[15] H.A.R. De Bruin, C.J. Moore, *Bound.-Layer Meteor.*, 42, 53-62 (1985).
[16] D.T. Mihailovic, B. Lalic, B. Rajkovic, I.D. Arsenic, *Bound.-Layer Meteor.*, 93, 425-451 (1999).
[17] B. Lalić, D.T. Mihailovic, B. Rajkovic, I.D. Arsenic, D. Radlovic, *Environ. Modell. Softw.*, 18 (10), 943-950 (2003).
[18] B. Lalic, D.T. Mihailovic, In *Fluid Mechanics of Environmental Interfaces*; C. Gualtieri and D.T. Mihailovic, Eds.; Taylor and Francis: London, GB, 2008, pp. 221-240.
[19] J. Kondo, T. Sato, *JMSJA*, 60, 461-470 (1982).
[20] A.K. Blackadar, H. Tennekes, *J. Atmos. Sci.*, 25, 1015 – 1025 (1968).
[21] A.S. Monin, A.M. Yaglom, *Statistical Fluid Mechanics*; MIT Press: Cambridge, USA, Vol. I: Mechanics of Turbulence, 1971, p. 766.
[22] H. H. Neumann, G. Den Hartog, R.H. Shaw, *Agr. Forest Meteorol.*, 45, 325-345 (1989).
[23] T.P. Meyers, D.D. Baldocchi, *Agr. Forest Meteorol.,* 53, 207-222 (1991).
[24] P. Kastner-Klein, M.W. Rotach, *Proceedings of the 3rd International Conference on Urban Air Quality*, Tempe, Arizona, USA (2001).
[25] H.R. Oliver, *Quart. J. Roy. Meteor. Soc.*, 97, 548-553 (1971).
[26] B. Lalić, PhD Thesis: Novi Sad, RS, 2006 (in Serbian) p. 139.
[27] J. R. Garratt, *The Atmospheric Boundary Layer*; Cambridge University Press: Cambridge, GB, 1992, p. 316.
[28] W. O. Pruitt, D. L. Morgan, E J. Lourence, *Quart. J. Roy. Meteor. Soc.*, 97, 370-386 (1973).
[29] B. B. Hicks, *Bound.–Layer Meteor.*, 10, 237-240 (1976).
[30] J. O. Hinze, *Turbulence*; McGraw Hill: New York, USA, 1975, p. 790.
[31] R. Chevray, J. Methieu, *Topics in Fluid Mechanics*; Cambridge University Press: Cambridge, GB, 1993, p. 320.
[32] C. Long, P.L. Wiberg, A.R. Nowell, *J. Hydr. Eng.*, 119, 1182-1190 (1993).
[33] P.A. Sheppard, *Proc. Roy. Soc.*, A 188, 208-222 (1947).
[34] F. Pasquill, *Proc. Roy. Soc.*, A 202, 143-153, (1950).
[35] S.S. Zilitinkevich, D.V. Chalikov, *Meteorologija i Gidrologija*, **2**, 8-25 (1968).
[36] W.B. Goddard, *J. Appl. Meteor.*, 9, 373-378 (1970).
[37] Dyer, *An Introduction to Liquid Scintilation Counting*; Heyden: New York, USA, 1974, p. 111.
[38] R.J. Francey, J.R. Garratt, *J. Appl. Meteor.*, 20, 603-618 (1981).
[39] A.J. Dyer, E.F. Bradley, *Bound.-Layer Meteor.,* 22, 3-19 (1982).
[40] U. Hogstrom, *J. Atmos. Sci.*, 42, 263-270 (1985).
[41] U. Hogstrom, *J. Atmos. Sci.*, 43, 2131-2134 (1986).
[42] U. Hogstrom, *Bound.-Layer Meteor.*, 42, 55-78 (1988).
[43] N.G. Mortensen, S.E. Larsen, L. Troen, T. Mikkelsen, *Proceedings of the 6th Symposium on Meteorology and Instrumentation*, 393-396 (1987).
[44] S.P. Oncley, J.A. Businger, C.A. Friehe, J.C. LaRue, E.C. Itsweire, S.S. Chang, *Proceedings of 9th Symposium on Turbulence and Diffusion*, Roskilde, DK, 237-240 (1990).
[45] J. R. Garratt, *Quart. J. Roy. Meteor. Soc.*, 106, 803-819 (1980).
[46] J.O. Hinze, *Mecanique de Turbulence*, 108, 129-165 (1962).

[47] J. Nikuradze, *Forschungsheft Verein Deutscher Ingenieure*, 356 (1932).
[48] P. Frenzen, C.A. Vogel, *Bound.–Layer Meteor.*, 68, 439-442 (1995).
[49] E. Andreas, K.J. Claffey, R.E. Jordan, C.W. Fairall, P.S. Guest, P.O. Persson, A.A. Grachev, *J. Fluid Mech.*, 559, 117-149 (2006).
[50] J. Wieringa, *Bound.–Layer Meteor*, 63, 323-364 (1993).
[51] E.L. Petersen, N.G. Mortensen, L. Landberg, J. Hojstrup, H.P. Frank, *Wind Energy,* 1, 55-72 (1998).
[52] Y. Shao, Y. Yang, *Atmos. Environ.*, 39, 7351-7361 (2005).
[53] P.J. Sellers, S.O. Los, C.J. Tucker, C.O. Justice, A.D. Dazlich, G.J. Collatz, D.A. Randal, *J. Climate*, 9, 706-737 (1996).
[54] D.T. Mihailovic, S.T. Rao, C. Hogefre, R. Clark, *Environ. Fluid Mech.,* 2, 315-337 (2002).
[55] D.T. Mihailovic, J. Lazic, J. Leśny, J. Olejnik, B. Lalic, D.V. Kapor, A. Cirisan, *Theor. Appl. Climatol.*, 100 (3-4), 299-323 (2009).
[56] Y. Xue, P.J. Sellers, J.L. Kinter, J. Shukla, *J. Climate*, 4, 345-364 (1991).
[57] Z.I. Janjic, J.P., Jr Gerrity, S. Nickovic, *Mon. Wea. Rev.*, 129, 1164-1178 (2001).
[58] Z.I. Janjic, *Meteor. Atmos. Phys.*, 82, 271-285 (2003).
[59] D.T. Mihailović, M. Ruml, *Computer Technics in Environmental Studies*; Computational Mechanics Publications: Southampton, GB, Vol. II: Environmental Systems, 1994; p. 398.
[60] P. Stangroom, *PhD Thesis*: Notthingam, GB, 2004; p. 298.

In: Essays on Fundamental and Applied Environmental Topics ISBN 978-1-61942-522-4
Editor: Dragutin Mihailovic

Chapter 6

NON-LOCAL EDDY DIFFUSION AND VERTICAL MIXING SCHEMES FOR USE IN CHEMICAL TRANSPORT MODELS

D. T. Mihailović[1,*]***, Z. Podraščanin***[2] ***and I. Arsenić***[1]
[1]Department for Field and Vegetable Crops, Faculty of Agriculture, University of Novi Sad, Novi Sad, Serbia
[2]Department of Physics, Faculty of Sciences, University of Novi Sad, Novi Sad, Serbia

ABSTRACT

The atmospheric boundary layer (ABL) parameterization is one of the most important components of air quality and chemical transport models. The easiest and most commonly used method for the parameterization of vertical turbulent mixing in ABL is one that is analogous to molecular diffusion. These types of schemes are known as K-schemes, in which mixing occurs only between the closest model layers. This approach is suitable only when the scale of turbulent motion is much smaller than the scale of mean motion, i.e., during stable and neutral atmospheric conditions.

K-schemes cannot describe the effects of large scale eddies, which are dominant in the convective boundary layer (CBL) and also cannot simulate counter gradient flows up to the gradient. Therefore, K-schemes are not recommended for the CBL. To avoid the disadvantages of K-schemes, a "non-local" turbulent kinetic energy (TKE) scheme can be used. However, the TKE scheme is able to represent the scale of the turbulence under any stability condition but the mixing is still local. It can also only represent small scales of turbulence, which reduces its reliability. During the convective condition, much of the mixing is caused by bounty plumes that originate in the surface layer and penetrate the capping inversion. In those circumstances, the ABL is characterized by narrow, fast-moving updrafts and wide, slow-moving downdrafts. This type of motion is usually parameterized using an asymmetrical convective mixing scheme (ACM). None of the above-mentioned schemes is able to represent mixing at all possible scales. However, a combination of the K-scheme and the ACM scheme can represent all of the spectrums of

* Corresponding author: E-mail address: guto@uns.ac.rs; Phone: +381 21 485 3203. Fax: +381 21 635 0552.

turbulence scales and is the most suitable to represent CBL mixing. In this chapter, a non-local TKE scheme, an ACM scheme and a linear combination of those schemes are comprehensively considered. To determine the features of the considered schemes, we performed numerical tests using Unified EMEP (European Monitoring and Evaluation Programme) chemical model and 1-D vertical diffusion models. The obtained outputs were compared with (i) the measured concentrations of the NO_2 pollutant and (ii) the Large Eddy Simulation (LES).

1. Introduction

Chemical transport models (CTMs) simulate the stocks and flows of one or more species, while related general circulation models (GCMs) simulate overall atmospheric dynamics (e.g., fluid and heat flows). CTMs solve the continuity equation only for the species of interest, whereas GCMs solve primitive equations for the atmosphere. CTMs are expected to accurately represent the entire cycle for the species of interest including fluxes (e.g., advection), chemical production/loss, and deposition. Because computing costs have declined in recent years, there has been an increasing tendency for GCMs to incorporate CTMs for climate-relevant species. CTMs are classified according to their simulation methodology and their species of interest. Jacob [1] classifies CTMs as either Eulerian ("box") or Lagrangian ("puff") models. Eulerian models simulate "boxes" through which fluxes pass and in which chemical production/loss and deposition occur over time. Lagrangian models simulate the production and motion of air parcels ("puffs") over time. An Eulerian CTM solves continuity equations using a global/fixed frame of reference, whereas a Lagrangian CTM uses a local/moving frame of reference. CTMs typically focus on one chemical species, but to realistically model the dynamics and chemistry, the CTM may be forced to account for many related species such as precursors or tracers.

The treatment of the atmospheric boundary layer (ABL) and the surface processes has important implications for simulations within air quality and chemical transport models, particularly for the pollutants that are greatly affected by them. There are two approaches, local and non-local, for solving the turbulence closure problem. Whereas the local closure approach assumes that turbulence is analogous to molecular diffusion, in the non-local closure approach, an unknown quantity at one point is parameterized by values of known quantities at many points in 3-D space. The K-scheme is the simplest (and very popular) local closure method for Eulerian air quality and chemical transport models and is used in both the ABL and the free troposphere. Because it uses local gradients at one point of the model grid, the K-scheme can be used only when the scale of turbulent motion is much smaller than that of the mean flow [2], i.e., in the case of stable and neutral conditions in the atmosphere when this scheme is consistent. However, it cannot (i) describe the effects of large-scale eddies, which are dominant in the convective boundary layer (CBL), or (ii) simulate counter-gradient flows in which a turbulent flux flows up to the gradient. Therefore, the K-scheme is not recommended for the CBL [2]. To avoid the disadvantages of the K-scheme, Alapaty [3] and Alapaty and Alapaty [4] recently suggested a non-local turbulent kinetic energy (TKE) scheme based on a K-scheme that was intensively tested using the Unified EMEP (European Monitoring and Evaluation Programme) chemical model [5, 6].

To quantify the transport of a passive tracer field in three-dimensional simulations of turbulent convection, non-local and non-diffusive behavior can be described by a transilient matrix whose elements contain the fractional tracer concentrations moving from one subvolume to another as a function of time. This approach, known as the turbulent transilient theory (T3; [7 - 9]), was originally developed for and applied to geophysical flows; however, this formalism can be extended and applied in an astrophysical context to three-dimensional simulations of turbulent compressible convection with an overshoot into convectively stable bounding regions [10]. The most frequently used non-local closure method is asymmetrical convective mixing suggested by Pleim and Chang [11]. The design of this model is based on Blackadar's scheme [12] but takes into account an important feature, namely, that in the CBL, the vertical transport is asymmetrical [13]. The buoyant plums are fast and narrow, whereas the downward streams are wide and slow. Accordingly, transport by upward streams should be simulated as non-local, and transport by downward streams should be simulated as local. The concept of this model is that the buoyant plume rises from the surface layer and transfers air and its properties directly into all of the layers above. Downward mixing only occurs between adjacent layers in the form of a slow subsidence. The asymmetrical convective mixing scheme (ACM) can only be used under convective conditions in the atmospheric boundary layer, whereas, for stable or neutral regimes, the K-scheme is used. Although, with this approach, a more realistic simulation of vertical transport within the CBL layer is obtained, it has some drawbacks that can be elaborated in a condensed form as follows: (i) because this method mixes the same amount of mass into every vertical layer in the atmospheric boundary layer, it has the potential to remove mass much too quickly out of the surface layer, and (ii) the method also fails to account for upward mixing in layers higher than the surface layer [14]. Wang [15] compared three different vertical transport methods including a semi-implicit K-scheme with local closure, and the ACM and T3 schemes with non-local closures. Of the three schemes, the ACM scheme moved mass more rapidly out of the surface layer into the other layers than the two other schemes in terms of the rate at which mass was mixed between different layers. Recently, the scientific community has attempted to demonstrate a model equation set with an elegant conservation of mass, total energy and others. Based on this result, there is strong interest in applying a Nambu formulation of the available model equation sets. Salmon [16] suggested a general discretization to the Nambu bracket in which its antisymmetry is preserved; therefore, the conservation quantities are also automatically preserved in the numerical scheme. This conservation property is of great importance in any numerical model [17] and generates a new class of numerical schemes including chemical and air pollution models. Recently, Pleim [18] suggested a scheme that includes the non-local ACM scheme and the K-scheme as a linear combination with a weighting factor (ACM2) determining their individual contributions to mixing the CBL.

In this chapter, we describe non-local eddy diffusion and vertical mixing schemes using chemical transport models. In Section 2, we consider the TKE scheme and compare the measured values of NO_2 concentrations and those obtained from the Unified EMEP chemical model outputs. Section 3 is devoted to the description of the ACM schemes, and in Section 4, we consider the combined non-local vertical mixing and eddy diffusion schemes. The performance of this scheme was tested using (Large Eddy Simulation) LES simulations. Section 5 contains our concluding remarks.

2. TKE Eddy Diffusion Scheme

2.1. Description of the TKE Eddy Diffusion Scheme

This approach begins by considering the general form of the vertical eddy diffusivity equation. For momentum, this equation can be written as

$$K_m = \frac{\overline{e}_* kz(1-\frac{z}{h})^p}{\Phi_m} \tag{1}$$

where K_m is the turbulent transfer coefficient, $\overline{e}_*$ is the mean turbulent velocity scale within the ABL that is to be determined (closure problem), k is the von Karman constant ($k = 0.41$), z is the vertical coordinate, h is the ABL height, p is the profile shape exponent taken from the similarity theory [19] (usually taken as 2), and Φ_m is the non-dimensional function of momentum. According to [20], we use the square root of the vertically averaged TKE in the ABL as a velocity scale in place of the single velocity scale to closure Eq. (1). Instead of using a prognostic approach to determine TKE, we use a diagnostic method. It is then logical to consider the diagnostic TKE to be a function of both friction velocity, u_*, and convective velocity, w_*. The square root of the TKE near the surface then serves as a closure to this problem [4].

However, it is more suitable to estimate $\overline{e}_*$ from the profile of the TKE through the entire ABL. According to [20], a linear combination of the TKE dissipation rates associated with shear and buoyancy can adequately approximate the vertical distribution of the TKE, $e(z)$, in a variety of boundary layers ranging from near neutral to free convection conditions. Following [20], the TKE profile can be expressed as

$$e(z) = \frac{1}{2}\left(\frac{L_E}{h}\right)^{2/3}\left[0.4w_*^3 + u_*^3\left(h-z\right)\left(\frac{\Phi_m}{kz}\right)\right]^{2/3} \tag{2}$$

where L_E characterizes the integral length scale of the dissipation rate. Here, $\Phi_m = (1-15z/L)^{-1/4}$, where L is Monin-Obukhov length, is an empirical function for the unstable atmospheric surface layer [21], which is applied to both the surface and mixed layers. We use $L_E = 2.6h$, which is in the range of $2.5h - 3.0h$ as suggested by [22]. For the stable atmospheric boundary layer, we modeled the TKE profile using an empirical function proposed by [23] based on aircraft observations

$$\frac{e(z)}{u_*^2} = 6\left(1-\frac{z}{h}\right)^{1.75} . \tag{3}$$

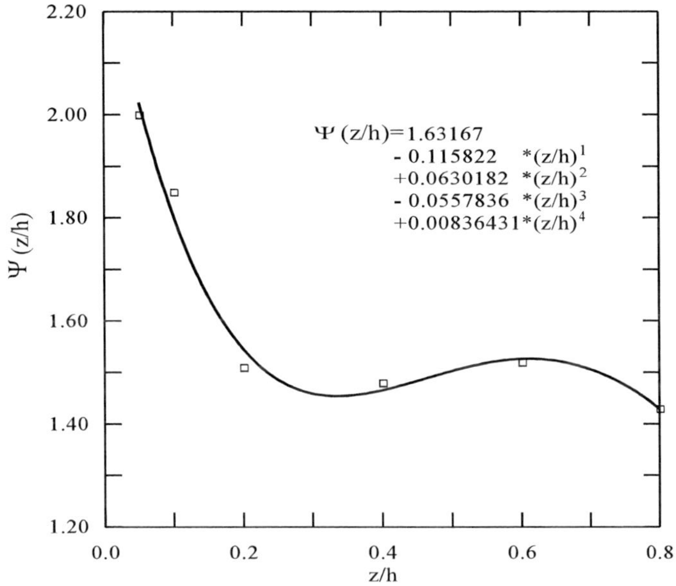

Figure 1. Dependence of the vertical profile function on z/h, which is used in the calculation of the mean turbulent velocity scale within the ABL in Eq. (4).

Based on works on the LES [20, 22], Alapaty [3] suggested a way to estimate the vertically integrated mean turbulent velocity scale $\overline{e}_*$, which, within the ABL, can be written as

$$\overline{e}_* = \frac{1}{h}\int_0^h \sqrt{e(z)}\psi(z)dz \tag{4}$$

where $\psi(z)$ is the vertical profile function for TKE as obtained by [20] based on LES studies, which was later modified by [24] (Figure 1), and dz is layer thickness.

The formulation of eddy-diffusivity by Eq. (1) depends on h. We follow [19] for the determination of h using

$$h = \frac{Ri_c\left\{u(h)^2 + v(h)^2\right\}}{\frac{g}{T_0}\{\theta_v(h) - \theta_s\}} \tag{5}$$

where Ri_c is a critical bulk Richardson number for the ABL, $u(h)$ and $v(h)$ are the horizontal velocity components at the ABL height h, g/T_0 is the buoyancy parameter, θ_s is the appropriate virtual potential temperature of air near the surface, and $\theta_v(h)$ is the virtual potential temperature at h. For unstable conditions $(L < 0)$, θ_s is given by [19], i.e.,

$$\theta_s = \theta_v(z_1) + C_0 \frac{\overline{w\theta_{v0}}}{w_s} \tag{6}$$

where $C_0 = 8.5$ [25], w_s is the velocity, and $\overline{w\theta_{v0}}$ is the kinematics surface heat flux. The velocity w_s is parameterized as

$$w_s = \left(u_*^3 + c_1 w_*^3\right)^{1/3} \tag{7}$$

and

$$w_* = ((g/\theta_0)\overline{w\theta_{v0}}h)^{1/3}, \tag{8}$$

using $c_1 = 0.6$. In Eq. (6), $\theta_v(z_1)$ is the virtual temperature at the first model level. The second term on the right-hand side of Eq. (6) represents a temperature excess, which is a measure in the lower part of the ABL. For stable conditions, we use $\theta_s = \theta_v(z_1)$ with $z_1 = 2$ m. Based on Eq. (5), the boundary layer height can be calculated by an iteration for all stability conditions when the surface fluxes and profiles of θ_v, u, and v are known. The computation begins by calculating the bulk Richardson number Ri between the level θ_s and the subsequent higher levels of the model. Once exceeds the critical value, the value of h is derived by linear interpolation between the level where $Ri > Ri_c$ and the level below. We use a minimum of 100 m for h. In Eq. (5), Ri_c is the value of the critical bulk Richardson number, which is 0.25 in this study.

In the free atmosphere, turbulent mixing is parameterized using the formulation suggested by [12] in which turbulent transfer coefficients are functions of the Richardson number and wind shear in the vertical. This formulation can be written as

$$K_m = K_0 + S(kl)^2 \frac{Rc - Ri}{Rc} \tag{9}$$

where K_0 is the background value (1 m^2 s^{-1}), S is the vertical wind shear [9], l is the characteristic turbulent length scale (100 m), Rc is the critical Richardson number, and Ri is the Richardson number that is defined as

$$Ri = \frac{g}{\theta_v S^2} \frac{\partial \theta_v}{\partial z}. \tag{10}$$

The critical Richardson number in Eq. (9) is determined as

$$Rc = 0.257(\Delta z)^{0.175} \tag{11}$$

where Δz is the layer thickness [26].

2.2. Parameterization of in-Canopy Resistance for Dry Deposition Calculation in Chemical Transport Models

Dry deposition, i.e., the transport of trace gases and particles to the earth's surface, is an important loss process for many reactive and soluble trace gases. The flux of a gas to the surface is commonly expressed as the product of a deposition velocity V_d and the concentration C of the gas in air

$$F = -V_d C \tag{12}$$

The deposition velocity V_d of a species i is calculated using an analogy to Ohm's law in electrical circuits as follows

$$V_d = \frac{1}{R_a + R_b^i + R_c^i} \tag{13}$$

where R_a is the aerodynamic resistance, which is the same for all gases, R_b^i is the quasi-laminar sublayer resistance, and R_c^i is the total surface resistance of the gas. The aerodynamic resistance in-canopy air space, R_{inc}, which is included in the parameterization of the surface resistance, can be written in the form

$$R_{inc} = \int_{z_g}^{h_c} \frac{1}{K_s} dz + \int_{h_c}^{h_a} \frac{1}{K_s} dz \tag{14}$$

where z_g is the effective ground roughness length, h_c is the canopy bottom height (the height of the base of the canopy), which is defined as given in [27], h_a is the canopy source height, which is parameterized according to [28], and K_s is the turbulent transfer coefficient inside the canopy. Using an expression for K_s as in [28], from (13), we obtain

$$R_{inc} = \frac{A_n}{u_*} \tag{15}$$

where A_n can be written in the form

$$A_n = \frac{H^2}{k\beta^2 h_c^2} \frac{e^{\beta\left(1-\frac{h_c}{H}\right)}}{\ln\frac{H-d}{z_0}} + \frac{H}{\beta\sigma_s} \frac{e^{\beta\left(1-\frac{h_c}{H}\right)} - e^{\beta\left(1-\frac{h_a}{H}\right)}}{\ln\frac{H-d}{z_0}} \tag{16}$$

where H is the canopy height, $\beta^2 = 2C_d\overline{L_d}(H-h_c)H/\sigma$, C_d is the leaf drag coefficient with a value of 0.2, $\overline{L_d}$ is the area-averaged stem and leaf area density, which is related to the leaf area index (LAI) as $LAI = \overline{L_d}(H-h_c)$, z_0 is the roughness length, d is the displacement height, and $\sigma = 2C_{dg}^2 H/[\overline{L_d}(H-h_c)]$, in which C_{dg} is the leaf drag coefficient that is estimated from the size of the roughness elements of the ground [27], that is, $C_{dg} = k^2/\left[\ln(h_c/z_g)\right]$. Expression (14) is formally the same as the expression used in chemical models ([27]), i.e. $R_{inc} = A/u_*$, where A is a constant with a value of $LAI \cdot H \cdot b$ where b = 14 m^{-1}.

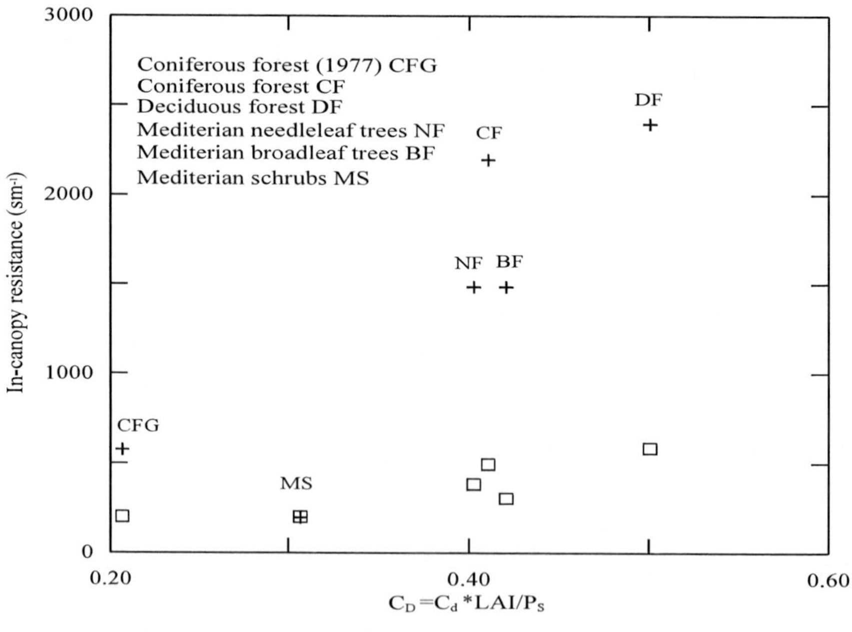

Figure 2. The values of the in-canopy resistance R_{inc} calculated by Eq. (14) (crosses) and using the expression in [29] (empty squares). The calculations were performed for the forest vegetation communities using data from the curve in [27]. The units in the x direction are the same as in this study, whereas, in the y direction, R_{inc} is used instead of conductance. Here, $C_D = C_d LAI/P_s$ where P_s is the shelter factor [28].

However, R_{inc}, which is given by Eq. (15), includes a more comprehensive architecture of the vegetation and its morphological characteristics. It can be obtained using Erisman's expression [29]. The values of R_{inc} obtained by these expressions approaches each to other for lower vegetation (short grass, crops, tall grass, shrubs and lower forest). However, their obtained values differ significantly for the forest (Figure 2). Therefore, it can be expected that the dry deposition of the calculated gases and particles, which follows the suggested parameterization of R_{inc}, will be closer to measured values. It is important to note that this issue is one of the most important in designing chemical models and performing pollution research [30 - 32].

2.3. Numerical Tests with the TKE Scheme Incorporated in the Unified EMEP Chemical Model

To compare the performance of the non-local TKE scheme [Eqs. (1) - (4)] and the local scheme, we used the Unified EMEP chemical model, where the vertical sub-grid turbulent transport is modeled as a diffusivity effect. It was designed following [33] (this scheme will be referred to as the OLD one). For an unstable case, K_m is determined as

$$K_m(z) = K_m(h) + \left(\frac{h-z}{h-h_s}\right)^2 \left\{\left[K_m(h_s) - K_m(h)\right] + (z-h_s)\left[\frac{\delta}{\delta_s} K_m(h_s) + 2\frac{K_m(h_s) - K_m(h)}{h-h_s}\right]\right\} \qquad h_s \le z < h \tag{17}$$

where h_s is the height of the surface boundary layer. In the model calculation, h_s is set as 4 % of the mixing height [34].

For the testing of both schemes, the reproduction of the vertical transport of pollutants in the ABL was performed using the Unified EMEP chemical model (UNIT-ACID, rv2_0_9). The basic physical formulation of the Unified EMEP chemical model is identical to that of Berge and Jacobsen [35]. A polar-stereographic projection, true at N and with a grid-size of 50×50 km^2, was used. The model domain used in the simulation had (101, 91) points covering the entire area of Europe and North Africa. The σ terrain-following coordinate was used with 20 levels in the vertical from the surface to 100 hPa and with the lowest level located close to 92 m. The Arakawa C grid was used as the horizontal grid in the model. All of the other details regarding this model can be found in [34]. The Unified EMEP chemical model uses 3-hourly resolution meteorological data from the dedicated version of the HIRLAM (High Resolution Limited Area Model) numerical weather prediction model with a parallel architecture [36]. The horizontal and vertical wind components are given on a staggered grid. All of the other variables are given in the center of the grid. Linear interpolation between the 3-hourly values is used to calculate the values of the meteorological input data at each advection step.

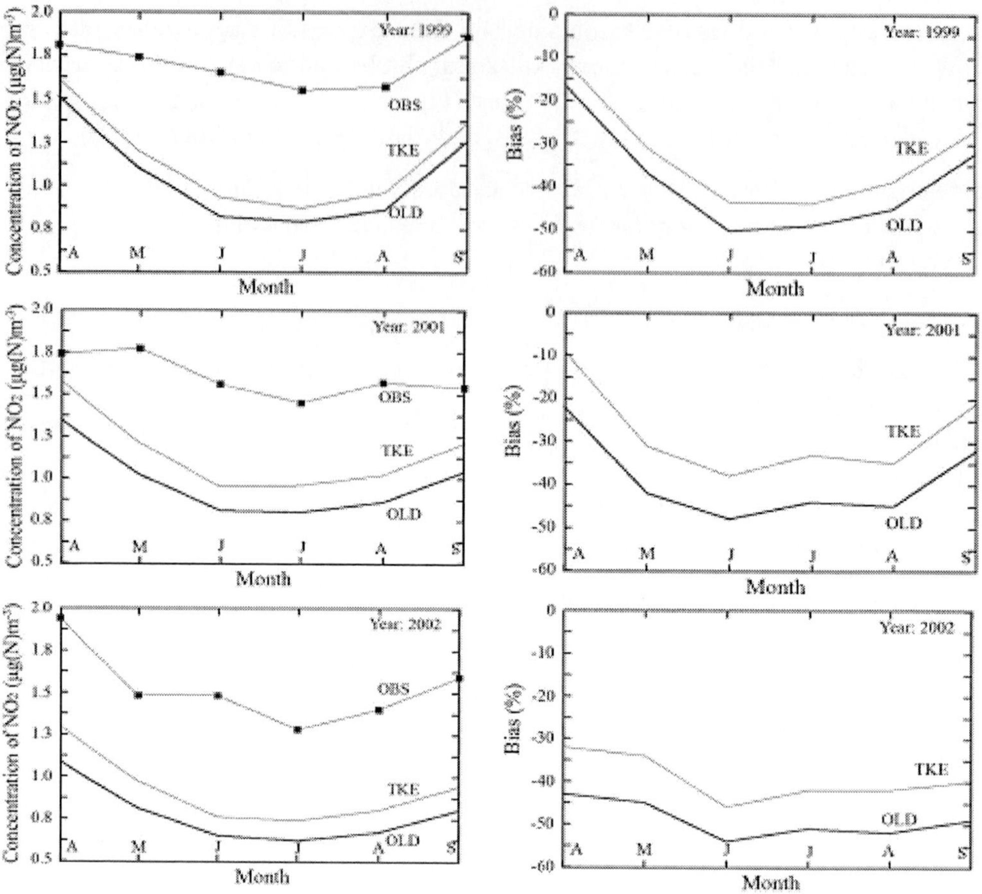

Figure 3. The OLD scheme versus the TKE scheme. Comparison of: (a) the modeled and observed NO_2 ($\mu g(N)\ m^{-3}$) concentrations in air and (b) their biases in the period of April-September for the years 1999, 2001 and 2002.

The time step used in the simulation was 600 s. In the Unified EMEP chemical model, the OLD scheme remarkably improved the vertical mixing in the ABL, particularly under stable conditions and conditions approaching free convection compared with the scheme previously used in the Unified EMEP chemical model. The improvement was particularly pronounced for NO_2 [37]. However, by reducing the horizontal grid-size and increasing the heterogeneity of the underlying surface in the Unified EMEP chemical model, there is a need for an eddy-diffusivity scheme with a higher level of sophistication for the simulation of turbulence in the ABL. The non-local eddy-diffusivity schemes should perform well for this task.

A comparison of the TKE and OLD schemes was performed using simulated and measured concentrations of the pollutant NO_2 because it is one of the most affected gases by the processes in the ABL layer. Simulations were performed for the years 1999, 2001 and 2002 for the months when the convective processes are dominant in the ABL (April - September). The station recording NO_2 ($\mu g(N)\ m^{-3}$) concentration in air was considered for comparison when measurements were available for at least 75 % of days in a year [1999 (80 stations), 2001 (78) and 2002 (82)]. We calculated the bias on the monthly basis as (M-

O)/O*100 % where M and O denote the modeled and observed values, respectively. The comparison of the modeled and observed NO_2 in air (μg(N) m^{-3}) concentrations and corresponding biases for both schemes are shown in Figure 3. The values used in the calculations were averaged over the entire domain of integration. Both schemes underestimated the observations. However, for all of the considered months, NO_2 concentrations calculated with the TKE scheme were generally higher and closer to the observations than those obtained by the OLD scheme (on the order of 10 %). Correspondingly, the bias of the TKE scheme was lower than that of the OLD scheme.

To quantify the simulated values of both schemes we performed an error analysis of the NO_2 concentration outputs. Following the study of Pielke [38], we computed the following statistical quantities:

$$\nu = [\sum_{i=1}^{N} (\Gamma_i - \hat{\Gamma}_i)^2 / N]^{1/2} , \quad (18)$$

$$\nu_{BR} = \{\sum_{i=1}^{N} [(\Gamma_i - \overline{\Gamma}) - (\hat{\Gamma}_i - \overline{\hat{\Gamma}})]^2 / N\}^{1/2} , \quad (19)$$

$$\eta = [\sum_{i=1}^{N} (\Gamma_i - \overline{\Gamma})^2 / N]^{1/2} \quad (20)$$

and

$$\hat{\eta} = [\sum_{i=1}^{N} (\hat{\Gamma}_i - \overline{\hat{\Gamma}})^2 / N]^{1/2} . \quad (21)$$

Here, Γ is the variable of interest (of the aforementioned variables in this study), and N is the total amount of data. An overbar indicates the arithmetic average, and a caret refers to an observation. The absence of a caret indicates a simulated value; ν is the root-mean-square error (RMSE), and ν_{BR} is the RMSE after the bias is removed. The RMSE provide a good overview of a dataset, in which large errors are weighted more than many small errors. The standard deviations in the simulations and the observations are given by η and $\hat{\eta}$. An RMSE that is less than the standard deviation of the observed value indicates skill in running the simulation. Moreover, the values of η and $\hat{\eta}$ should be close if the prediction is to be considered realistic. The statistics gave the following values: TKE (ν=0.548, ν_{BR}=0.293, $\eta = 0.211$, $\hat{\eta} = 0.147$) and OLD ($\nu = 0.802$, $\nu_{BR} = 0.433$, $\eta = 0.303$, $\hat{\eta} = 0.147$). A comparison of η and $\hat{\eta}$ shows that difference between them was smaller with the TKE scheme. We realize that the results depend on the different interactions of the two diffusion schemes with the other part of the model. Because no attempt was made to calibrate the eddy diffusion schemes to give the best possible simulations, the present comparisons can only serve as illustrations of the impact.

3. Non-Local Vertical Mixing Schemes

3.1. General

As mentioned above, the non-local vertical mixing schemes were designed to describe the effects of large scale eddies that are dominant in the CBL and to simulate counter-gradient flows where a turbulent flux flows up to the gradient. During convective conditions in the atmosphere, both small-scale, sub-grid and large-scale, super-grid eddies are important to vertical transport. In this section, we describe three different non-local vertical mixing schemes including Blackadar's scheme [39], the asymmetrical convective model [11] and a scheme with varying upward mixing rates [40 - 42].

Turbulent eddies that exist in the ABL can transport heat, mass, and momentum over large vertical distances that are often comparable to the height of the ABL. Therefore, at least during convective conditions, turbulent vertical mixing processes can be non-local. Stull and Driedonks [8] use "*transilient*" (from the Latin for "jump over") to indicate non-local vertical mixing in their parameterization of the ABL. Let S be any variable such as potential temperature, horizontal wind components, mixing ratio, or the concentration of a trace gas species in the atmosphere. The new value of S due to sub-grid-scale turbulent vertical mixing for a grid cell i at a future time ($t+\Delta t$) can be written as

$$S_i(t+\Delta t) = \sum_{j=1}^{N} C_{ij}(t,\Delta t) S_j(t) \tag{22}$$

where C is the transilient matrix, and i and j refer to two different grid cells in an atmospheric column [9]. In the turbulent mixing between grid cells i and j, C_{ij} represents the fraction of the air mass ending in grid cell i that came from grid cell j, grid cell i is considered the "destination" cell, and grid cell j is considered the "source" cell. Therefore, the change in the variable S due to sub-grid-scale vertical mixing for grid cell i after a time interval Δt is a simple matrix multiplication with the value of S in the source cell. Estimating the transilient matrix is the closure problem in this parameterization. This can be solved by considering the TKE equation in a non-local form. After the parameterization of the unknown terms using the horizontal wind and temperature parameters, we obtain the mixing potentials (Y). Next, the transilient matrix C_{ij} can be written as

$$C_{ij} = \frac{m_j Y_{ij}}{\|\mathbf{Y}\|_\infty} \quad i \neq j \tag{23}$$

where m_j is the mass fraction of air in cell j [9], and $\|\mathbf{Y}\|_\infty$ is the L_∞ norm of matrix $\mathbf{Y}$. The diagonal elements (for $i=j$) of the transilient matrix can be written as

$$C_{ii} = 1 - \sum_{j=1}^{N} C_{ij} \quad i \neq j \tag{24}$$

Turbulent mixing in the ABL for any dynamic or thermodynamic variable can be written as

$$\frac{\partial S_i}{\partial t} = \sum_{j=1}^{N} M_{ij}(t, \Delta t) S_j \tag{25}$$

where the elements in the matrix $\mathbf{M}$ represent mass mixing rates. This equation is similar to the transilient turbulence parameterization [Eq. (22)]. However, only a few pathways that represent the dominant mixing scales in the CBL are considered resulting in a very sparse transilient matrix that can be numerically solved much faster [9].

3.2. Overview of Non-Local Vertical Mixing Schemes

Blackadar's scheme [39] is a simple non-local closure scheme that is designed to describe convective vertical transport by eddies of varying sizes. The effect of convective plumes is simulated by mixing material directly from the surface layer with every other layer in the convective layer. The schematic representation of vertical mixing simulated by Blackadar's scheme (BLACK) is presented in Figure 4a. The mixing algorithm can be written for the surface and for every other layer as

$$\frac{\partial c_1}{\partial t} = -Muc_1 \sum_{i=2}^{N} \frac{\Delta \xi_i}{\Delta \xi_1} + Mu \sum_{i=2}^{N} c_i \frac{\Delta \xi_1}{\Delta \xi_i}, \tag{26}$$

and

$$\frac{\partial c_k}{\partial t} = Muc_1 \frac{\Delta \xi_k}{\Delta \xi_1} - Muc_k \frac{\Delta \xi_1}{\Delta \xi_k} \quad (2 < k \leq N), \tag{27}$$

respectively, where Mu represents the mixing rate, ξ is the vertical coordinate, and $\Delta\xi$ denotes the layer thickness. The mixing matrix that controls this model is non-zero only for the top row, the left-most column, and the diagonal.

The asymmetrical convective model [11] is a non-local, vertical mixing scheme based on the assumption of the vertical asymmetry of buoyancy-driven turbulence. The concept of this model is that buoyant plumes, according to Blackadar's scheme, rise from the surface layer to all levels in the CBL, but downward mixing occurs between adjacent levels only in a cascading manner. The schematic representation of the vertical mixing simulated by the ACM is presented in Figure 4b. The mixing algorithm is driven by equations

$$\frac{\partial c_1}{\partial t} = Md_2 c_2 \frac{\Delta\xi_1}{\Delta\xi_2} - Muc_1 \sum_{i=2}^{N} \frac{\Delta\xi_i}{\Delta\xi_1}, \tag{28}$$

$$\frac{\partial c_k}{\partial t} = Muc_1 - Md_k c_k + Md_{k+1} c_{k+1} \frac{\Delta\xi_{k+1}}{\Delta\xi_k} \quad (2 < k < N), \tag{29}$$

and

$$\frac{\partial c_N}{\partial t} = Muc_1 - Md_N c_N, \tag{30}$$

where Mu and Md are the upward and downward mixing rates, respectively. The downward mixing rate from level k to level $k-1$ is calculated as

$$Md_k = \frac{\xi_N - \xi_k}{\Delta\xi_k} Mu \quad (k = 1, 2, \ldots, N-1). \tag{31}$$

The mixing matrix controlling this model is non-zero only for the leftmost column, the diagonal and superdiagonal.

The scheme with varying upward mixing rates (VUR) [40] is a modified version of the ACM, where the upward mixing rate Mu_k ($k = 2, \ldots, N$) changes with the height (position of the box, $\Delta\xi_k$), providing slower mixing.

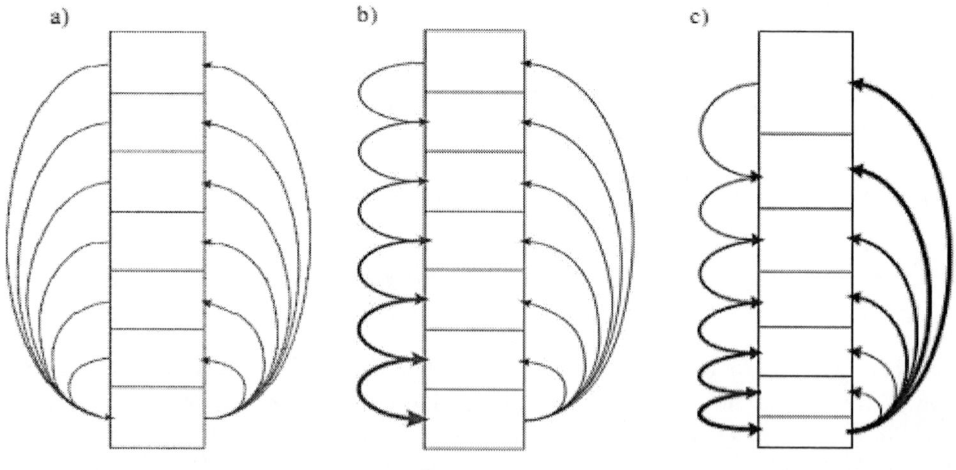

Figure 4. Schematic representation of vertical mixing in a one-dimensional column as simulated by the (a) BLACK, (b) ACM scheme and (c) VUR. Line thicknesses are proportional to mixing rates.

The schematic representation of vertical mixing simulated by the scheme is shown in Figure 4c. The upward mixing rates are scaled with the amount of TKE in the layer as

$$Mu_k = Mu_1 \frac{e_k \Delta\xi_k}{\sum_{i=1}^{N} e_i \Delta\xi_i} \tag{32}$$

where Mu_1 is the upward mixing rate from surface layer to layer above, and e_k denotes the TKE in the considered layer. The upward mixing rate from surface layer to layer above is parameterized as

$$Mu_1 = \frac{\rho u_*^4 + H_f w_*}{h(\rho u_*^3 + H_f)} \tag{33}$$

where ρ is the air density, and H_f represents the sensible heat flux. Using the VUR scheme, the mixing algorithm for the lowest layer can be written in the form

$$\frac{\partial c_1}{\partial t} = Md_2 c_2 \frac{\Delta\xi_1}{\Delta\xi_2} - Mu_1 c_1 - \sum_{k=3}^{N} Mu_k c_1 . \tag{34}$$

The algorithm for the other layers is very similar to the ACM algorithm [Eqs. (28) - (29)], with the upward mixing rate *Mu* substituted with varying upward mixing rates Mu_k.

4. Combined Non-Local Vertical Mixing and Eddy-Diffusion Schemes

4.1. Description of the Combined Scheme

In this section, we suggest a scheme that includes the non-local ACM and TKE schemes as a linear combination with a weighting factor determining their individual contributions in mixing the CBL. As indicated, none of the above-described schemes can represent mixing at all scales. The first group captures only local scale mixing, whereas the second one describes non-local scale mixing. Therefore, the combination of non-local TKE and ACM schemes can represent all of the possible spectrums of turbulent scales. However, the contribution of each scheme is not identical and depends on the weighting factor f. Following [18] (Figure 5), the partitioning of the upward convective mixing rate Mu' and turbulent transfer coefficient K' is calculated as $Mu' = fMu$ and $K' = (1-f)K$. Then, the turbulent mixing of quantity φ in model layer i is calculated as

$$\frac{\partial \varphi_i}{\partial t} = Mu'\varphi_1 - Md'_i\varphi_i + Md'_{i+1}\varphi_{i+1}\frac{\Delta z_{i+1}}{\Delta z_i} + \frac{1}{\Delta z_i}\left[\frac{K'_{i+1/2}(\varphi_{i+1}-\varphi_i)}{\Delta z_{i+1/2}} - \frac{K'_{i-1/2}(\varphi_i-\varphi_{i-1})}{\Delta z_{i-1/2}}\right] \tag{35}$$

where the half index indicates the location between layers i and $i-1$. In the combined scheme (ACMTKE from now on), for momentum, the turbulent transfer coefficient given by Eq. (1) ($p=2$) is used, and for the heat

$$K_h = K_m \left\{ 1.35 \frac{(1-9z/L)^{1/2}}{(1-15z/L)^{1/4}} \right\}. \tag{36}$$

Let us note that $Mu = K_h(z_{1+1/2})/[\Delta z_{i+1/2}(h - z_{1+1/2})]$ is parameterized similarly to [18] but with K_h calculated using Eq. (33). We believe that this is a better choice because the friction velocity for moderate to strong convective conditions is not a representative scale [4, 24]. The key parameter $f \in [0,1]$ controls the degree of the contributions of the terms in the ACMTKE scheme. It is estimated as suggested by [18], i.e.

$$f = \left[1 + \frac{k^{-\frac{2}{3}}}{0.1a} \left(-\frac{h}{L} \right)^{-\frac{1}{3}} \right]^{-1} \tag{37}$$

where a is set to 7.2. For $f=0$, (i.e., stable ABL), the TKE profile is modeled using an empirical function proposed by [23] based on aircraft observations. Next, $\overline{e}_*$ is calculated from Eq. (4).

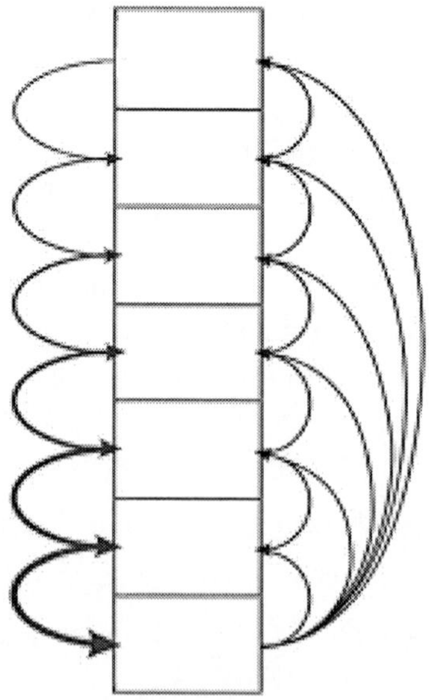

Figure 5. Schematic representation of vertical mixing in a one-dimensional column as simulated by the ACM2 scheme. Line thicknesses are proportional to mixing rates.

4.2. Numerical Tests Using the Combined Non-Local Vertical Mixing and Eddy Diffusion Scheme

To validate the performance of the proposed ACMTKE scheme, we compared it to the OLD (Eq. 17) scheme by (i) using the observations derived from the LES studies (LESs) and (ii) calculating the bias because the comparison with the LESs is a common approach for testing ABL schemes. Although LESs are high-resolution, time-dependent model simulations, quasi-steady ABL profiles can be obtained by horizontally averaging across the entire computational grid, using periodic boundary conditions, and averaging in time over many large-eddy turn-over time periods ($\tau = z_i / w_*$). The experiments represent a cloud-free ABL driven by shear and buoyancy. Various stability and wind shear conditions were generated using constant prescribed forcing by surface sensible heat flux and geostrophic wind profiles. After a period of $\sim 5\tau$ to spin up the turbulent dynamics, passive tracers were added to the simulations and the LES were run for another $\sim 10\tau$ of which the last $\sim 6\tau$ were averaged to produce the final profiles. The ABL schemes were initialized from the LES profiles after the $\sim 5\tau$ spin up, run for 10τ, and averaged over the final $\sim 6\tau$ for a comparison with the LES final profiles.

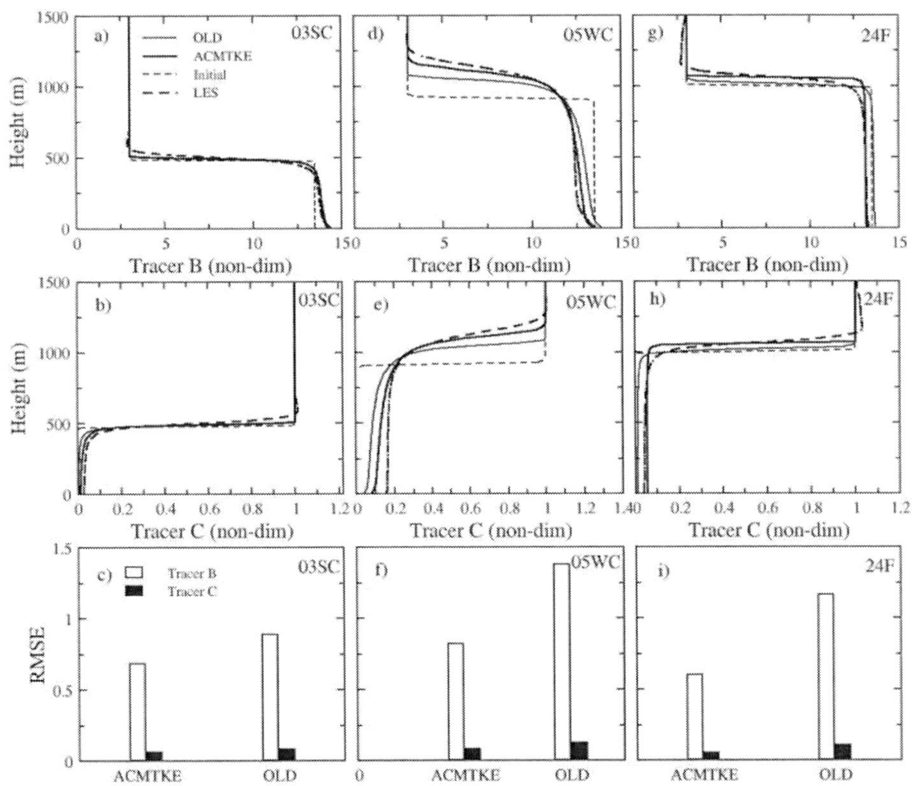

Figure 6. ACMTKE versus OLD scheme. Comparison of calculated concentrations of the tracers B and C with the LES and the corresponding RMSE for profiles: 03SC (a - c), 05WC (d - f) and 24F (g - i).

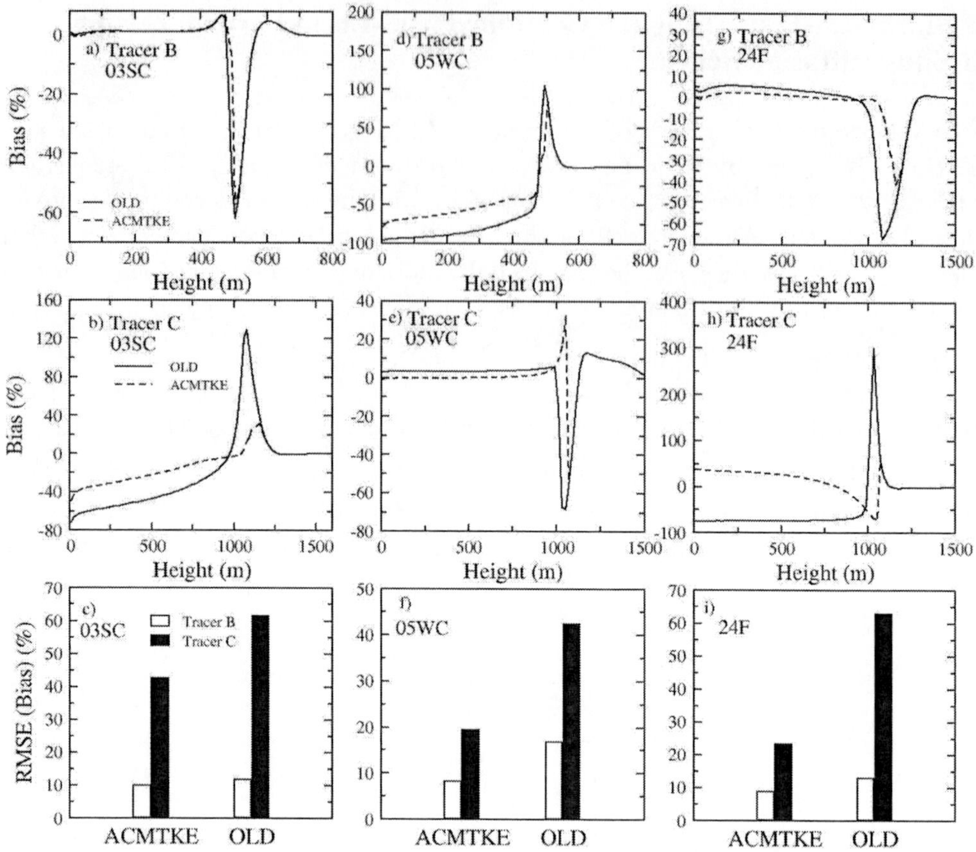

Figure 7. The ACMTKE versus OLD scheme. Comparison of biases of tracers B and C with the LES and the corresponding the RMSE for profiles: 03SC (a - c), 05WC (d - f) and 24F (g - i).

See [43] for a detailed description of the LES model and the simulation experiments. Three cases were chosen to test the ACMTKE and OLD schemes - one had a weak capping inversion and a low value of surface sensible heat flux, which was designated as 05WC (0.05 K m s^{-1}), and two had strong capping inversions with low and high values of surface sensible heat flux, which were labeled 03SC (0.03 K m s^{-1}) and 24F (0.24 K m s^{-1}), respectively. All three cases were under dry conditions and had a roughness length of 0.16 m with geostrophic wind components m s^{-1} and m s^{-1} at all levels. The vertical grid used for the ACMTKE simulation was identical to the LES vertical grid. The comparisons were made for two concentrations of tracers B and C using a 1-D vertical diffusion model based on Eq. (35).

Figures 6 and 7 show the comparisons of the ACMTKE and OLD schemes. Figure 6 shows that for both tracers (B and C), proposed combined scheme gave (1) better (Figs. 6a - 6c and 6g - 6i) and (2) much better (Figs. 6d - 6f) results than the OLD scheme. This is more evident in the panels in Figure 7. This figure, which depicts the calculated biases (see subsection 3) shows that the bias for both tracers was much better for the ACMTKE scheme, particularly for both tracers in the case of the 05WC profile.

It is important to note that the cadence was set based its modeling aspect rather than on the statistical analysis of the obtained results. This analysis could be more comprehensive in the case of its application to the so-called detrended fluctuation analysis (DFA). Therefore, we offer a short overview of the potential uses of this method in this work's field of study. In stochastic processes, chaos theory and time series analysis, DFA is a method for determining the statistical self-affinity of a signal. It was introduced in [44] and represents an extension of the (ordinary) fluctuation analysis, which is affected by nonstationarities. Most of the geophysical quantities (e.g., air temperature, humidity, precipitation, ozone concentration, etc.) obey nonlinear laws, which usually generate spatio-temporal nonstationarities. Currently, the estimation of the scaling exponents of the power-law in the air turbulence field can be achieved by the employment of the aforementioned method, which has already proved its usefulness in several complex systems such as surface air pollutants [45], aerosol index [46], total ozone content and global tropospheric temperature [47].

CONCLUSIONS

In this chapter, we considered the non-local eddy diffusion and vertical mixing schemes for use in chemical transport models during the convective period. All of these were comprehensively described, and one (TKE) was tested using NO_2 concentration outputs from the Unified EMEP chemical model. The simulated concentrations of NO_2 by the TKE scheme were closer to the observations when compared to those using the OLD scheme [33], which is used in the current Unified EMEP chemical model. Further, we developed a new combined non-local vertical mixing and eddy diffusion scheme for use in the chemical transport model. This was accomplished by their linear combination, which included weighting factor. The performance of the scheme was tested by a one-dimensional boundary layer model and compared with (i) a LES and (ii) a pure local diffusion scheme. The proposed scheme gave results that were closer to the observations when compared to those obtained using a local eddy diffusion scheme [33] that is often incorporated into air quality and chemical transport models.

ACKNOWLEDGEMENTS

This chapter was realized as a part of the project "Studying climate change and its influence on the environment: impacts, adaptation and mitigation" (No. 43007), which is financed by the Ministry of Education and Science of the Republic of Serbia within the framework of integrated and interdisciplinary research over the period 2011 - 2014. The authors would like to thank to Dr. Keith Ayotte and Dr. Peter Sullivan for providing the LES data.

REFERENCES

[1] D.J. Jacob, *Introduction to Atmospheric Chemistry; 1st Edition*; Princeton University Press: Princeton, US, 1999; p. 271.

[2] R.B. Stull, *An Introduction to Boundary Layer Meteorology*; Kluwer Academic Publishers: Norwell, US, 1988; p. 667.

[3] K. Alapaty, *The 4th WRF Users' Workshop*, Boulder, CO, US (2003).

[4] K. Alapaty, K.M. Alapaty, *Research Note,* MCNC-North Carolina Supercomputing Center, Research Triangle Park, US, 2001; p. 5.
[5] D.T. Mihailovic, J.E. Jonson, *Air Pollution Report 5/2005,* Norwegian Meteorological Institute, Oslo, NO (2005).
[6] D.T. Mihailovic, K. Alapaty, *Environ. Modell. Softw.*, 22, 1685-1689 (2007).
[7] R.B. Stull, *J. Atmos. Sci.*, 41, 3351-3367 (1984).
[8] R.B. Stull, *Bound.-Lay. Meteorol.*, 40, 209-239 (1987).
[9] K. Alapaty, J.E. Pleim, S. Raman, D.S. Niyogi, D.W. Byun, *J. Appl. Meteorol.*, 36, 214-233 (1997).
[10] M.S. Miesch, A. Brandenburg, A. Zweibel, E.G. Zweibel, *Phys. Rev. E.,* 61, 457-467 (2000).
[11] J.E. Pleim, J.S. Chang, *Atmos. Environ.,* 26A, 965-981 (1992).
[12] A.K. Blackadar, *4th Symposium on Atmospheric Turbulence Diffusion and Air Quality*, Reno, NV, US, 443-447 (1979).
[13] J.C. Wyngaard, R.A. Brost, *J. Atmos. Sci.*, 41, 102-112 (1984).
[14] G. Tonnesen, J. Olaguer, M. Bergin, T. Russell, A. Hanna, P. Makar, D. Derwent, Z. Wang, *Relative Reactivity Working Group*, NASCO, Pasco, US, 1998; p. 55.
[15] Z. Wang, *Proceedings of the Photochemical Reactivity Workshop. U.S. Environmental Protection Agency*, Durham, NC, USA, 4-29 (1998).
[16] R. Salmon, *Nonlinearity*, 18, R1-R16 (2005).
[17] P. Névir, R. Blender, *J. Phys. A.-Math. Gen.*, 26, L1189-L1193 (1993).
[18] J.E. Pleim, *J. Appl. Meteorol. Clim.*, 46, 1383-1395 (2007).
[19] Troen, L. Mahrt, *Bound.-Lay. Meteorol.*, 37, 129-148.(1986).
[20] C. Zhang, D.A. Randall, C.H. Moeng, M. Branson, M. Moyer, Q. Wang, *Mon. Weather Rev.*, 124, 2521-2536 (1996).
[21] J.A. Businger, Y. Izumi, E.F. Bradley, *J. Atmos. Sci.*, 28, 181-189 (1971).
[22] C.H. Moeng, P.P. Sullivan, *J. Atmos. Sci.*, 51, 999-1022 (1994).
[23] D.H. Lenschow, X.S. Li, C.J. Zhu, B.B. Stankov, *Bound.-Lay. Meteorol.*, 42, 95-121 (1988).
[24] D.T. Mihailović, Z. Podrascanin, K. Alapaty, *Environ. Sci. Pollut. R.*, 15, 296-302 (2009).
[25] A.A.M. Holtslag, B.A. Boville, *J. Climate*, 6, 1825-1842 (1993).
[26] D.L. Zhang, R.A. Anthes, *J. Appl. Meteorol.,* 21, 1594-1609 (1982).
[27] P.J. Sellers, J.L. Dorman, *J. Clim. Appl. Meteorol.*, 26, 622-651 (1987).
[28] D.T. Mihailovic, K. Alapaty, B. Lalic, I. Arsenic, B. Rajkovic, S. Malinovic, *J. Appl. Meteorol.,* 43, 1498-1514 (2004).
[29] J.W. Erisman, A. van Pul, P. Wyers, *Atmos. Environ.*, 28, 2595-2607 (1994).
[30] M.J. Sanz, A. Carratala, C. Gimeno, M.M. Millan, *Environ. Pollut.*, 118, 259-272 (2002).
[31] H.M. Chang, F.W. Lee, F.T Jeng, *Environ. Sci. Pollut. R.*, 9, 385-391 (2002).
[32] M. Sofiev, P. Siljamo, I. Valkama, M. Ilvonen, J. Kukkonen, *Atmos. Environ.,* 40, 674-685 (2006).
[33] J.J. O'Brien, *J. Atmos. Sci.*, 27, 213-1215 (1970).
[34] D. Simpson, H. Fagerli, J.E. Jonson, S. Tsyro, P. Wind, J.P. Tuovinen, *Europe. Part I: Unified EMEP Model Description. EMEP Status Report 2003,* Norwegian Meteorological Institute, Oslo, NO, 2003; p. 74.

[35] E. Berge, H.A. Jacobsen, *Tellus B*, 50, 205-223 (1998).
[36] D. Bjorge, R. Skalin, *Research Report No. 27,* Norwegian Meteorological Institute, Oslo, NO (1995).
[37] H. Fagerli, A. Eliassen, *Europe. EMEP Summary Status Report, Research Report No. 141*, Norwegian Meteorological Institute, Oslo, NO, p. 74 (2002).
[38] R.A. Pielke, *Mesoscale Meteorological Modeling; 2nd Edition;* Academic Press: New York, US, 2002; p. 676.
[39] A.K. Blackadar, *4th Symposium Atmospheric Turbulence, Diffusion and Air Quality*, Boston, MA, US, 46-49 (1976).
[40] D.T. Mihailović, K. Alapaty, M. Sakradžija, *Environ. Sci. Pollut. R.,* 15, 296-302 (2008).
[41] D.T. Mihailović, A. Firanj, In *Air Quality*; A. Kumar Ed.; Sciyo: Rijeka, HR, 2010; pp. 233-246.
[42] D.T. Mihailović, Á. Bordás, K. Alapaty, In *Advances in Environmental Modeling and Measurements*; D.T. Mihailovic, B. Lalic Eds.; Nova Science Publishers, Inc.: New York, US, 2010; pp. 121-132.
[43] K.W. Ayotte, P.P. Sullivan, A. Andren, S C. Doney, A. Holtslag, W.G. Large, J.C. McWilliams, C.-H. Moeng, M. Otte, J. Tribbia, J.C. Wyngaard, *Bound.-Lay. Meteorol.,* 79, 131-175 (1996).
[44] C-K, Peng, S.V. Buldyrev, S. Havlin, M. Simons, H.E. Stanley, A.L. Goldberger, *Phys. Rev. E.*, 49, 1685-1689 (1994).
[45] C. Varotsos, J. Ondov, M. Efstathiou, *Atmos. Environ.*, 39, 4041-4047 (2005).
[46] C. Varotsos, J. Ondov, A. Cracknell, M. Efstathiou, M Assimakopoulos, *Int. J. Remote Sens.*, 27, 3593-3603 (2006).
[47] C. Varotsos C, D. Kirk-Davidoff, *Atmos. Chem. Phys.*, 6, 4093-4100 (2006).

In: Essays on Fundamental and Applied Environmental Topics ISBN 978-1-61942-522-4
Editor: Dragutin Mihailovic

Chapter 7

EFFECT OF PERMEABILITY ON HYPORHEIC FLOWS ACROSS A TRIANGULAR DUNE

C. Gualtieri[*]
Hydraulic, Geotechnical and Environmental Engineering Department (DIGA),
University of Napoli "Federico II", Via Claudio, Napoli, Italy

ABSTRACT

Despite the traditional separation between the studies of surface water and groundwater flows, for several decades, it has been recognized that rivers and aquifers are strongly connected and that their interaction gives rise to a continuous exchange of water and solutes, which exerts a significant influence on water quality. The analysis of this exchange has led to the identification of a zone, termed the *hyporheic zone*, where groundwater and stream water mix. The hyporheic zone has hydrodynamic, physiochemical and biotic characteristics different from those of both the river and the subsurface environments and controls the distribution of solutes, colloids, dissolved gases and biogeochemical reactions from ripple to global scales. Thus, this zone affects the distribution of benthic flora and fauna in lakes, oceans, bays and estuaries, as well as hyporheic and riparian organisms in fluvial systems.

This chapter presents some results of a numerical study carried out to investigate the influence of the permeability of a porous medium on a coupled free fluid-porous medium flow. Laminar flow in the water column above triangular bedforms and Darcian flow in the underlying permeable sediments were considered. Numerical simulations were carried out under steady-state conditions using Comsol Multiphysics 3.5a in a range of bedforms height-based Reynolds number Re_H from 6 to 4442 and for three different values of the permeability of the porous medium. First, numerical results are shown to confirm the close relationship between the characteristics of the separation region in the water column downstream of the bedform crest and those of the hyporheic zone in the porous medium. Second, numerical results highlight the influence of the permeability on the development of the hyporheic zone. Finally, the issues that arise in the application of these results to turbulent flows and to large scales are discussed.

[*] Corresponding author: E-mail address:carlo.gualtieri@unina.it; Phone: + 39 081 768 3433. Fax: + 39 081 5938936.

1. Introduction

Despite the traditional separation between the studies of surface water and groundwater flows, for several decades, it has been recognized that rivers and aquifers are strongly connected and that their interaction gives rise to a continuous exchange of water and solutes, which exerts a significant influence on water quality [1]. The analysis of this exchange has led to the identification of a zone, known as the *hyporheic zone*, where groundwater and stream water mix. The hyporheic zone has hydrodynamic, physiochemical and biotic characteristics different from those of both the river and the subsurface environments [2, 3] and controls the distribution of solutes, colloids, dissolved gases and biogeochemical reactions from ripple to global scales [4 - 6]. Thus, this zone affects the distribution of benthic flora and fauna in lakes, oceans, bays and estuaries, as well as hyporheic and riparian organisms in fluvial systems [7, 8]. The effects of the hyporheic exchange processes on both the river and the subterranean environments are twofold. First, the hyporheic zone acts as a storage zone or a dead zone, which temporarily traps stream-transported solutes and subsequently releases them after some time. Due to the low filtration velocities in the subsurface, the residence time of solutes in the hyporheic zone is usually much longer than the travel time in the main stream. This difference between the characteristic timescales of in-stream and hyporheic flow produces long tails in the time-concentration curves [1, 9]. Second, the metabolic activity of the hyporheic microorganisms significantly alters the in-stream concentration of chemicals, both at the reach scale [10] and at the basin scale [11]. The surface of the hyporheic sediments is covered by biofilms of aerobic microorganisms, which use the dissolved oxygen from the surface water to oxidize nitrogen and phosphorus for their metabolism [3, 12]. The hyporheic exchange is also important for riparian vegetation because the oxygen flux from the stream regulates the redox conditions in the aquifer and, therefore, the dynamics of nutrients and trace metals [8]. Examples of vertical, lateral and longitudinal hyporheic exchange processes between a river and its surrounding groundwater are presented in Figure 1. Hyporheic exchange flows are indicated by blue arrows.

Different approaches have been proposed to study surface water-groundwater interactions and the hyporheic zone, as pointed out by Boano [13]. First, a number of studies have relied on detailed field measurements of the topographical and hydraulic features of specific sites to calibrate 2D or 3D numerical groundwater flow models. This approach allows hyporheic paths to be modeled and the interactions between the hyporheic zone and the surrounding aquifer to be described at scales on the order of hundreds of meters [14 - 17]. However, this approach remains site-specific, and its results cannot easily be transferred to different locations, but this method allows one to consider the reciprocal influence of all the key factors affecting surface water-groundwater exchange.

A second approach studies how the transport of chemicals in streams is influenced by the exchange with hyporheic zones [19 - 21]. Thus, the in-stream concentration of a tracer injected into a river has been observed at certain sampling points, and these data have been used to calibrate a one-dimensional stream transport model, such as the transient storage model (TSM) [1]. The calibrated parameters can subsequently be related to the characteristics of the hyporheic zone. However, this approach can identify only the shorter and faster hyporheic paths in the shallower streambed, whereas no information regarding the slower transport to the deeper part of the bed is provided.

Finally, the hyporheic exchange can be analyzed both experimentally and numerically on smaller scales. The exchange due to ripples and dunes has particularly been thoroughly studied because such bedforms are frequently present in rivers [22 - 34]. A physically based model (advective pumping model, APM) has been proposed [35, 24, 25]; this model identifies two basic processes as being responsible for the exchange of solutes with the hyporheic zone: *pumping* and *turnover*. Pumping indicates the water exchange due to gradients of head on the bed surface caused by the separation of the flow behind the bedforms, while turnover denotes the entrapment and release of water in the sediments, due to the deposition and scour of the bed. Furthermore, other models consider the exchange of solutes due to turbulent dispersion [36], temperature gradients [37], wave motion [38] and meander sinuosity [39 - 41]. Despite those many efforts, the study of hyporheic zones is far from being complete and requires further research.

This chapter presents some results of a numerical study carried out to investigate the influence of the permeability of a porous medium on the hyporheic flow in a coupled free fluid-porous medium domain. Laminar flow in the water column above triangular bedforms and Darcian flow in the underlying permeable sediments were considered. The interaction between free-fluid flow and the flow within the porous medium was studied by considering the characteristics of the separation region in the free fluid, the depth of the interfacial exchange zone and the fluxes across the bedforms at the sediment-water interface.

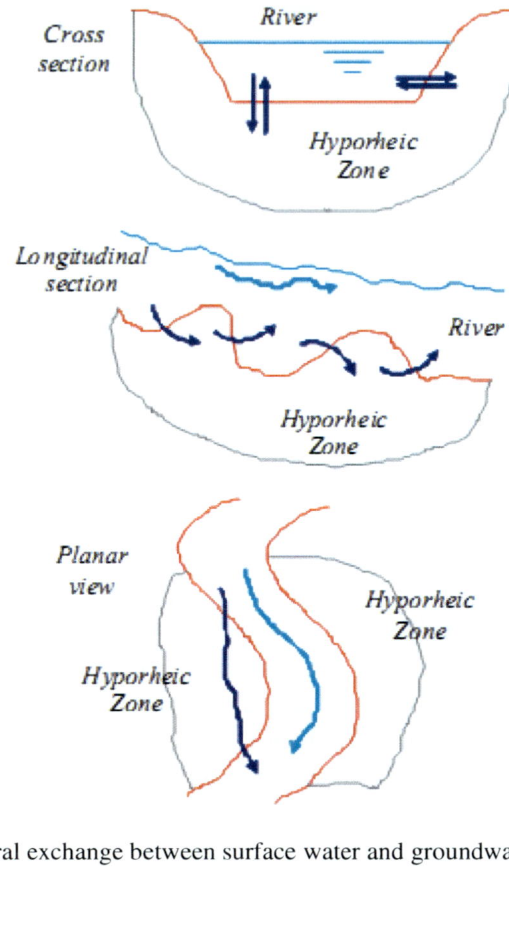

Figure 1. Vertical and lateral exchange between surface water and groundwater (after Findlay, 1995).

2. Numerical Simulations

The analysis of the flow field in a free fluid-porous medium domain was carried out using numerical simulations. Free-fluid flow was simulated using the well-known mass conservation and momentum balance equations. For a planar, steady-state and incompressible flow, these equations are

$$\frac{\partial u}{\partial x} + \frac{\partial v}{\partial y} = 0 \tag{1}$$

$$\rho\left(u\frac{\partial u}{\partial x} + v\frac{\partial u}{\partial y}\right) = -\frac{\partial p}{\partial x} + \mu\nabla^2 u + F_x$$
$$\rho\left(u\frac{\partial v}{\partial x} + v\frac{\partial v}{\partial y}\right) = -\frac{\partial p}{\partial y} + \mu\nabla^2 v + F_y \tag{2}$$

where in the Navier-Stokes equations, ρ is fluid density, p is fluid pressure, F_x and F_y are force terms accounting for gravity or other body forces, and u, v are velocity components in the x and y directions, respectively.

In the porous medium, the governing equations again are continuity and momentum balance equations. For a planar, steady-state and incompressible flow, the continuity equation is Eq. (1), while momentum balance equations in a porous medium are the empirical Darcy's law, which states that the volume-averaged velocity field is determined by the pressure gradient, the fluid viscosity and the structure of the porous medium

$$\frac{\mu}{k}u = -\frac{\partial p}{\partial x} + F_x$$
$$\frac{\mu}{k}v = -\frac{\partial p}{\partial y} + F_y \tag{3}$$

where k is porous medium permeability [L^2], which is a scalar for an isotropic porous medium.

These equations were solved using the Multiphysics 3.5a™ modeling package, which is a commercial multiphysics modeling environment [42]. Sequential coupling was implemented. Multiphysics first solved Eqs. (1) and (2) for the pressure p and the velocity vector components u and v within the free fluid domain. Next, the calculated pressure distribution along the bed surface was assigned as a Dirichlet boundary condition for the groundwater flow equation, and the flow field in the porous medium was solved.

Multiphysics 3.5a™ was applied to the 2D geometry presented in Figure 2, which represents a free fluid-porous medium domain with free fluid and porous medium depth of 0.45 m and 2.0 m, respectively. These sub-domains are separated by a series of five triangular bedforms 1.0 m long and 0.05 m high, with the crest being located at 0.9 m. Thus, if L is the wavelength of the dune, the bedform aspect ratio $H_{bed\ form}/L$ is 0.05. This value is within the typical range for dunes, i.e. 0.04÷0.21 [43].

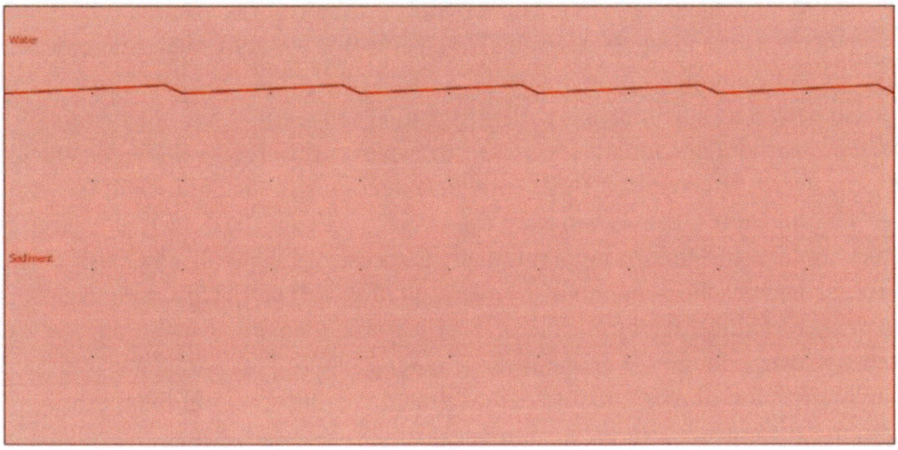

Figure 2. The investigated geometry.

This bedform geometry is the same as that used as the base case by Bayani-Cardenas and Wilson in their numerical study [31]. For the simulations, water at 20 °C with density ρ = 998.16 kg m^{-3} and dynamic viscosity $\mu = 1.00\times10^{-3}$ was selected as the fluid, while the intrinsic permeability of the porous medium K ranged from 1×10^{-8} to 1×10^{-12} m². These values correspond to gravel, well-sorted coarse sand and fine sand. Flow conditions were characterized using a bedform height-based Reynolds number, defined as

$$Re_H = \frac{u_{avg} \, H_{bed\ form}}{\nu} \tag{4}$$

where u_{avg} is the average velocity in the free fluid flow.

Twenty-two numerical simulations were carried out using Multiphysics 3.5a™ to investigate the influence of the bedform height-based Reynolds number Re_H on the flow field. Table 1 lists all the operative conditions.

Boundary conditions were assigned in both the free-fluid and porous medium sub-domains:

- at the top and the bottom of the water column, a no-flow *symmetry* boundary condition and a *no-slip* condition were assigned, respectively;
- at the top boundary of the porous domain, a prescribed *pressure* condition was assigned. This pressure was derived by solving the Navier-Stokes equations in the above free-fluid domain;
- the lower boundary of the porous domain was considered *impermeable*;
- to approximate an infinite horizontal domain solution, a spatially *periodic pressure* and *velocity* boundary condition were assigned along the vertical sides for both the water column and the porous medium. The same mean pressure drop, *dp*, was prescribed between the two vertical boundaries for both domains, thereby resulting in an ambient flow that always ran from left to right. This finding is consistent with simulations reported in the literature [31].

The stability and accuracy of the Navier-Stokes solution was ensured by using Lagrange $p2 - p1$ elements (second-order Lagrange elements for velocity and linear for pressure). Similarly, second-order Lagrange elements were used for the Darcian domain. Note that a Lagrange element is a finite element with polynomial shape functions of a certain order.

Different mesh characteristics were tested. Next, different values for the maximum element size were selected for the free fluid, the porous medium and the fluid-porous interface. The maximum element sizes were 0.025, 0.05 and 0.01 m for the free fluid, the porous medium and the fluid-porous interface, respectively. Overall, the mesh that was used had 47 160 elements with a minimum element quality of 0.664. This parameter is related to the element aspect ratio, which means that anisotropic elements can produce a low-quality measurements, even though the element shape is reasonable. This parameter is a scalar from 0 to 1. The visualization of mesh quality demonstrated the uniformity of the mesh.

Regarding the solver settings, a stationary solver was used, where the relative tolerance and the maximum number of segregated iterations were set to 1.0×10-6 and 50, respectively.

3. Numerical Results: Discussion

3.1. General Characteristics of the Flow Field: Comparison with the Results from Bayani-Cardenas and Wilson [31]

Numerical simulations provided velocity field and pressure values throughout the flow domain. The analysis of the results was first focused on the characterization of the flow field close to the sediment-water interface.

Table 1. Key parameters of the numerical simulations

Run	$K - m^2$	$dp - Pa$	$u_{avg} - m/s$	Re_H
1	1×10^{-10}	0.00001	0.00011	6
2	1×10^{-10}	0.00011	0.00120	60
3	1×10^{-10}	0.00033	0.00353	176
4	1×10^{-10}	0.00107	0.01139	569
5	1×10^{-10}	0.00213	0.02250	1123
6	1×10^{-10}	0.00319	0.03344	1669
7	1×10^{-10}	0.00425	0.04455	2224
8	1×10^{-10}	0.00529	0.05532	2761
9	1×10^{-10}	0.00614	0.06411	3200
10	1×10^{-10}	0.00740	0.07718	3852
11	1×10^{-10}	0.00850	0.08850	4417
12	1×10^{-12}	0.00011	0.00120	60
13	1×10^{-12}	0.00107	0.01140	569
14	1×10^{-12}	0.00213	0.02250	1123
15	1×10^{-12}	0.00319	0.03355	1674
16	1×10^{-12}	0.00614	0.06411	3200
17	1×10^{-8}	0.00011	0.00120	60
18	1×10^{-8}	0.00107	0.01139	569
19	1×10^{-8}	0.00213	0.02258	1127
20	1×10^{-8}	0.00319	0.03372	1683

21	1×10^{-8}	0.00614	0.06450	3219
22	1×10^{-8}	0.00850	0.08900	4442

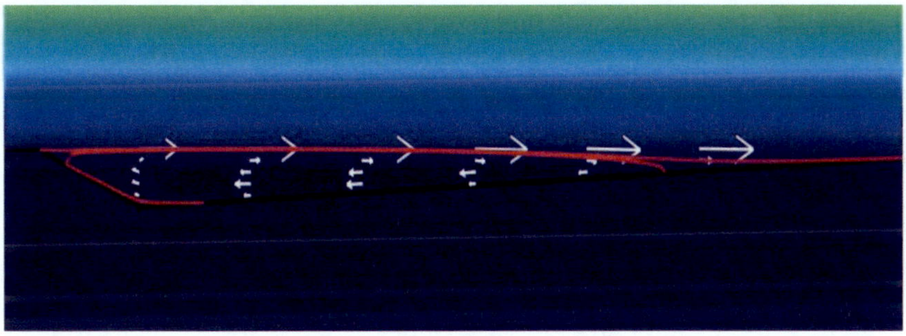

Figure 3. The separation region downstream of the dune crest for Run 7 with $Re_H = 2224$.

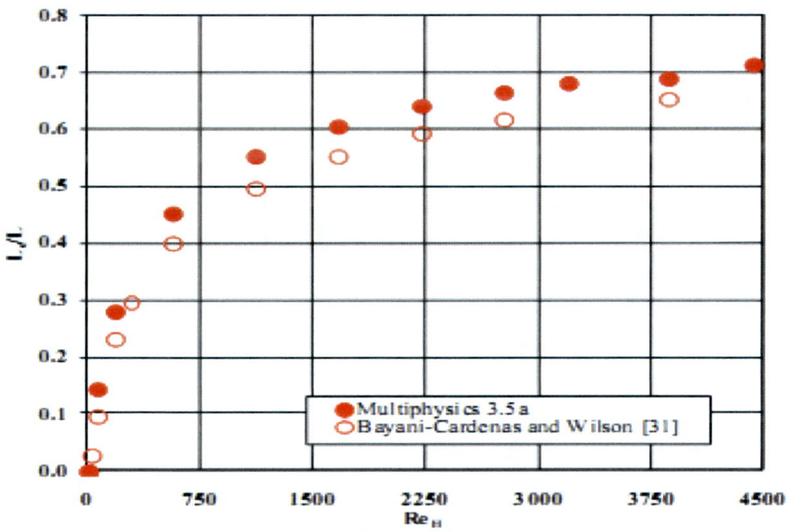

Figure 4. Non-dimensional separation length L_e/L vs. Re_H for Runs 1 - 11.

The flow in the water column accelerates along the stoss side and decelerates along the lee side of the bedform. A separation region can be observed downstream of the bedform crest with backflow to the crest (Figure 3). Low velocities (blue) were observed in the porous medium and in the separation region.

The separation region became larger with increasing Re_H. The eddy detachment point, x_d, migrates upstream and up the lee face toward the crest, while the reattachment point, x_r, migrates downstream and up the stoss face.

Figure 4 presents the values of L_e/L against the bedform height-based Reynolds number Re_H for Runs 1 - 11. The numerical results from Multiphysics 3.5a were successfully compared with those calculated by Bayani-Cardenas and Wilson [31]. At the lowest Re_H, no separation region was observed. The eddy size, measured by length $L_e = x_r - x_d$, was particularly sensitive to the current at low Re_H. Over $Re_H = 1000$, the increase in separation length was slow up to an asymptotic value.

The value and the location of the separation region was closely related to the distribution of the pressure from the Navier-Stokes equations at the sediment-water interface because the detachment and reattachment points were located close to the points of minimum and maximum pressure along the bedform. In the bulk water column, the mean pressure gradient is dominant over local pressure gradients generated by the current–bedform interaction. Note that the pressure was continuous across the sediment–water interface. Figure 5 presents the non-dimensional bottom pressure p^* along the bedform for different values of Reynolds number Re_H. The non-dimensional pressure was calculated as follows [31]:

$$p^* = \frac{p - p_{mid}}{p_{amp}} \quad (5)$$

where p_{mid} and p_{amp} are the midpoint pressure and pressure amplitude, respectively, which are defined as

$$p_{mid} = \frac{p_{max} + p_{min}}{2}$$
$$p_{amp} = \frac{p_{max} - p_{min}}{2} \quad (6)$$

where p_{max} and p_{min} are maximum and minimum pressure along the bedform, respectively.

Note that in Figure 5 the trough is located at $x/L = 0$ and $x/L = 1$, while the crest is at $x/L = 0.9$. For the lowest Re_H, the maximum value of p^* was located near the trough. As Re_H increased, the position of maximum pressure, which is close to the reattachment point x_r, moved downstream. For all values of Re_H, the minimum value of p^* was located close to the dune crest where the eddy detachment point x_d was located. Thus, around the crest, there is a relevant change in pressure. Overall, the trend shown in Figure 5 is consistent with that presented in [31]. Below the interface, a backflow region was observed within the porous medium. The area of this region increased with increasing Re_H, following the pattern observed for L_e/L. Over $Re_H = 1000$, two backflow regions were observed: one close to the crest and another downstream of the dune trough along the stoss face.

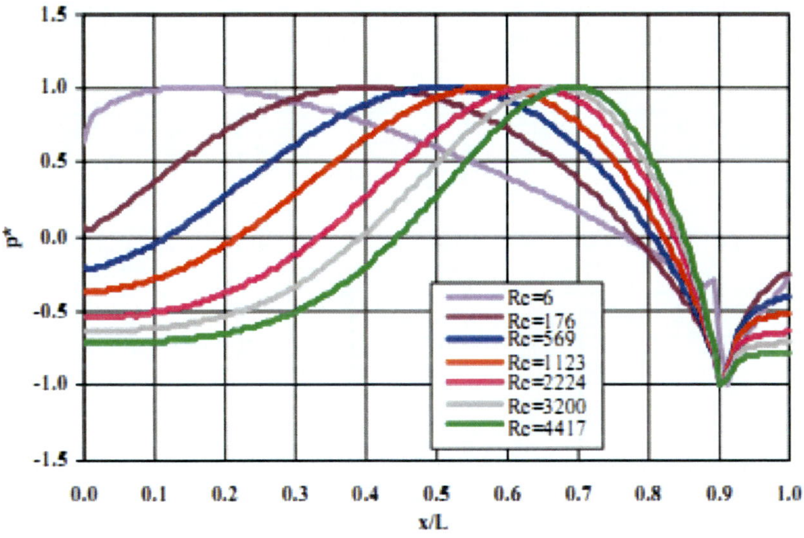

Figure 5. Non-dimensional pressure along the bedform for Runs 1, 3, 4, 5, 7, 9 and 11.

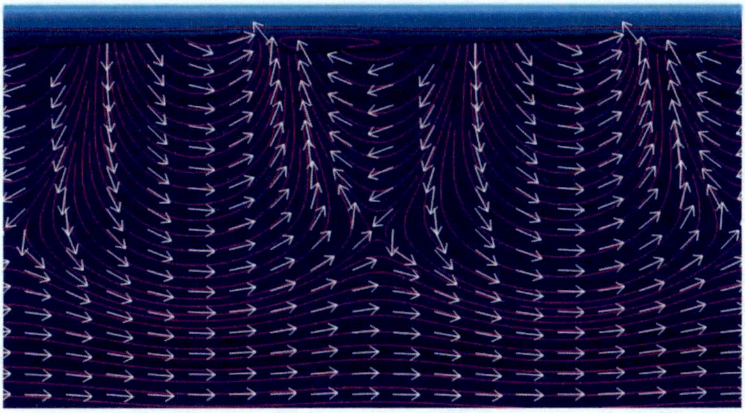

Figure 6. Streamline and velocity vectors in the porous medium for Run 4 with $Re_H = 569$.

Second, the characteristics of the flow below the sediment-water interface were considered. Figure 6 presents the flow field in the porous medium for $Re_H = 569$. Note that the velocity vectors (arrows) show only the flow direction (not magnitude). Water entered the bed along the upper part of the bedform's stoss face and back up into the water column on the lee face and the lower part of the stoss face. This area is the hyporheic zone or the interfacial exchange zone [31]. Below the hyporheic zone, the mean pressure gradient was dominant over local gradients. This results in *underflow*, or ambient flow [31], within the sediments and was present in all the simulations. Note that the mean pressure gradient is the same in both domains. As performed by Bayani-Cardenas and Wilson [31], the hyporheic zone was visually identified on a high-density plot of streamlines, selecting the streamline that separated the sediments region affected by exchange flows with the water column from the unaffected area dominated by underflow. The maximum vertical depth of this zone, d_z, was defined as the vertical distance from the deepest point of this streamline to the trough of the third bedform. Furthermore, in the hyporheic zone, a sub-horizontal exchange flow between the sediments region below adjacent bedforms was observed.

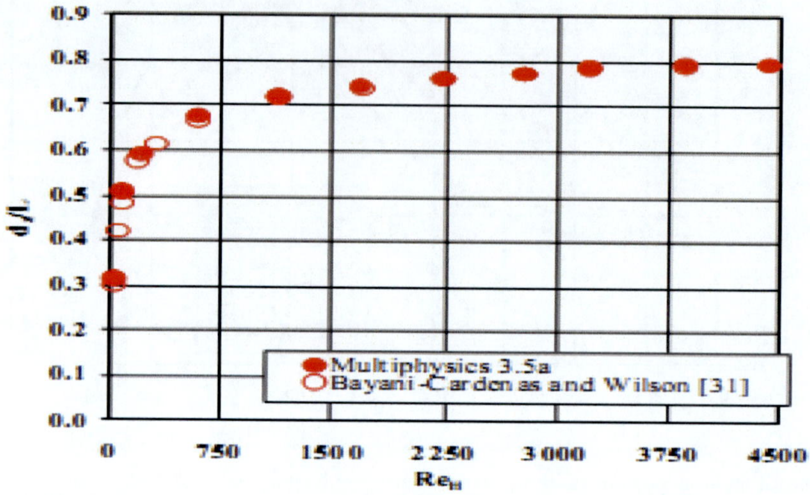

Figure 7. Non-dimensional hyporheic zone depth d_z/L vs Re_H for Runs 1 - 11.

Figure 7 presents d_z/L data against Re_H for Runs 1 - 11. As the bedform height-based Reynolds number Re_H increased, the depth of the hyporheic zone rapidly increased up to $Re_H = 1000$. After that observation, d_z increased gently toward an asymptotic limit (Figure 7). Ultimately, the depth of the hyporheic zone exhibited the same trend with respect to the Re_H value along the length of the separation region downstream of the bedform crest. As has been pointed out by Bayani-Cardenas and Wilson [31], this result confirmed that the separation region and the related bottom pressures controlled the flow field in the upper part of the porous medium and the size of the hyporheic zone. The d_z/L data were in close agreement with those from [31]. It was demonstrated that the d_z/L data fall on one curve that can be described by a Michaelis–Menten model [31].

Third, the water fluxes exchanged between the free fluid and the porous medium were considered. In particular, the non-dimensional effective flux density q^*_{int} was calculated following the approach proposed in [31], that is, by integrating the magnitude of the normal flux over a bedform surface and dividing it by L and by two because the integration does not discriminate between fluxes going into and out of the bed, which are equal. This effective flux density based on the bedform length q_{int} was non-dimensionalized by dividing by the hydraulic conductivity of the porous medium as follows

$$q^*_{int} = \frac{q_{int}\, v}{K\, g} \tag{7}$$

where g is the gravitational constant and v is the kinematic viscosity of water. Figure 8 presents the flux density against Reynolds number for Runs 1 - 11. It is noted that the flux density grows with Re_H and with the difference between the maximum and minimum bottom pressures, both of which are without limit. As the pressure drops, dp and Re_H increase, as the variation of the pressure at the interface increases. Therefore, the flux density increases. As has been pointed out by Bayani-Cardenas and Wilson [31], the relationship between q^*_{int} and *Re* becomes almost linear for $Re_H > 1500$ and can be described by a power function.

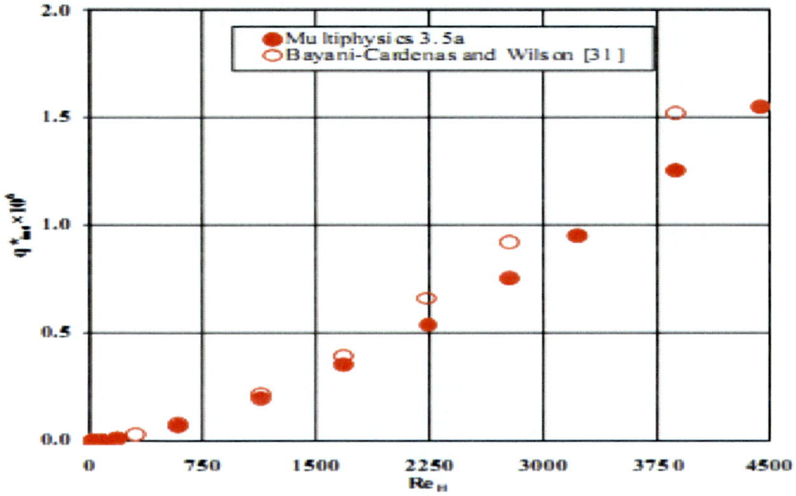

Figure 8. Non-dimensional flux density q^*_{int} vs Re_H for Runs 1 - 11.

Overall, Runs 1 to 11 confirmed the previous findings from Bayani-Cardenas and Wilson [31] regarding the laminar flow above a triangular bedform and the resulting hyporheic flows across the bedform.

3.2. Effect of Permeability of the Porous Medium on the Hyporheic Exchange

Runs 12 - 22 were carried out to investigate the effect of changes in permeability on the structure of hyporheic flows. As in Runs 1 - 11, the permeability of the porous medium was 1×10^{-12} m²; Runs 12 - 16 refer to a lower permeability, i.e., 1×10^{-12} m², and Runs 17 - 22 to a higher permeability, i.e., 1×10^{-8} m².

First, the analysis of the separation region showed insignificant changes caused by the reduction of the permeability, whereas for the larger permeability, i.e., 1×10^{-8} m², it was noted that the dimensionless separation length L_e/L increased with Re_H until $Re_H \approx 600$ and subsequently decreased. Figure 9 shows the numerical data for all of the Runs performed.

This different trend could be explained by considering how x_d and x_r migrated along the dune with the increasing Re_H at the largest permeability.

Figure 10 presents the flow field with streamline and velocity vectors close to the sediment-water interface for Run 21 with $K = 1\times10^{-8}$ m². It is noted that due to the larger value of the permeability, the velocity vectors are notably sub-horizontal below the crest and the lee face. These vectors enter the water column downstream of the dune trough along the stoss face. Also, large values of the Darcian velocity were observed. Thus, while for the lower permeabilities, the location of the detachment point moved upstream of the dune crest with increasing Re_H for $K = 1\times10^{-8}$ m², x_d migrated downstream of the dune trough for $Re_H > 600$. Therefore, despite the fact that even x_r migrated downstream, the reattachment length L_e/L

decreased with increasing Re_H, and the separation length for the largest permeability was located entirely along the stoss face of the dune.

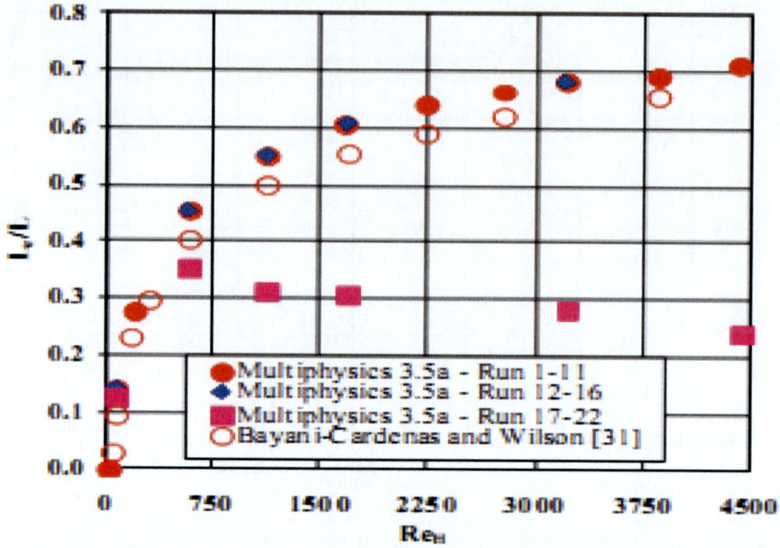

Figure 9. Non-dimensional separation length L_e / L vs Re_H for Runs 1 - 22.

Figure 10. Streamline and velocity vectors in the porous medium for Run 21 with $Re_H = 3219$.

Second, this trend also affected the pressure field along the dune, which interacts with the pressure field in the porous medium and, in turn, results in the hyporheic zone. Figure 11 presents the distribution of the non-dimensional pressure p^* along the bedform for two series of data, which are characterized by similar values of Re_H but different permeability values. The full lines refer to $K = 1 \times 10^{-10}$ (Runs 4, 5, 9 and 11) and the dashed lines to $K = 1 \times 10^{-8}$ m² (Runs 18, 19, 21 and 22). It is noted that the maximum p^* for the larger permeability holds for any Re_H located downstream of that for the smaller permeability. Also, for $K = 1 \times 10^{-8}$ m² at the larger Re_H, the minimum non-dimensional pressure p^* is located close to the dune trough, i.e., $x / L = 1.0$. For $K = 1 \times 10^{-8}$ m², the local gradients at the sediment-water interface were less important, thereby resulting in a weaker pumping effect on the porous medium immediately below the interface. Therefore, as expected, the non-dimensional hyporheic zone depth d_z / L increased with Re_H up to $Re_H \approx 600$ after which it decreased (Figure 11).

Finally, and again, the different trend regarding the growth of separation length with Re_H resulted in smaller non-dimensional density fluxes across the sediment-water interface (Figure 12) at $K = 1\times10^{-8}$ m², though the numerical data followed a power law, as indicated by the numerical results for the larger permeabilities [31].

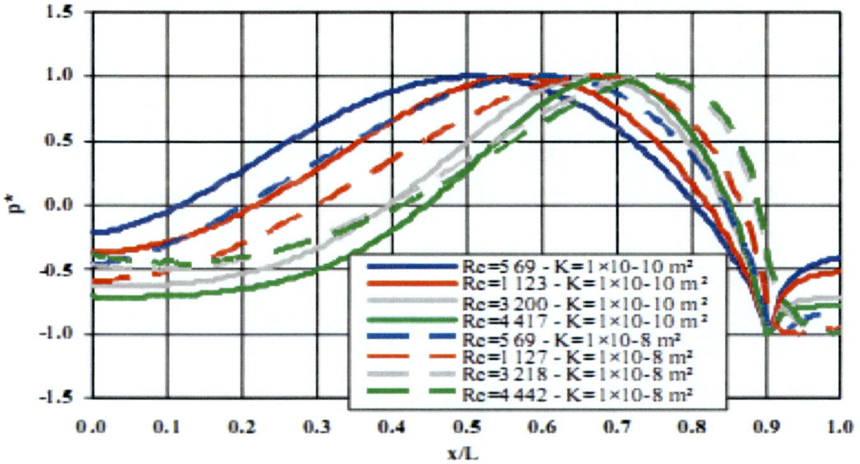

Figure 11. Non-dimensional pressure along the bedform for Runs 4, 5, 9, 11, 18, 19, 21 and 22.

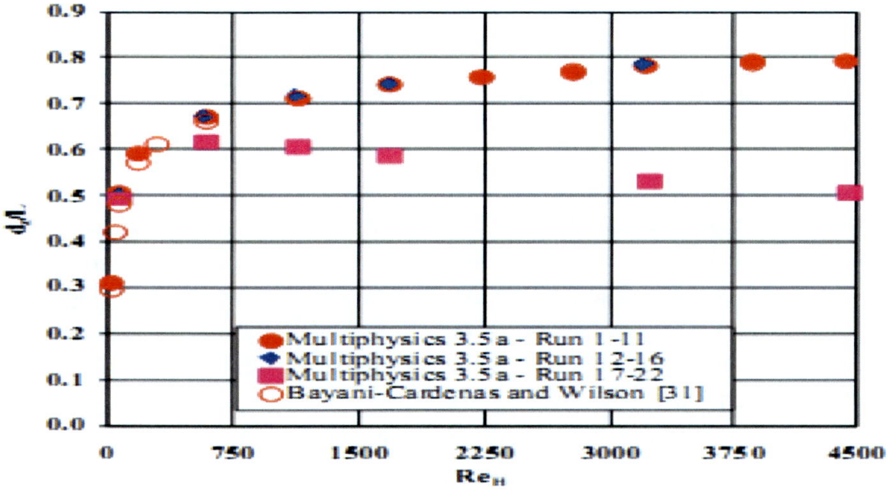

Figure 11. Non-dimensional hyporheic zone depth d_z / L vs Re_H for Runs 1 - 22.

Overall, the numerical results demonstrated the effect of large increases in the permeability of the porous medium on the characteristics of the separation region downstream of the dune crest as well as of the hyporheic exchange flows.

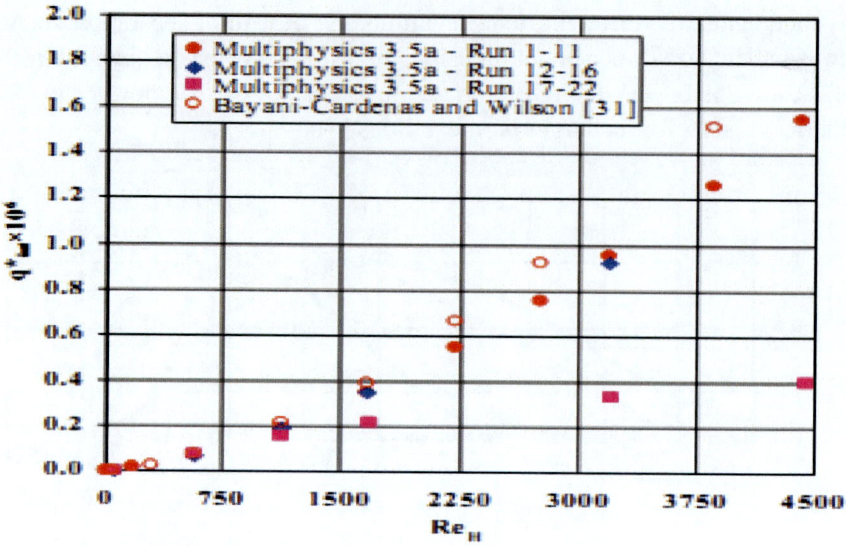

Figure 12. Non-dimensional flux density q^*_{int} vs Re_H for Runs 1 - 22.

3.3. Effect of Turbulence and Large Scales on the Results: Discussion

Finally two issues of more general interest must be carefully discussed.

The first issue is related to the assumption of laminar flow in the water column and pertains to the possibility of extending or not extending the above results to turbulent flows. Turbulence is ubiquitous in natural fluid flows because of the large scales that these flows typically occupy. Only in a limited number of cases where natural flows are characterized by small currents, such as in quiescent lakes, wetlands, or ocean bottoms, is the flow laminar. In these cases, the presented results could be used to predict flow behavior. Note that there are no published results for laminar or transitional flows to be compared with the numerical results presented in this study. Their extension to turbulent flows should consider the findings from experimental and numerical works about the turbulent flow over bedforms and from flows with similar features, such as the classical flow in a channel with a sudden expansion (backward-facing step flow). Consider, for example, the separation region observed downstream of the bedform crest (Figure 3). Experiments for fully turbulent flow over two-dimensional triangular bedforms [44, 45] resulted in separation lengths shorter than those observed in the laminar flow presented in this study. This finding was also confirmed by preliminary numerical simulations carried out on the same geometry using a k-ω turbulence model [46]. Finally, it is well known that in the backward-facing step flow, if laminar flow conditions hold, the length of the separation region, which is termed reattachment length, increases with increasing Reynolds number, which controls the flow. Beyond that, the length irregularly decreases in the transitional region with a sharp initial decrease followed by a continued gradual but irregular decrease to a minimum at approximately $Re_H = 2075$. Beyond $Re_H = 2475$, the flow is turbulent, and the primary reattachment length ceases to be a function of the Reynolds number Re_H and remains constant [47]. It is believed that in the backward-facing step flow, the laminar regime occurs for $Re_H < 450$. On the other hand, the transitional regime falls within the range from $Re_H > 450$ to $Re_h < 2475$, and the turbulent regime begins

CONCLUSIONS

In recent years, an increasing number of studies have addressed the analysis of the hyporheic zone where groundwater and stream water mix. Despite considerable efforts and significant findings, the study of hyporheic zones is far from being complete and requires further research.

This chapter presented the results of a numerical study of the flow in a coupled free fluid-porous medium domain. Laminar flow was assumed in the free-fluid sub-domain. Two series of simulations were carried out. The first series (Runs 1 - 11) was aimed at performing a comparison with previous literature results [31], whereas the second (Runs 12 - 22) was aimed at investigating the influence of the permeability of the porous medium on the flow field across the entire domain. The numerical results from Runs 1 - 11 confirmed previous numerical studies that have indicated that the separation region and the related bottom pressures are a key parameter that control the flow field in the upper part of the porous medium and the size of the hyporheic zone. Second, these numerical results confirmed the role of the bedform height-based Reynolds number of the water column in the evolution of the characteristics of both the separation region and the hyporheic zone.

Runs 12 - 22 were carried out for the same domain but with different values of the permeability of the porous medium to explore the influence of this parameter on the hyporheic exchange. These numerical results showed that a large decrease in the permeability produces no significant changes in the characteristics of either the separation region or the hyporheic zone. On the contrary, a large increase in the permeability produced an increase in the separation length with Re_H until $Re_H \approx 600$, whereas above this value, L_e decreased. Also, local gradients at the interface were less important, and the depth of the hyporheic zone also decreased with increasing Re_H above $Re_H \approx 600$. Finally, weaker dimensionless flux densities were observed for the largest permeability. Overall, the numerical study demonstrated that the permeability of porous media may be another key parameter in hyporheic flows.

Finally, it should be noted that the presented results were obtained under the assumption of laminar flow in the water column, which is rarely true in natural flow. Therefore, their extension to turbulent flows is not straightforward, despite several arguments suggesting that the separation region downstream of the bedform in turbulent flow is shorter than in laminar flow. Moreover, the results presented in this study, together with other numerical results, suggest that the aspect ratio of the bedform affects both the length of the separation region and the depth of the hyporheic zone, despite the fact that there are no numerical simulations available for low aspect ratios, such as those observed in large rivers.

ACKNOWLEDGEMENTS

The author acknowledges the preliminary work carried out by his students F. Salzano, P. Naviglio, A. Crocamo, R. Di Giovanni, F. Addelio, G. Lombardi, M. R. Mattei, V. Porzio and F. Sorrentino. The author thanks Prof. Fabián A. Bombardelli for his helpful comments.

REFERENCES

[1] K.E. Bencala, R.A. Walters, *Water Resour. Res.*, 19(3), 718-724 (1983).

[2] W.B. Dade, *Limnol. Oceanogr.*, 38(1), 52-69 (1993).

[3] Boulton, S. Findlay, P. Marmonier, E. Stanley, H. Valett, *Annu. Rev. Ecol. Syst.*, 29, 59-81 (1998).

[4] M. Huettel, W. Ziebis, S. Forster, G.W. Luther, *Geochimica et Cosmochimica Acta*, 62(4), 613-631 (1998).

[5] M. Huettel, H. Røy, E. Precht, S. Ehrenhauss, *Hydrobiologia*, 494(3), 231-236, (2003).

[6] G.N. Nishihara, J.D. Ackerman, In *Fluid Mechanics of Environmental Interfaces;* C. Gualtieri, D.T. Mihailović, Eds.; TaylorandFrancis Ltd: London, GB, 2008; pp. 299-326.

[7] S. Findlay, *Limnol. Oceanogr.*, 40(l), 159-164 (1995).

[8] M. Brunke, T. Gonser, *Freshwater Biol.*, 37, 1-33 (1997).

[9] Wörman, A.I. Packman, K. Jonsson, *Water Resour. Res.*, 38(1), 1001-1016 (2002).

[10] S. Findlay, D. Strayer, C. Goumbala, K. Gould, *Limnol. Oceanogr.*, 38(7), 1493-1499 (1993)

[11] J.W. Harvey, C.C. Fuller, *Water Resour. Res.*, 34(4), 623-636 (1998).

[12] P. Hancock, A. Boulton, W. Humphreys, *Hydrogeol. J.*, 13, 98-111 (2005).

[13] F. Boano, *Ph.D. Thesis* (2006)

[14] J.W. Harvey, K.E. Bencala, *Water Resour. Res.*, 29(1), 89-98 (1993).

[15] J.K. Anderson, S.M. Wondzell, M.N. Gooseff, R. Haggerty, *Hydrol. Process.,* 19, 2931-2949 (2005).

[16] M.N. Gooseff, J.K. Anderson, S.M. Wondzell, J. LaNier, R. Haggerty, *Hydrol. Process.,* 20, 2443-2457 (2006).

[17] A.G. Fernald, D.H. Landers, P.J. Jr. Wigington, *River Res. Applic.*, 22, 1111-1124 (2006).

[18] T. Kasahara, A.R. Hill, *J. Am. Water Resour. Ass.*, 43(5), 1257-1269 (2007).

[19] A.G. Fernald, P.J. Jr. Wigington, D.H. Landers, *Water Resour. Res.*, 37(6), 1681-1694 (2001).

[20] F. De Smedt, *J. Hydrol.*, 339, 29-38 (2007).

[21] L.L. Lautz, D.I. Siegel, *Hydrol. Process.,* 21, 3533-3548 (2007).

[22] L.J. Thibodeaux, J.D. Boyle, *Nature*, 325(6102), 341-343 (1987).

[23] S.A. Savant, D.D. Reible, L.J. Thibodeaux, *Water Resour. Res.*, 23(9), 1763-1768 (1987).

[24] A.H. Elliott, N.H. Brooks, *Water Resour. Res.*, 33(1), 123-136 (1997).

[25] A.H. Elliott, N.H. Brooks, *Water Resour. Res.*, 33(1), 137-151 (1997).

[26] Marion, M. Bellinello, I. Guymer, A. Packman, *Water Resour. Res.*, 38(10). 1209-1221 (2002).

[27] A.I. Packman, M. Salehin, M. Zaramella, *J. Hydraul. Eng.*, 130(7), 647-656 (2004).

[28] F.J.R. Meysman, O.S. Galaktionov, P.L.M. Cook, F. Janssen, M. Huettel, J.J. Middleburg., *Biogeosciences Discuss.*, 3, 1809-1858 (2006).

[29] F. Boano, R. Revelli, L. Ridolfi, *Adv. Water Resour.*, 30(1), 148-156 (2007).

[30] M. Bayani-Cardenas, J.L. Wilson, *J. Hydrol.*, 331, 103-109 (2006).

[31] M. Bayani-Cardenas, J.L. Wilson, *Adv. Water Resour.*, 30(1), 301-313 (2007).

[32] M. Bayani-Cardenas, J.L. Wilson, *Water Resour. Res.*, 43(3), (2007).
[33] M. Bayani-Cardenas, J.L. Wilson, *J. Hydrol.*, 346, 69-80 (2007).
[34] Z.G. Feng, E.E. Michaelides, *Environ. Fluid Mech.*, 9(6), 617-634 (2009).
[35] A.H. Elliott, Rep. No. KH-R-52, W. M. Keck Laboratory of Hydraulics and Water Resources, California Institute of Technology, Pasadena, Calif. (1990)
[36] F. Habel, C. Mendoza, A.C. Bagtzoglou, *Adv. Water Resour.*, 25, 455-469 (2002).
[37] M. Carr, B. Straughan, *Adv. Water Resour.*, 26, 263-276 (2003).
[38] F. Habel, A.C. Bagtzoglou, *J. Am. Water Resour. Ass.*, 41(2), 461-476 (2005).
[39] F. Boano, C. Camporeale, R. Revelli, L. Ridolfi, *Geophys. Res. Lett.*, 33 (2006).
[40] M. Bayani-Cardenas, *J. Hydrol.*, 362, 134-141 (2008).
[41] M. Bayani-Cardenas, *Water Resour. Res.*, 45(1), (2009)
[42] Multiphysics 3.5a, User's Guide, ComSol AB, Sweden, 2009.
[43] A.J. Paarlberg, C.M. Dohmen-Janssen, S.J.M.H. Hulscher, P. Termes, *Water Resour. Res.*, 43 (2007).
[44] M.C.M.L. Van Mierlo, J.C.C. De Ruiter, *Delft Hydr: Rep. TOW, A55, Q789,* Delft, The Netherlands (1988).
[45] S.J. Bennett, J.L. Best, *Sedimentology*, 42 (1995).
[46] C. Gualtieri, Unpublished Numerical Simulations.
[47] B.F. Armaly, F. Durst, J.C.F. Pereira, B. Schönung, *J. Fluid Mech.*, 127 (1983).
[48] C. Gualtieri, *Proceedings of ASME Fluids Engineering Summer Conference (FEDSM2005)* (2005).

Part Two:
Application Issues

In: Essays on Fundamental and Applied Environmental Topics ISBN 978-1-61942-522-4
Editor: Dragutin Mihailovic

Chapter 8

DYNAMICAL DOWNSCALING: MONTHLY, SEASONAL AND CLIMATE CASE STUDIES

B. Rajković[1,*], K. Veljović[1] and V. Djurdjević[1]
[1]Faculty of Physics, University of Belgrade, Belgrade, Serbia

ABSTRACT

At the present state of modeling skills and computer resources, monthly forecasts are the shortest integrations falling in the long-term integrations category. What are the possible applications and the general meanings of such results relative to standard 5 to 10 day forecasts? Another issue is the ability to conduct forecasts without resorting to the nudging of large scales during integration. Finally, we discuss the quantitative scores for such forecasts.

The other types of integrations/simulations are climate time scale integrations. The need for the assessment of their quality is paramount. The first step in that direction involves downscaling the ERA-interim data set, which can be achieved by performing integrations in both the non-coupled and coupled modes.

1. INTRODUCTION

The simulation of regional climates allows the representation of phenomena occurring on a wide range of spatial scales, from the large-scale to the regional scale. "Large-scale" refers to scales that range from a few hundred kilometers to the planetary level. "Regional scale" refers to scales that range from a few hundred kilometers to the local level. Regional Climate Models (RCMs) have been developed with the intent to increase the resolution of longtime (climate) predictions in a specific area. The use of RCMs dates to the late 1980s. This type of model currently dominates in the field of high resolution, multi-decade climate simulations and, especially, climate change projections and seasonal prediction. Today, over twenty RCMs are available worldwide offering a comprehensive specification of regional climate

[*] Corresponding author: E-mail address: bora@ff.bg.ac.rs; Phone: +381 11 26 25 981. Fax: +381 11 32 82 619.

modeling genesis. Coupled Atmosphere-Ocean General Circulation Models (AOGCMs), which are defined by grid points with resolutions of about 150 to 300 km and involve objective global analysis/re-analysis with low resolution, can adequately simulate phenomena on continental scales only (main characteristics of global atmospheric circulation). They are not capable of achieving real regional scale processes. Regional climate simulation performed by coupling a Limited Area Model (LAM) with "global" information is a standard method for obtaining climate details. On the one hand, climate is a global phenomenon, and strictly speaking, only a General Circulation Model (GCM) with some representation of the ocean can simulate climate as a response of the atmosphere-ocean-land system to solar forcing. Global modeling must also take into account the very large-scale structure of the forcing resulting from the air-sea distribution, large-scale mountain ranges and large-scale vegetation structures. On the other hand, forcing also occurs on a very local scale to create local climates, or more precisely, local perturbation of the climate. At present, the only way to model local climates is through regional scale simulations by RCMs. This concept has been established in weather forecasting, which uses large-scale, time changing fields and limited area models over a sub-domain with much higher resolution. This procedure is known as the nesting technique (Figure 1).

There are two types of nesting: one-way and two-way. As a limited area model, the RCM requires the provision of both initial and boundary (lateral and in some cases surface) conditions, which are delivered from GCM simulation. In one-way nesting, the RCM does not return information to the global model, while in two-way nesting, it does return information.

The one-way nesting approach is more prevalent because of it is simpler to implement. An implicit assumption is that a nested RCM should produce high-resolution climate information even from low-resolution lateral boundary conditions (LBC) in a computationally efficient manner. Nested RCMs have been used in Numerical Weather Prediction (NWP) for a long time. The climatology of every RCM is determined by two factors: (i) the regional characteristics produced by internal forcing (e.g., by complex topography, air-land-sea interaction, coastlines, etc.) of the RCM [1] and (ii) the large-scale information provided by the LBCs.

Figure 1. Nesting of a regional model inside the global domain.

In addition, simulation inherits the errors existing in the driving GCMs, which opposes the assumptions made in theory. Thus, the RCM should not significantly interfere with the large-scale information from the GCM but should instead only add regional details derived from the regional scale forcing ([2], among others). Hence, RCM nesting has also been called "dynamical downscaling" because the objective is to obtain fields with realistic regional details as possible, but without contradicting the large-scale driving fields. The resolution in RCMs has improved over time. Today, models run at 30 - 50 km horizontal resolution, and some RCM experiments are performed with even higher horizontal resolutions up to 10 km. If there is a very large-scale separation between the driver model and the nested one, another (or more) nesting might be necessary. The major technical issues in designing an RCM run are as follows: the model resolution and the LBC update interval. Recent studies based on so-called Big-Brother experiments introduced by Denis et al. [3] have addressed these technical matters and determined that the LBC update interval (at first it was 12 h, now it is 6 h) is important. Another conclusion of these studies was that the ratio between the space resolution in GCM and RCM is relevant [4].

The other technical considerations, which are also important, are as follows: type of boundary conditions and choice of model domain (domain location and domain size). The RCM solution varies with the computational domain size (e.g., [5, 6]) as well as location and season (e.g., [7, 8]). These and other issues of boundary conditions as well as their relevance in limited area modeling have been further addressed in several papers [9 - 15], among others. With the continuing advances in regional climate modeling, these problems drew even greater attention.

2. Lateral Boundary Conditions

The first analysis of LBCs used in the meteorology and oceanography was that of Charney [16]. Modeling requires the defining of "well-posed" boundary conditions so that the partial differential equations have a unique solution that depends continuously on the provided data. In the context of linearized, one-dimensional shallow water equations, Charney [16] provides the following rule: "At least two conditions have to be specified at inflow points, and one condition at outflow." The discretization of equations of motion poses additional problems for LBCs, which differ depending on grid configurations (the A, B, C, D and E) in Arakwa-Winninghoff notation [17]. Each of these grids has its particular variant of LBCs. The problem arises from the fact that numerical schemes create numerical modes in addition to the natural one. If we have a single wave with the single-phase speed, we could devise an outflow boundary condition, but with dispersion, we have multiple phase speeds, which make it very impractical to devise separate outflow boundary conditions for each one. A clear example of the problem coming from the exit boundary is presented by Miguez-Macho et al. [7]. By showing the difference between the analysis and simulations for a particular month, they claim that the error due to the length of their integration propagates inward and creates degradation even on the largest resolvable scales.

The most often used numerical scheme for lateral boundaries was proposed by Davies [9, 10]; its main goal is to mitigate large-scale differences between the driving fields and internal model fields by employing a relaxation procedure.

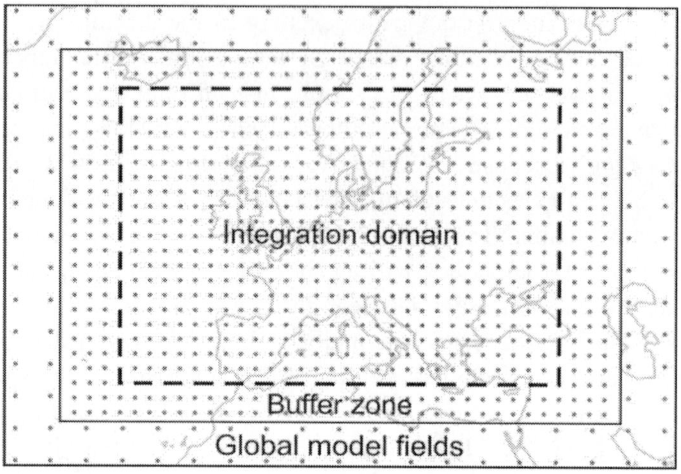

Figure 2. Schematics of the Davies boundary scheme. Relaxation is done over points in the buffer zone.

The scheme prescribes the introduction of LBCs over several (5 - 8) points in a wide buffer zone (Figure 2) where the variables of the regional model are gradually constrained to match the corresponding values of the driver model. In mathematical terms, the scheme adds the Newtonian relaxation term to the internal model tendency for each LAM LBC variable, $a = a(x, y, z, t)$

$$\frac{da}{dt} = \ldots - N\left(a - a_{dm}\right), \tag{1}$$

where epresents the Newtonian relaxation coefficient as a function of distance from the boundaries x, which takes the largest value at the lateral boundary and decreases toward the interior of the integration domain. The is a value given by the driver model. $N(x)$ can be quantified by different functional forms, such as linear or exponential, which are used to provide the gradual driving of the regional model solution toward specified large-scale fields [18].

3. Separate Treatment of the Large Scale in the Regional Models

Despite the implicit optimism of the "dynamical downscaling" approach, there are researchers who claim that limited area models are not able to maintain large-scale circulation (e.g., [3]). Based on a number of experiments in which the Regional Atmospheric Modeling System (RAMS) model was driven using the NCEP (National Centers for Environmental Prediction) Global Reanalysis, Castro et al. [5] made even stronger statements about the shortcomings of limited area models. They write about "the failure of the RCM to correctly retain value of the large-scale," stating that "the utility of the RAMS RCM, then, is not to add increased skill to the large-scale, rather the value added is to resolve the smaller scale features which have a greater dependence on the surface boundary." Separovic et al. [6] advocated this interpretation, pointing out the lower forecasting capacity of large-scales in their RCM simulation against global reanalysis. To circumvent this RCM "ineffectiveness," some researchers have suggested the use of a "spectral nudging" [19 - 21]. The practice of nudging large-scales in an LAM seems to have started with Waldron et al. [22]; they, in turn, described their inspiration as the NCEP's Regional Spectral Model (RSM) of Juang and Kanamitsu (e.g., [23]). Waldron et al. supported this method by stating that "the largest scale features resolved by the LAM may be better predicted by the global model." The spectral nudging method consists of large-scale forcing throughout the whole model domain. When considering that the regional scale climate statistics are determined by the consistency between continental scale atmospheric conditions and regional characteristics, such as mountain ranges or coastlines, in the spectral nudging approach, the regional model is forced to satisfy large-scale flow conditions inside the integration area. According to some researchers, spectral nudging may successfully keep the simulated state close to the driving state on larger scales, while smaller scale features are generated [19].

4. Downscaling Experiments Using a Regional Model

4.1. Models

The first operational version of our atmospheric model was created at the Hydrometeorology Institute and Belgrade University (HIBU) in the 1970s. It was still a dry model, but it was also very advanced because of the dynamical core that featured an energy and enstrophy conserving advection scheme, pressure gradient force and the accompanying omega-alpha term formulation [24, 25] choice of lateral boundary condition [26].

Because of efficient time marching, the model was extremely fast. The vertical coordinate was the sigma coordinate. Later, an additional vertical coordinate was introduced, the Eta coordinate [27]. In 1985, Janjic formulated [28] the so-called physics of the model as a combination of the convection [29], turbulence (Mellor-Yamad 2.5) and radiation scheme (GFDL, [30]). The surface scheme was the NOAH scheme [31].

In the second round, Janjic [32] made substantial revisions of both the convection (Betts-Miller-Janjic) and turbulence (Mellor-Ymada-Janjic) schemes until they arrived at what is more or less their present forms. The final stage of the model was the introduction of the non-hydrostatic mode so that the model could cover scales from several hundreds of meters to the synoptic one in the single code [33]. In this form, it was used as the operational model of the National Center for Environmental Predictions (NCEP), i.e., the Mezzo-NCEP model (Eta).

The philosophy behind these choices was that all components of a model should be of comparable complexity. Janjic (personal communication) always maintained that the weather forecasting model should be very efficient. It turns out that his models performed equally well as research tools. In that mode, computing efficiency is not a critical issue, but it is very helpful if the most advanced computers are not required and if many experiments can be performed.

For the LBCs, we have used various global fields: the ECMWF global runs or analysis, the ERA-interim fields and the SINTEX climate simulations of the CMCC in Bologna. The

SINTEX model is a coupled global general circulation model that combines the atmospheric model [34] and Ocean Parallelise (OPA, [35]) with the ORCA2 configuration.

At Belgrade University's Institute for Meteorology, we have been conducting dynamical downscaling for several years. For that purpose, we have used the EBU coupled atmosphere-ocean model as our regional model (RCM); the atmospheric portion used a slightly older version of the Mezzo-NCEP (Eta), while the POM (Princeton Ocean Model) was used to model the ocean component. Coupling is performed at every physical time step of the atmospheric model. For further information about the POM model, please see POM's user guide at http://www.aos.princeton.edu/. Because the issue of dynamical downscaling has been connected with LBCs and the solution of the reflection problem at the outgoing boundary points, we will explain those features of the EBU in some detail. Values from a global model that have been linearly interpolated are used at the first row of points in the regional model. For the inflow points, both normal and tangential velocity components are specified, while for the outgoing points, only the normal component is specified. In the second row, values are obtained as the average of the values from the first and third rows. This must be done to avoid sub-grid separation on two C sub-grids. However, we believe that a more important design feature for preventing the false reflection of the outgoing parts of the boundaries is the implementation of the upstream advection scheme for the outflow. This idea was introduced by Elvius and Sundstrom's [36] paper and is applied at several (4 - 5) points next to the boundary. The strong damping characteristics of upstream advection effectively cushion the down velocity at the outgoing points and thus reduce the numerical reflection. We have also tested Davies scheme, which, as we noted earlier, reduces false reflection by forcing the model values toward the large-scale values over several boundary points.

4.2. The Verification Method

A novel categorical statistics method that measures the correspondence between the estimated and observed occurrence of events and that focuses on spatial verification is used to verify wind estimates. A method has been proposed by Mesinger [37] to adjust the Equitable Threat Score (ETS) to arrive at a value that corresponds to unit bias to show the model's or forecaster's accuracy in placing precipitation. We will attempt to use it for wind forecast verification to assess not only wind speed but also the correspondence between the areas with similar wind intensities.

The difference between global and regional simulations for wind speeds that are 250 hPa greater than chosen wind speed thresholds at a given time is obtained through the calculation of two verification measures: the bias adjusted equitable threat score (ETS_a) and the bias score (BS). These are computed as

$$ETS_a = \frac{H_a - \frac{O^2}{N}}{2O - H_a - \frac{O^2}{N}}, \qquad (2)$$

$$Bias = \frac{F}{O} \quad (3)$$

with N denoting the total number of verification points, or events.

The bias adjusted value of the number of hits, H_a in the above equation is obtained as

$$H_a = O - \frac{F - H}{\ln\left(\frac{O}{O - H}\right)} \text{lambertw}\left(\frac{O}{F - H} \ln\left(\frac{O}{O - H}\right)\right) . \quad (4)$$

A triplet of F , H and O represents the forecast area, the corrected forecast area ("hits") and the observed area, respectively. The Lambertw function, also known as the Omega function or product log, is the inverse function of $f(w) = we^w$, where e^w is the natural exponential function, and w is any complex number.

Notice that the bias score, also known as the frequency bias index (FBI), measures the ratio of the estimated area (frequency) to the observed area (frequency). Thus, a bias equal to one implies no bias. If we want to compute these two scores such that they are representative for the whole ensemble, the standard approach is to calculate scores for each ensemble member separately and then formulate the arithmetic mean over all individual scores. However, we have formed these aggregate scores in a different way. Our sets of events (H = hits, O = observations and F = forecasts) are counted from the union of the sets of events of all ensemble members. The motivation for this approach is the basic premise of the ensemble idea that all members are equally probable. Beside the ETSa and the FBI scores, we have calculated and presented the standard area averaged standard bias (SB) and root mean square error (RMSE) scores. The time interval at which all scores were calculated was set at 6 hours. When we perform longer integrations, e.g., on a monthly, seasonal or 30 year scale, in the case of climate simulations, we should express the results in additional ways. Another important interpretation of these results is a quantitative measure of our confidence in them, which can be achieved by the ensemble approach. Our monthly and seasonal runs were represented in this way. These results are presented by the spaghetti diagrams of the "standard" scores of the four already mentioned scores, but we have also presented the EPSgrams of the wind fields at 250 hPa that are stronger than 45 m s^{-1}. In the case of the seasonal forecasts, the monthly averages of a given meteorological variable are calculated, followed by the respective EPSgrams. In the case of time slices, because of the lack of adequate computing recourses, we have performed a single integration, and those results are given in terms of the SB and/or RMSE scores. Apart from these quantitative measures, these results are also verified through visual inspection of the respective fields.

5. Results

This section contains some of the results obtained with our regional model covering almost all time scales, all of which are of interest today with the exception of the decadal forecast. The lack of computational resources prevented the formulation of the decadal forecast; because the other time scales overlapped that particular time scale, we do not anticipate serious problems. The interpretation of results, however, might prove to be more involved. Through these results, we claim that it is both possible and advantageous to use higher resolution regional models nested inside a global model. In particular, we will address whether the regional model is capable of reproducing the large-scale features present in the global runs and passed to the regional model only through LBCs. These results will be verified using previously defined quantitative scores: ETSa, FBI, RMSE and SB.

5.1. NWP, Mid and Long Range Runs

Before we present results from the ensemble integrations, we would like to mention one result from our previous study [38], which represents our first experience in prolonged forecasting. Using ECMWF analysis, we were able to create the so-called "perfect boundary conditions" for a period of 17 days. The study was part of the ADRICOSM-EXT project, under the coordination of IOC-UNESCO; the project leader was Dr. Nadja Pinardi, whose goal was to examine the feasibility of operational forecasts for the Adriatic Sea. Our job was to produce ocean forcing and to investigate into real time control using satellite data for surface winds. In Figure 3, we present the scores for 10 m winds (direction and speed) against the satellite observations over the Adriatic Sea. We can see that the results are quite good with a speed bias of - 8 m s^{-1}.

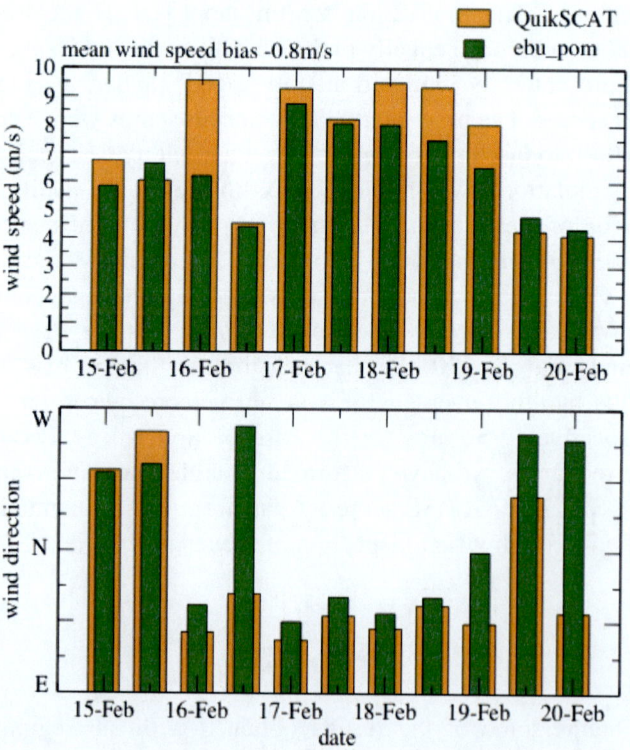

Figure 3. Scores for 10 meter winds for the domain over the Adriatic Sea.

Encouraged by these results, we started our work on monthly forecasts with an ensemble of 26 members (control + 25), which represented half of what is available at the ECMWF (European Centre for Medium-range Weather Forecasts). The first results we present are for the domain over the USA encompassing the Rocky Mountains, which comprise a sufficiently large obstacle that would have a significant influence on the large-scale flow structure. In that case, the RCM is allowed "enough time" to generate its own circulation inside the domain. The length of the runs was 32 days with LBCs from the global model with a resolution of T399/62 for the first 10 days, while for the rest of the run, the resolution was reduced to T255/62. The frequency of LBC updating was also changed: for the first 10 days, the LBCs were updated every 6 hours, but after this period, they were updated every 12 hours. The 6 hour gaps in LBCs were obtained through linear-interpolation of the available fields. The run started on January 1st at 0000 UTC. We chose winter as the starting point because we expected the jet stream activity to be enhanced during the cold season. The model time step was chosen according to the numerical stability criteria and was set at 90 s. Verification of both ECMWF forecasts and ours was made against the ECMWF analyses over regional model domain. The results reported here are for runs with 31 km horizontal resolution and 45 vertical layers.

As mentioned before, there was some concern about the reliability of the largest scales of regional models, so we have begun our verification analysis at that level and examined the polar jet stream, an excellent large-scale feature for this purpose. Forecast areas of 250 hPa wind speeds greater than 45 m s^{-1} were used as the proxy, and the placement of these areas was verified.

To satisfy ourselves that we had extended the jet stream scenario, we looked for the appearance of points with winds stronger than 45 m s^{-1} during the integration period. That information is given in Figure 4, which shows that a substantial portion of the domain (never less than 10 %) had strong winds at all times. The abscissa is the time in days, and the ordinate is the normalized frequency of the occurrence of winds stronger than 45 m s^{-1} anywhere in the domain.

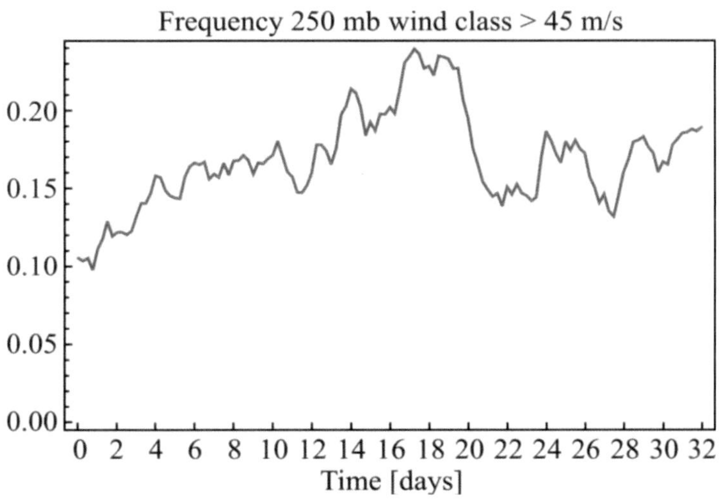

Figure 4. Frequency of winds stronger than 45 m s^{-1} over the whole domain during the integration time [in days].

We now present the results from our simulation. First, we examined the ability to forecast winds at 250 hPa; these results are presented in Figure 5. The left panel contains the 250 hPa wind field for the last day of the integration, as well as the top wind vectors and bottom wind intensity. The right panels display the standard scores for the global and regional models in red and black, respectively. The upper panel has the RMSE scores, while the lower one has the SB scores.

The main point is that there is no visible difference in the scores for both models. While they both degrade with time, they degrade roughly by the same amount. The same conclusion holds true for the ETSs and the FBI scores shown in Figure 6. Please remember that these scores are calculated from the union of forecasts over all ensembles. The similar ETS scores permit us to claim that not only the strength of the winds is equally well forecasted but also their position. This finding implies that large-scale simulations are equally well-forecasted by both models.

Two of the most important characteristics of an ensemble of forecasts are its spread and eventual clustering, which allows for grouping of the subsets of forecasts. Inspection of these aspects is usually performed in the form of a "spaghetti" plot with all members represented by their history curves for a given variable/parameter/score. In Figure 7, we present scores in such a format; the top portion displays the RMSE and SB scores, while the bottom four panels display the ETSa and Bias (FBI) scores for the regional run (left) and global run (right). The different colors are automatically set by the graphical package for each ensemble member.

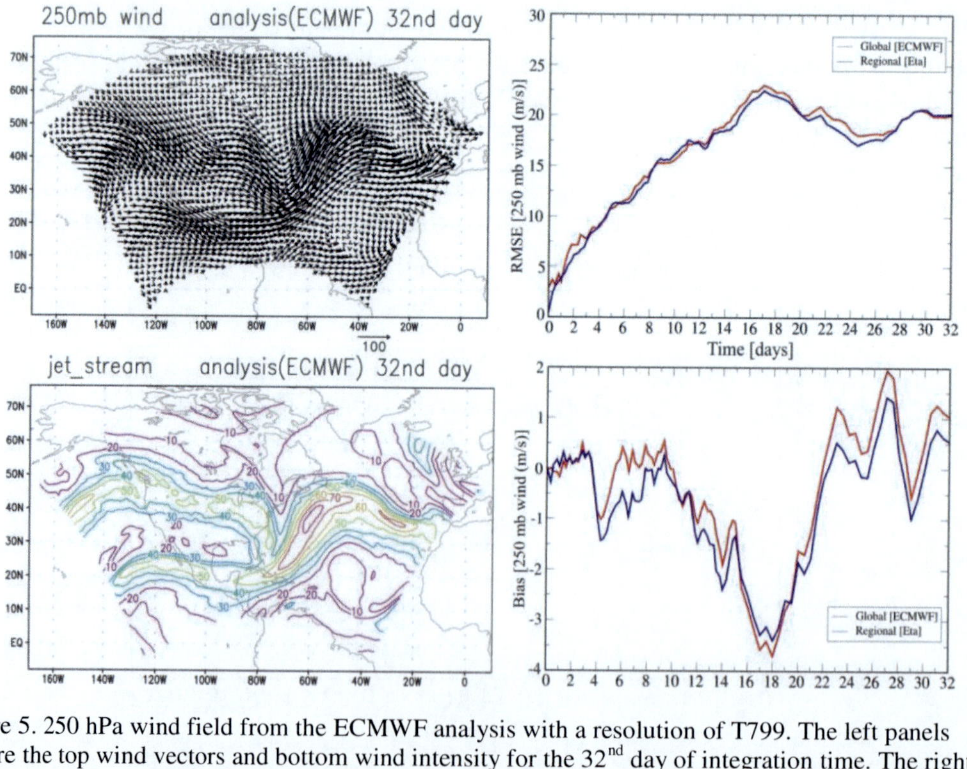

Figure 5. 250 hPa wind field from the ECMWF analysis with a resolution of T799. The left panels feature the top wind vectors and bottom wind intensity for the 32nd day of integration time. The right panels show the scores for both the regional and global models; the upper panel contains the RMSE, and the lower panel has the SB for 250 hPa winds as a function of lead days.

Finally, we present pairs of RMSE and SB scores member-by-member in Figure 8 as they change over time. From the inspection of Figs. 7 and 8, it is apparent that there are differences among the different members and a large spread for the ensemble, which is a desirable quality of every ensemble. The individual pairs of scores closely mirror each other.

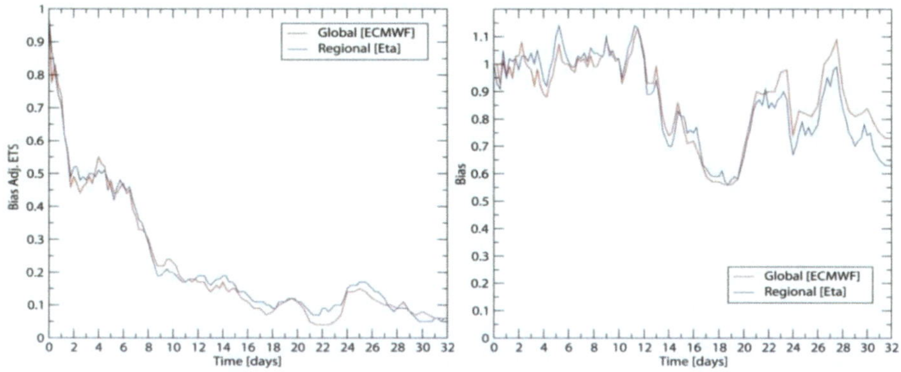

Figure 6. The ETS (Bias Adjusted) score (left panel) and FBI score (right panel) for the 250 hPa wind field as a function of lead days.

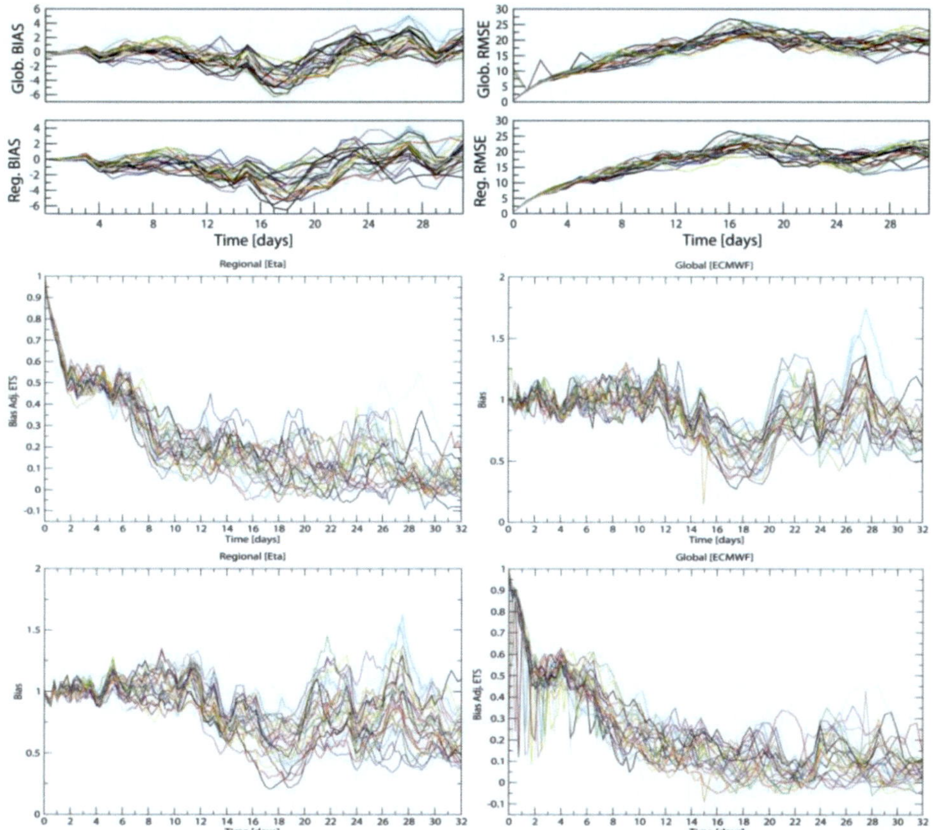

Figure 7. The "spaghetti" curves, top panel, for the RMSE and SB for the global and regional runs. The bottom left panels have the ETSa and Bias (FBI) scores for the regional run, while the bottom right panels have the same information for the global run.

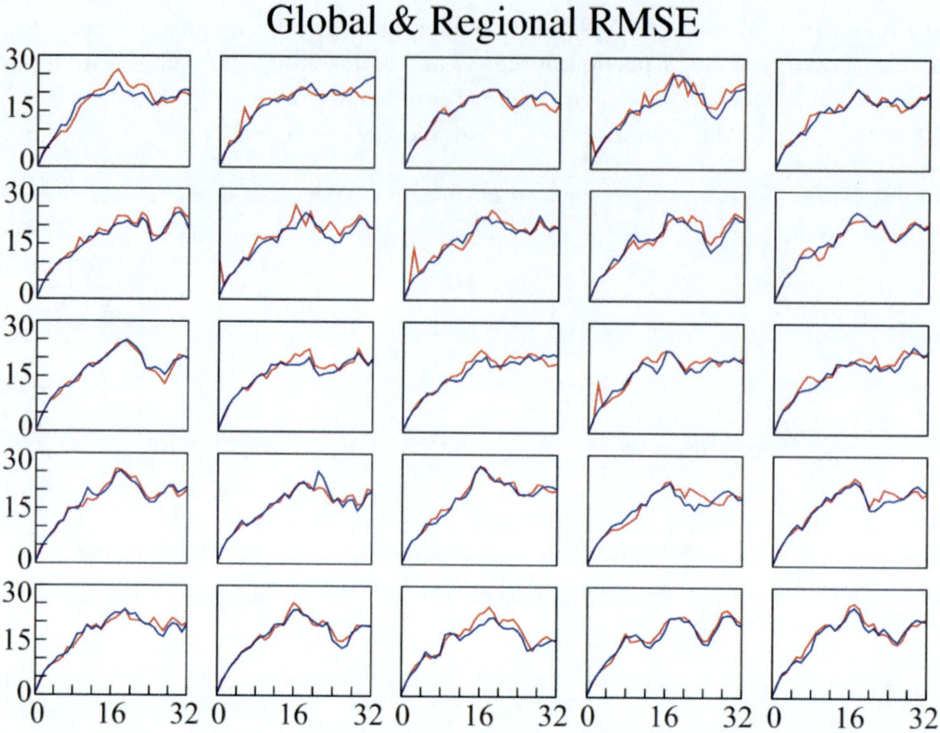

Figure 8a. The individual pairs of the RMSE for the 250 hPa wind forecasts using the global (red line) and the regional (blue line) model, for all ensemble members as a function of lead days.

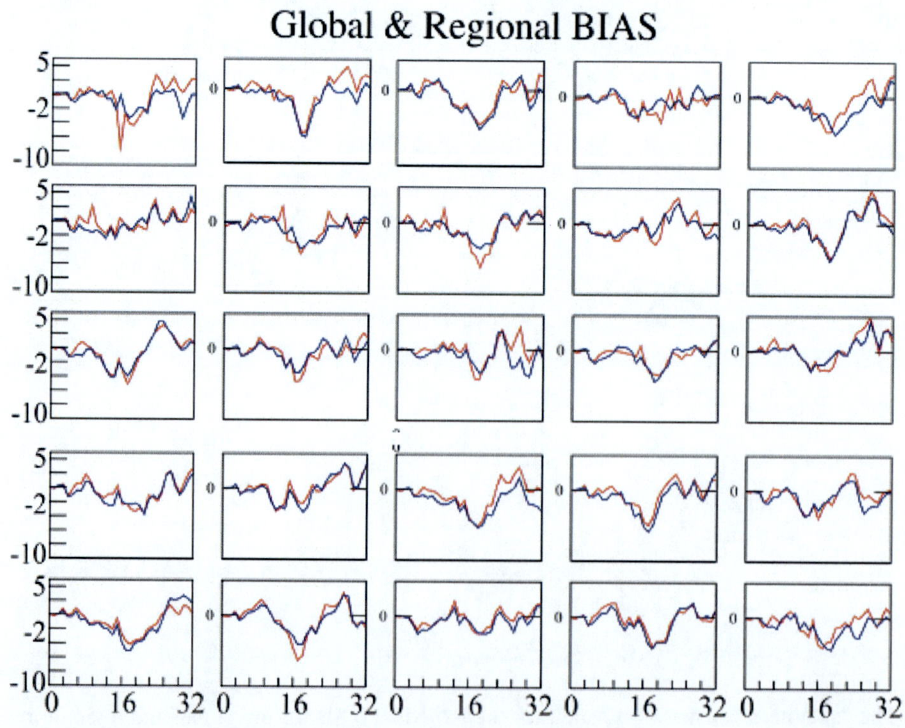

Figure 8b. The same as in Figure 8a, except for the Bias score.

The last piece of evidence presented for visual inspection shows 5 members at different moments. They are as follows: the control run, the two "best" runs (members no. 21 and 22) and the two "worst" runs (members no. 16 and 24), as judged by the "spaghetti" ETSa scores. The time moments are days 4, 8, 16 and 32. They are presented in Figure 9a with four panels in the horizontal with the first two representing the analysis (green contours), the third representing the analysis plus the global data (black contours) and the fourth representing the analysis plus the regional data. From top to bottom, the control is represented, followed by members 16, 21, 22 and 24. In Figure 9b, the information is presented for the same members as in Figure 9a, except for days 16 and 32.

The control and the two "best" cases showed very good agreement between the global and regional runs, except for the 16th and 24th members; however, these were still close enough as to maintain the large-scale flow. It is worth noting that these cases did not agree with the analysis, but this is a separate issue.

The findings from the verification of two 32 day 250 hPa wind forecasts, two of which were performed using the regional model and only differed with respect to the formulation of the LBC scheme (blue for standard and green for Davies, on scores panels, Figure 10) and one which was performed using ECMWF's global model (red line on scores panels, Figure 10), were compared for the first two ensemble members. The difference in the regional model skills for the different LBC schemes for these two members was quite small.

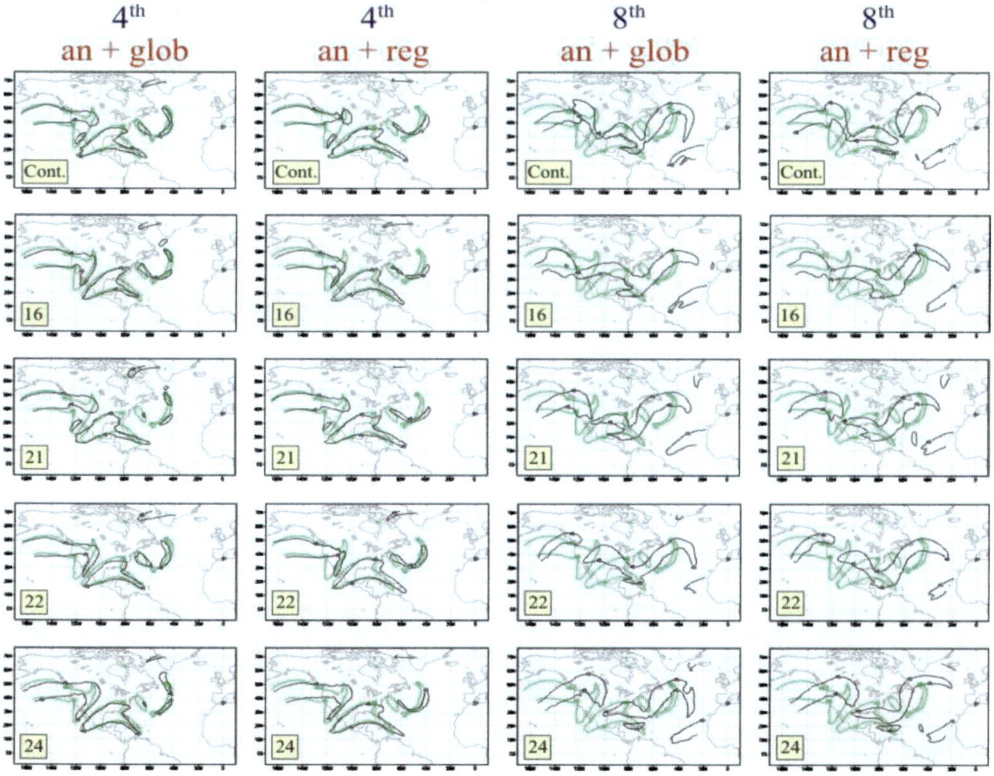

Figure 9a. From left to right: the first two panels reflect the analysis (green contours); the third, the analysis plus the global data (black contour); and the fourth, the analysis plus the regional data. From top to bottom: the control, 16th, 21st, 22nd and 24th ensemble members.

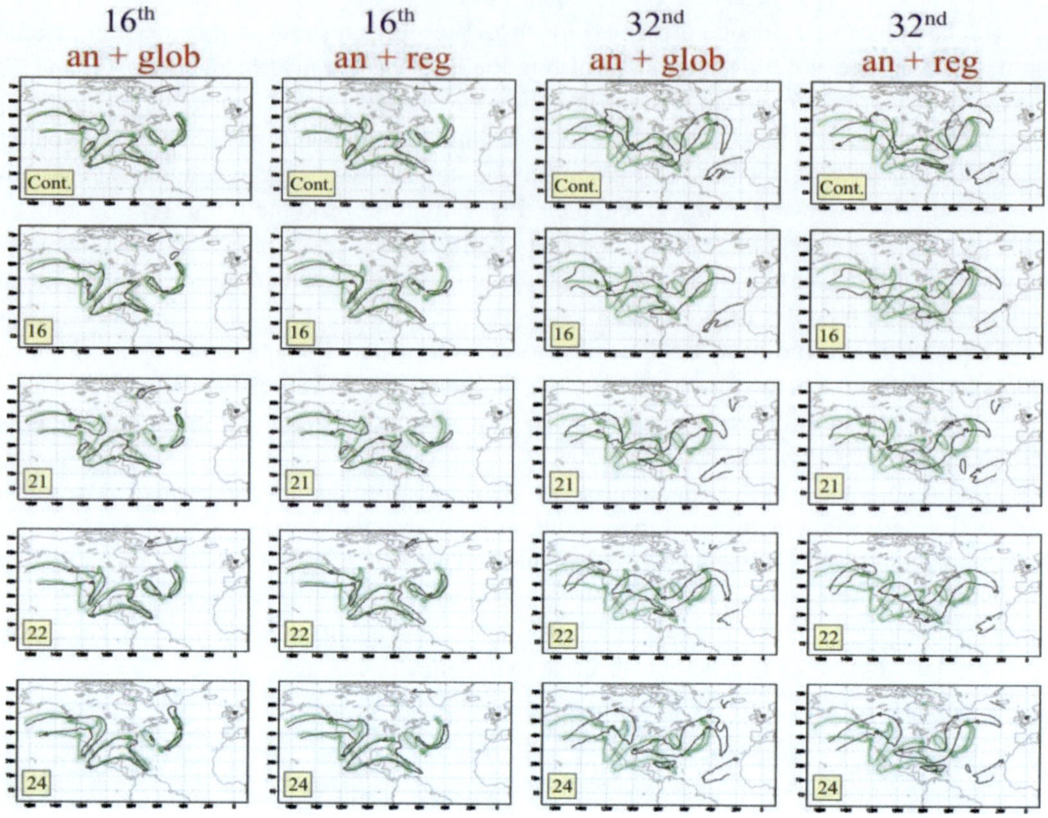

Figure 9b. The same as in Figure 9a, except for days 16 and 32.

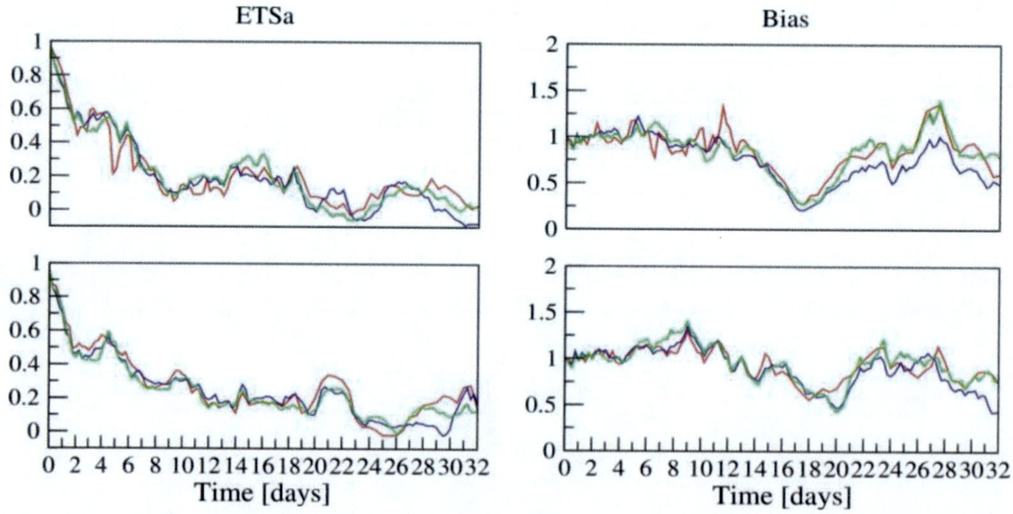

Figure 10. ETSa, left panels, and FBI, right panels, of 250 hPa wind speeds ≥ 45 m s^{-1}, of two 32 day integrations as a function of forecast time (days).

Usually, the starting point in the evaluation of a model skill is the anomaly correlation coefficient, ACC, which quantifies the spatial correlation between the forecast and observed deviations from a climatology value. We have used the following definition of ACC:

$$ACC = \frac{\sum_{m=1} f'_m o'_m}{\left[\sum_{m=1}^{M} (f'_m)^2 \sum_{m=1}^{M} (o'_m)^2\right]^{1/2}}$$

$$f'_m = f_m - c_m \qquad o'_m = o_m - c_m \tag{5}$$

where f are forecasts, o are observations and c are climatology values for all points (M) in the regional domain. The accepted threshold for skill is a value of 0.6. Below this value, a forecast has no skill. Climatology should be the model's climatology in relation to which forecast results can be expressed. However, its calculation requires substantial computing resources. Thus, our first approximation used the climatology of the global model (ECMWF) as a driver model. In this way, we slightly favored the ECMWF simulations. When we have an ensemble of forecasts, a synthetic, and yet informative, way to assess the reliability of forecast is an EPSgram with the 50th (median), 25th, 75th, 10th and 90th percentiles. We have computed the ACCs for both models and compared them. In Figure 11a, we present an EPSgram for the ACC of the 500 hPa height forecasts of the ECMWF T399 global ensemble (control plus 25 members) as a function of forecast time (days). The red line with diamonds represents the correlation coefficient of the control forecast. The box-and-whisker plot represents the median with a short horizontal line, the 25th and 75th percentiles with a wide, red vertical box, the 10th and 90th percentiles with narrow gray boxes and the minimum and maximum values with vertical lines.

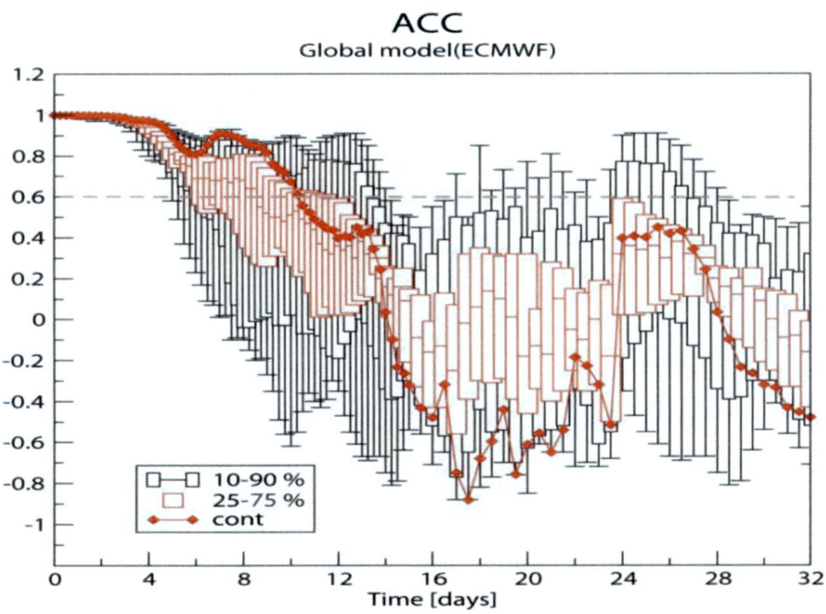

Figure 11a. The EPS forecast distribution – EPSgram for anomaly correlation coefficient (ACC) of the 500 mb height forecasts of the ECMWF T399 global ensemble (control plus 25 members) as a function of forecast time (days). Initial time is 0000 UTC 1 January 2009. Red line with diamonds: the control forecast. The EPSgram is represented by a box-and whiskers plot showing the median (short horizontal line), the 25th and 75th percentiles (wide red vertical box), 10th and 90th.

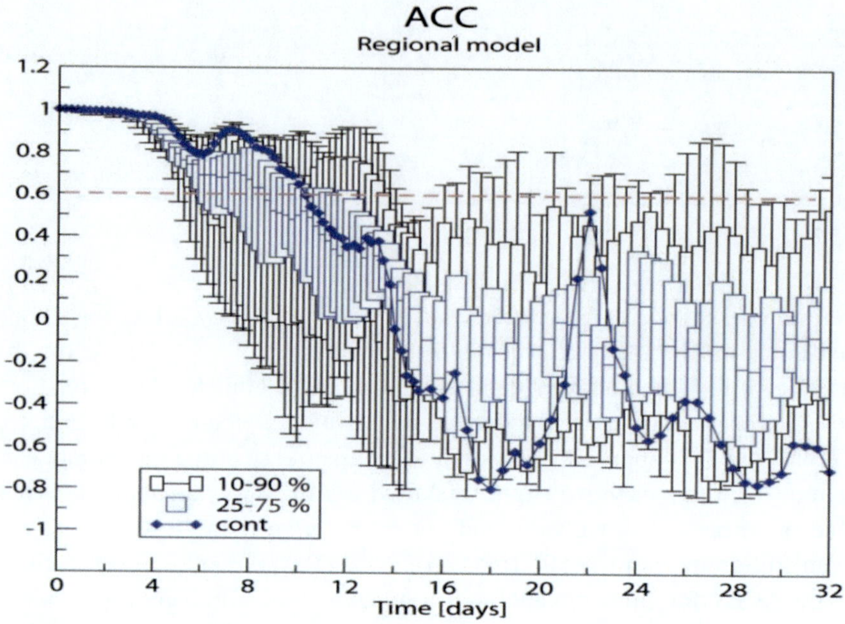

Figure 11b. The same as Figure 11a, except for the regional model.

Figure 11b is the same as Figure 11a, except for the regional model. Again, we mention only how close the scores are; thus, we can conclude that these figures show that the skills of both models are similar up to about 24 days. After that, the spread in the regional model stays the same, while the spread in the global model is narrower and features medians that become smaller at the end of the forecasts. Generally, both models exhibit similar evolution in the ACC score during the whole forecast period, and there is not a systematic error in the downscaled fields of the regional model.

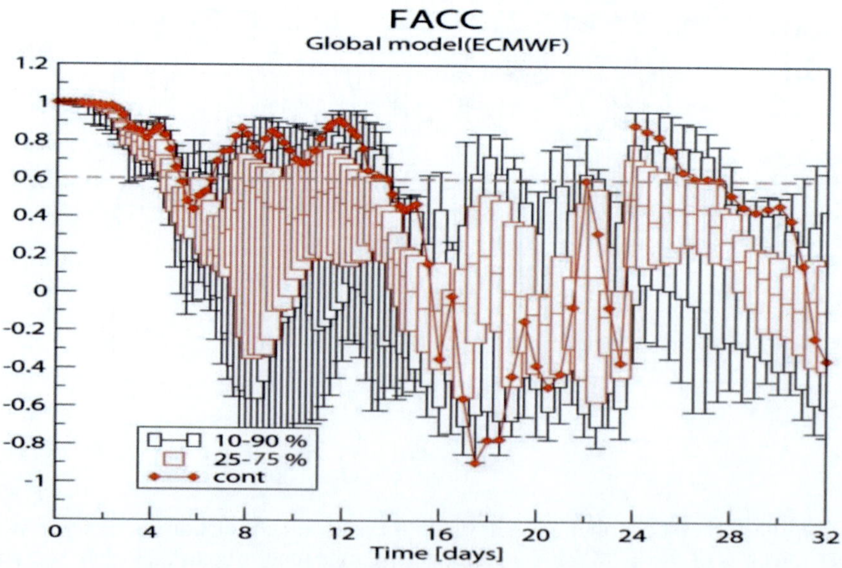

Figure 12a. The same as Figure 11a, except for a low-pass filtered fields.

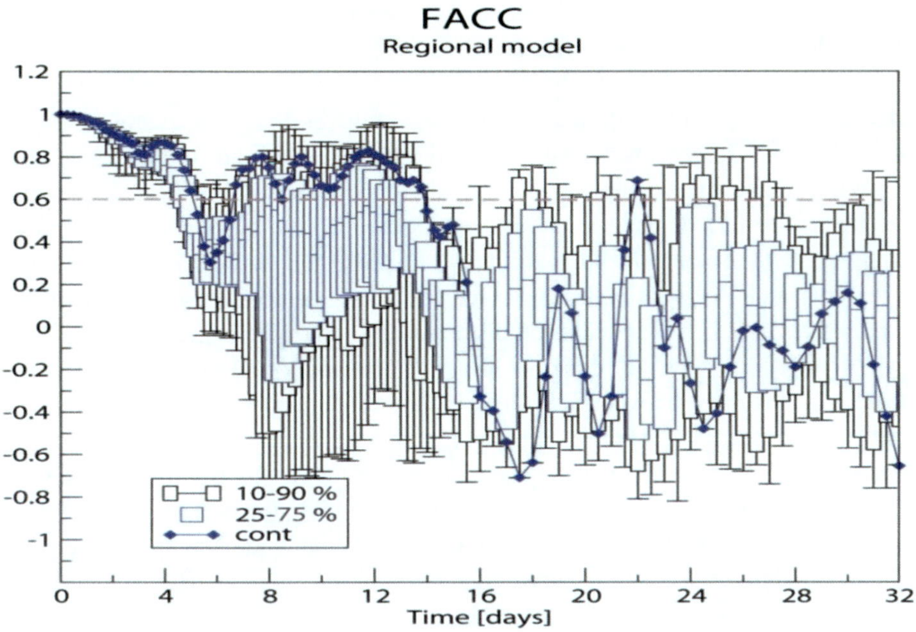

Figure 12b. The same as Figure 11b, except for a low-pass filtered fields.

The last part of our analysis of our monthly forecast relates to the idea of nudging the largest resolvable scales, as was mentioned in the introduction. The clearest way of forming part of the field with the largest resolvable waves is through Fourier decomposition. In the field of communication, this is referred to as low pass filtering. We performed a re-synthesis with the first 3 waves defined for the regional domain. The Fourier decomposition was conducted as described in Errico's paper with de-trending in both directions so that the remaining fields were periodic. The low wave number composition was then obtained using the phase shifted cosine function. For the obtained low passed fields, the ACC was calculated and presented as an EPSgram (Figure 12a and 12b). The most interesting, and perhaps expected, observation was the extension of the interval needed for the ACC score to fall below the 0.6 level. In the case of the full fields, this interval was about 10 days; for the filtered fields, this interval was 14 - 15 days. The two models were once again quite similar, indicating that the forecasts have similar statistics as well.

5.2. Seasonal Forecasts

The regional seasonal forecast started within the SECCOF concept in several parts of Africa and has been favorably received there. The same level of success is not possible at middle latitudes, but it has still become operational at several forecast centers. At the South East European Virtual for Climate Change Center (SEE-VCCC), in collaboration with the Institute for Meteorology, seasonal forecasts are run operationally once a month. The model is the same as the CRU data, i.e., the EBU-POM, and is yet another example of global to regional information transformation. The global model is again the ECMWF with its seasonal ensemble forecast from which an ensemble of BCs is formed in the same manner as for the standard 5 day forecasts. But the results have to be presented in a format that is very different from that of a deterministic 5 day forecast.

First, we do not think in terms of the atmospheric state for some moment in time and some particular location. Instead, we compute monthly or 3 month averages. Because the ensemble consists of 25 members, we calculate the 50th, 25th and 75th percentiles together with the minimum and maximum values. The same is performed for the global runs and for the climatology (normal). These values are verified against the measured T2m, shown as red dots, at those particular stations. These figures are presented in Figure 13. We can see that both the regional and global models reproduce the transition from the summer to the winter with acceptable accuracy in the prediction of the mean monthly temperature. For the Belgrade station (right panel), for 5 of 7 months, the observed mean monthly temperature is inside the 50th percentile boxes (box between 25th and 75th percentile) of the regional model forecast. The global model forecast for this station was less successful, but this situation was reversed for the Nis station, where the global model captured, 5 of the 7 monthly mean temperatures within the 50th percentile box. Overall, it is apparent that downscaling with the regional model did not deprecate the forecast skill of the global model and even shows the potential to introduce added value as a consequence of higher resolution.

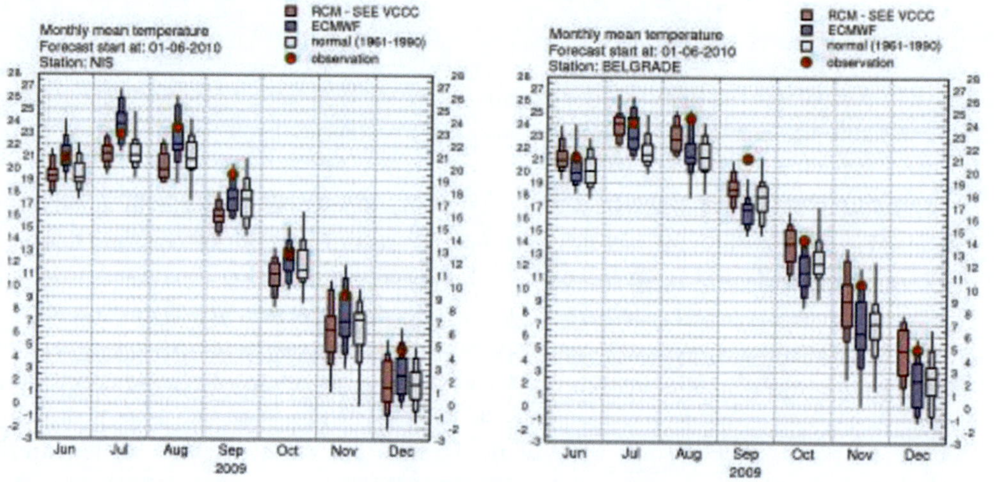

Figure 13. Seasonal forecast in the form of monthly means of T2m temperatures for two stations in Serbia, Belgrade and Nis from the regional model, global (ECMWF) model, observed climatology and the observed temperatures for that station.

5.3. Downscaling of the Present Climate Simulations

Today's frontier in global to regional interplay involves dynamical downscaling for a chosen time period, called time slices. Here, we give a brief overview of some of our findings in that area. The model is again the same RCM. The first step consists of downscaling some of the existing reanalysis data sets. Because the boundary conditions are perfect, one should simulate the observed climate; we performed this simulation using the ERA-interim data set. From this simulation, in Figure 14, we present the annual accumulated precipitation for one year (1989) from the reanalysis data and from our regional simulation.

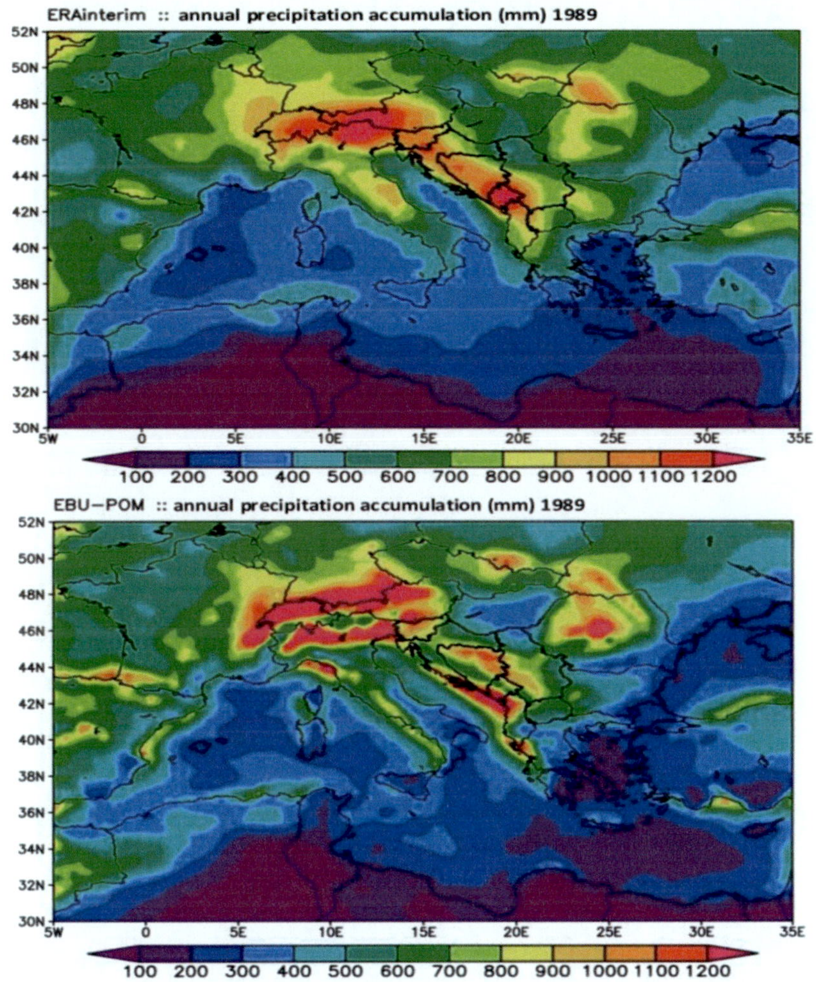

Figure 14. Climatology of the precipitation fields for the ERA-interim data set, upper panel, and the same field from the regional down-scaling, lower panel.

Upon visual inspection, it becomes clear that the spatial structure of the down-scaled precipitation field follows the spatial structure from the reanalysis dataset; the maxima and gradient coincide between arid African regions and Europe. The absolute values at the maximum locations are also similar. As a consequence of a higher resolution, the regional model results in more details being visible in the field, especially in the mountain and coastal regions.

The next step comprised the downscaling of the "present" climate from a global climate model. The previously mentioned SINTEXG model was used for this step. The two meter temperature and precipitation are the variables that are first examined in such simulations. Our T2m temperatures are presented in Figure 15 together with the global field and the observed CRU data.

Those temperatures are for the winter (djf) season over the domain covering most of the Europe. The scores for these simulations are well within the range of scores presented in the papers.

Table 1. BIAS, Mean Absolute Error and the Root Mean Square Error for all three fields

	BIAS	MAE	RMSE
CRU	-0.21	1.88	2.15
WM	-0.06	1.97	2.28

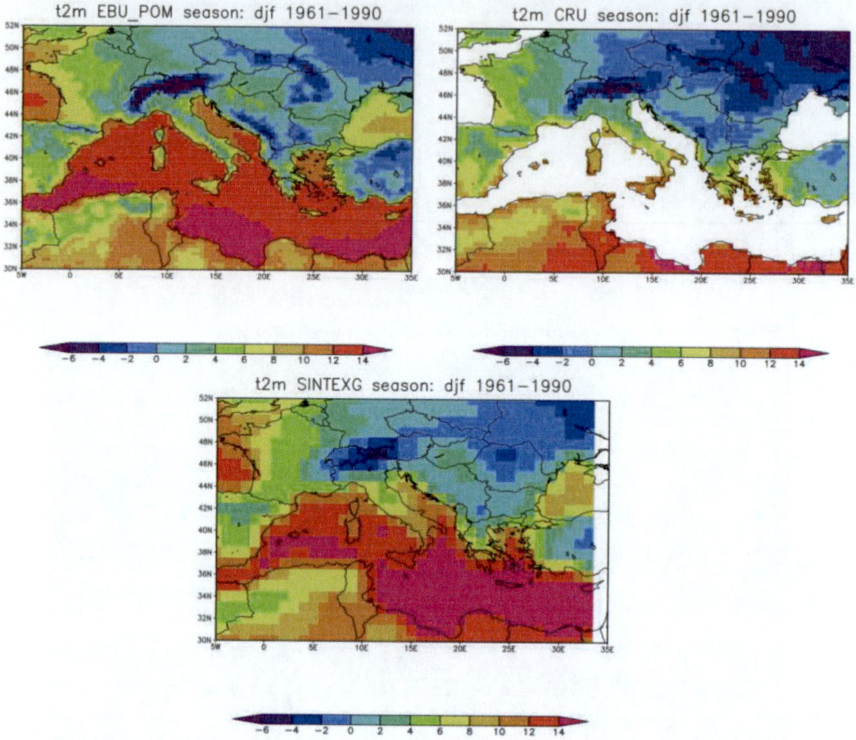

Figure 15. EBU-POM T2m field for the winter (djf) season, upper left panel. The upper right panel has the observed T2m data from the CRU data set (horizontal resolution of 0.5 deg. lat-lon). The lower left panel has the SINTEXG T2m for the same season.

Visual inspection of this figure reveals all expected benefits of the downscaling process, greater variability and in the “right” places. An even clearer example of improvement in the regional fields over global ones is shown in the next figure (Figure 16).

An inspection of the figure shows that the global model field has two obvious deficiencies. The wave-like form of the isolines (highlighted in green) is incorrect, and the horizontal gradient in the North-South direction is too large. This gradient is the consequence of erroneous overheating. The first problem is absent in the regional simulation, while the second one, although reduced, is still present. The final findings, as shown in Figure 17, concern the modeling of surface precipitation. As in the case of the ERA-interim data, the downscaled fields are closer to the observations of the CRU data set. A closer inspection of the precipitation in the region of the Monte-Negro coast shows that the downscaled fields performed better than the CRU data. However, it is important to keep in mind that the CRU data have only a 0.5 degree resolution, while the topography has much higher variability. From the observations, it is well known that that area has local extremes for precipitation that are well above those of the CRU dataset. For a full report of the down-scaling results, please refer to the SINTA final scientific report [39].

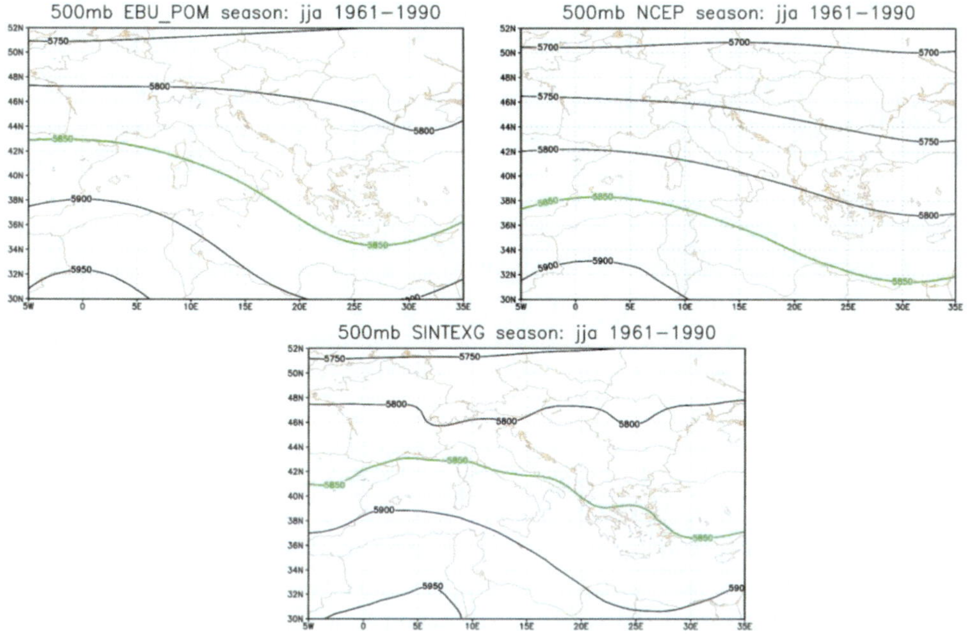

Figure 16. Geopotential isolines at 500 mb for the EBU-POM, NCEP-Reanalysis and SINTEXG model, all for the summer (jja) season. Note the unphysical shape of the 5 025 isoline (in green) and the very weak gradient for the SINTEXG filed. The EBU-POM field is closer to the NCEP-reanalysis in both respects.

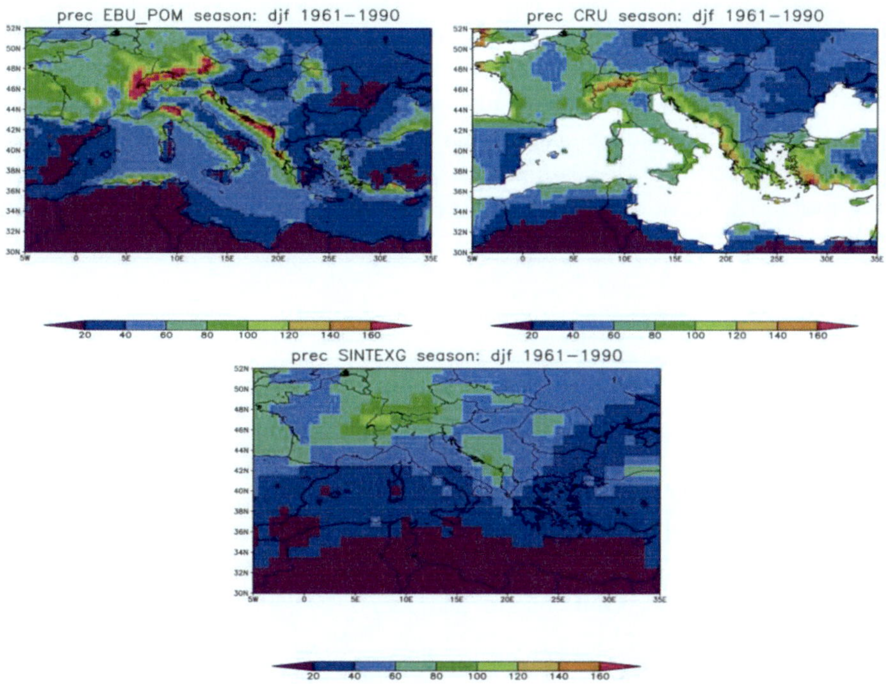

Figure 17. The EBU-POM precipitation field for the winter (djf) season (upper left panel), the EBU-POM, the CRU dataset (upper right panel) and the SINTEXG precipitation (lower left panel) for the same season.

Conclusion

Nesting of a regional model is a powerful tool for adding local information into global fields, which are generated by a global model. For a much smaller region, in comparison with the whole globe, there is the possibility of a substantial increase in resolution. In practice, this ratio is 4:10, depending mostly on the local computing resources. This conclusion holds on all time scales, including 5 - 10 day, monthly and seasonal forecasts as well as climate time slices (~ 30 years and longer). The medium time scales, i.e., the monthly and seasonal forecasts, are the most difficult to perform, and additional efforts are needed to achieve a proper understanding and extract useful information from these processes. Another challenge is presenting these forecasts in a manner that can be understood by non-specialist potential users. A technical point should be noted here: the statistical and probabilistic nature of their results requires climatology models that are very demanding of computing time and resources, especially in terms of memory storage. This point needs to be considered every time we "touch" the model, whether by improving the resolution, the dynamics, the physics, etc. The concept of the decadal forecast will be even more difficult to interpret, but the logistics of transferring global field information to a regional model is exactly the same as for all other time scales. Climate downscaling when forced completely dominates over the initial state.

Acknowledgements

This chapter was realized as a part of the project "Studying climate change and its influence on the environment: impacts, adaptation and mitigation" (No. 43007) and "Meteorological extremes and climate change in Serbia" (No. 176013), which are financed by the Ministry of Education and Science of the Republic of Serbia within the framework of integrated and interdisciplinary research and basic research, respectively, for the period 2011 - 2014. We gratefully acknowledge Mr. Dragomir Bulatović for technical preparing of the text and figures.

References

[1] F. Giorgi, L.O. Mearns, *Rev. Geophys.*, 29, 191-216 (1991).

[2] P. Lucas-Picher, D. Caya, S. Biner, *Research Activities in Atmospheric and Oceanic Modelling*, WMO/TD, 1220 (34), 2004, 7.27-7.28.

[3] B. Denis, R. Laprise, D. Caya, J. Côté, *Clim. Dyn.*, 18, 627-648 (2002).

[4] B. Denis, R. Laprise, D. Caya, *Clim. Dyn.*, 20, 107-26 (2003).

[5] L. Castro, R.A. Pielke, G. Leoncini, *J. Geophys. Res.*, 110, D05108, doi:10.1029/2004JD004721 (2005).

[6] L. Separovic, R. de Elia, R. Laprise, *Mon. Weather Rev.*, 136, 4942-4961 (2008).

[7] G. Miguez-Macho, G.L. Stenchikov, A. Robock, *J. Geophys. Res.*, 109, D13104, doi:10.1029/2003JD004495, 200-210 (2004).

[8] D. Caya, S. Biner, *Clim. Dynam.*, 22, 33-46 (2004).

[9] H.C. Davies, *Q. J. Roy. Meteor. Soc.*, 102, 405-418 (1976).
[10] H.C. Davies, *Mon. Weather Rev.*, 111, 1002-1012 (1983).
[11] Staniforth, *Meteorol. Atmos. Phys.*, 63, 15-29 (1997).
[12] T.T. Warner, R.A. Peterson, R.E. Treadon, *Bull. Amer. Meteor. Soc.*, 78, 2599-2617, (1997).
[13] R. Errico, *Mon. Weather Rev.*, 113, 1554-1562 (1985).
[14] R.M. Errico, D.P. Baumhefner, *Mon. Weather Rev.*, 115, 488-504 (1987).
[15] T. Vukicevic, J. Paegle, *Mon. Weather Rev.*, 117, 340-350 (1989).
[16] J. Charney, *Proceedings of the International Symposium of Numerical Weather Prediction*, Tokyo, JP, 131-152 (1962).
[17] F. Winninghoff, *Ph.D. Thesis*, University of California, Los Angeles, US, 1968; p. 161.
[18] F. Giorgi, M.G. Marinucci, *Mon. Weather Rev.*, 121, 2814-2832 (1993).
[19] H. von Storch, H. Langenberg, F. Feser, *Mon. Weather Rev.*, 128, 3664-3673 (2000).
[20] B. Rockel, C.L. Castro, R.A. Pielke, H.von Storch, G. Leoncini, *J. Geophys. Res.*, 113, D21107, doi: 10.1029/2007JD009461 (2008).
[21] J.C-F. Lo, Z-L. Yang, R.A. Pielke, *J. Geophys. Res*, 113, D09112, doi:10.1029/2007 JD009216 (2008).
[22] K.M. Waldron, J. Paegle, J.D. Horel, *Mon. Weather Rev.*, 124, 529-547 (1996).
[23] H-M.H. Juang, S-Y. Hong, M. Kanamitsu, *Bull. Amer. Meteor. Soc.*, 78, 2125-2143 (1997).
[24] Z.I. Janjic, *Contr. Atmospheric Phys.*, 50, 186-199 (1977).
[25] Z.I. Janjic, *Mon. Weather Rev.*, 112, 1234-1245 (1984).
[26] F. Mesinger, *Contr. Atmospheric. Phys.*, 50, 200-210 (1977).
[27] F. Mesinger, Z.I. Janjic, S. Nickovic, D. Gavrilov, D.G. Deaven, *Mon. Weather Rev.*, 116, 1493-1518 (1988).
[28] Z.I. Janjic, *Mon. Weather Rev.*, 118, 1429-1443 (1990).
[29] A.K. Betts, M.J. Miller, *Q. J. R. Meteorol. Soc.*, 112, 693-710 (1986).
[30] S.B. Fels, M.D. Schwarzkopf, *J. Atmos. Sci.*, 32, 1475-1488 (1975).
[31] M.B. Ek, K.E. Mitchell, Y. Lin, E. Rogers, P. Grunmann, V. Koren, G. Gayno, J.D. Tarpley, *J. Geophys. Res.*, 108, 8851, 16 PP., doi:10.1029/2002JD003296 (2003).
[32] Z.I. Janjic, *Mon. Weather Rev.*, 122, 927-945 (1994).
[33] Z.I. Janjic, J.P. Gerrity, S. Nickovic, *Mon. Weather Rev.*, 129, 1164-1178 (2001).
[34] S. Gualdi, A. Navarra, E. Guilyardi, P. Delecluse, *Ann. Geophys.*, 46, 1-26 (2003).
[35] G. Madec, P. Delecluse, M. Imbard, C. Levy, *Internal Report*, 11, Institute Pierre-Simon Laplace, Paris, FR (1999).
[36] T. Elvius, A. Sundstrom, *Tellus*, 25(2), 132-156 (1973).
[37] F. Mesinger, *Adv. Geosci.*, 16, 137-143 (2008).
[38] V. Djurdjevic, B. Rajkovic, *Ann. Geophys.*, 26, 1935-1954 (2008).
[39] S. Gualdi, B. Rajkovic, V. Djurdjevic, S. Castellari, E. Scoccimarro, A. Navarra, M. Dacic (2008). Final Scient Report [http://www.earth-prints.org/handle/2122/4675].

In: Essays on Fundamental and Applied Environmental Topics ISBN 978-1-61942-522-4
Editor: Dragutin Mihailovic

Chapter 9

Moss Biomonitoring Technique Used to Study the Spatial and Temporal Atmospheric Deposition of Heavy Metals and Airborne Radionuclides in Serbia

***M. Krmar*[1,*], *D. Radnović*[2] and *M. V. Frontasyeva*[3]**

[1]Department of Physics, Faculty of Sciences, University of Novi Sad, Novi Sad, Serbia
[2]Department of Biology and Ecology, Faculty of Sciences, University of Novi Sad, Novi Sad, Serbia
[3]Department of Neutron Activation Analysis and Applied Research, Division of Nuclear Physics, Frank Laboratory of Neutron Physics, Joint Institute for Nuclear Research, Dubna, Russia

Abstract

Biomonitoring techniques, first introduced in Scandinavia with the use of terrestrial mosses, are very suitable for studying the atmospheric deposition of heavy metals as well as other trace elements, including airborne radionuclides. Mosses do not have a rooting system and obtain water and most nutrients directly from the air via precipitation and dry deposition, accumulating these elements in measurable concentrations. Measurements of the concentrations of heavy elements in mosses, which have been performed on a regular basis every 5 years since 1990 in most European countries, can be used to determine both the spatial distribution of atmospheric deposition as well as the temporal changes in depositional patterns. The deposition of radionuclides from the atmosphere can be studied by using biomonitors. Airborne radionuclides, such as ^{7}Be and ^{210}Pb, can be used as tracer elements for the study of the atmospheric transport and deposition of aerosols. Recent studies have shown that mosses are an excellent sampling medium for spatial and temporal studies of the atmospheric deposition of airborne radionuclides.

---

* Corresponding author: E-mail address: krmar@df.uns.ac.rs; Phone: +381 21 459 368. Fax: +381 21 459 367.

1. Introduction

Detailed knowledge of the locations of alarmingly high emissions of heavy elements and their atmospheric deposition patterns requires a relatively high-density sampling network. Studies of atmospheric contamination based on standard air sampling techniques that use vacuum pumps or precipitation collectors can be limited by their high cost and the difficulty of deploying a large number of sampling devices for a long-term and large-scale program requiring frequent sampling and sample analysis. Hence, the use bioaccumulators for monitoring air pollution is a topic of great interest.

Terrestrial mosses (carpet forming) grow in different climatic zones. Mosses do not have a rooting system. Therefore, they obtain water and most nutrients directly from the air via precipitation and dry deposition, and their uptake of nutrients from the substrate on which they are growing is insignificant. The absence or strong reduction of the cuticle and their thin leaves allow for easy uptake of water and nutrients from the atmosphere. Their accumulation of elements acquired from the air in measurable concentrations makes mosses a robust sampling medium for heavy elements. Additional advantages of sampling mosses are the simplicity of sample collection and the possibility that a much higher sampling density can be achieved than with conventional precipitation collections. Results from moss surveys allow for the examination of both the spatial and temporal trends in the deposition of heavy elements and the identification of areas exposed to high deposition levels of pollutants.

The study of heavy metal deposition by using moss biomonitoring, originally established in Scandinavia [1, 2], has been shown to be very suitable for studying the atmospheric deposition of heavy metals as well as other trace elements. In the last three decades, the moss biomonitoring technique has been used a large number of times. The first moss survey conducted at the European scale was performed in 1990, and it has been repeated at five-year intervals in an increasing number of countries. The 2005/2006 moss survey was conducted in 32 countries, and moss samples from over 7000 sites across Europe were analyzed [3]. A pilot study of air pollution using the moss technique was carried out in the northern part of Serbia in 2000 for the first time [4]. When the study was repeated in 2005, almost all areas of the Republic of Serbia were covered.

Mosses are a very promising medium for the investigation and monitoring of the deposition of airborne radionuclides because an inexpensive sampling procedure can be combined with a simple gamma spectroscopy measurement, without any chemical pre-treatment of the sample. Air sampling provides a good determination of the amount of radionuclides in the air [5], and because sampling stations are usually located near the ground surface, the temporal variations in the activity concentration of radionuclides in the surface air (in Bq m^{-3}) can be measured by using this technique. Aerosols are removed from the atmosphere via dry deposition or wet deposition. Deposition samplers usually consist of different size vessels collecting rainwater samples or dry deposition samples. Using this technique [6], the activity concentration of the radionuclides in the depositional fallout (in Bq m^{-2}) for a selected time interval can be measured. However, a limited number of precipitation and air sampling stations cannot provide a satisfactory resolution for the spatial analysis of the atmospheric deposition of airborne radionuclides. On the other hand, mosses, which can collect airborne radionuclides transported and deposited by the same mechanisms as heavy elements, can be easily collected across large areas. The activity concentrations of some radionuclides can reach measurable levels. Mosses and lichens are already being used as biological indicators of environmental contamination by long-lived artificial radionuclides [7, 8]. The contents of airborne radionuclides near nuclear installations [9] as well as the amount of radioactivity deposited around coal-fired power stations [10] have occasionally been measured in mosses. Mosses in the Northern Hemisphere usually have a ^{137}Cs activity that is higher than the detection threshold of a standard low-background High Purity Germanium HPGe gamma spectroscopy system. The great majority of ^{137}Cs that has been released into the atmosphere is from the nuclear accident in Chernobyl and the more than 400 atmospheric nuclear weapons tests that were performed in the Northern Hemisphere. A study concerning the deposition of ^{137}Cs after the Chernobyl accident [11] is one of the most comprehensive attempts at using terrestrial mosses as a sampling medium for the measurement of the atmospheric transport and deposition of radionuclides.

Berylium-7 (^{7}Be), formed by the spallation reaction between cosmic rays (mostly high energy protons) and the nuclei of oxygen and nitrogen in the stratosphere and upper troposphere, can be used in investigations of atmospheric deposition. By following and monitoring the deposition of atmospheric ^{7}Be, it is possible to investigate the intrusions of stratospheric air masses into the troposphere, the thickness of the tropopause, vertical transport in the troposphere, the horizontal movement of air masses and the residence times of aerosol particles in the troposphere, among other atmospheric phenomena. Lead-210 (^{210}Pb) is a natural radionuclide and a member of the ^{238}U chain. In contrast to the ^{7}Be that is produced in the higher atmosphere, ^{210}Pb is introduced into the atmosphere from the soil. After the decay of ^{226}Ra, a member of the ^{238}Y chain, the daughter nuclei ^{222}Rn (a noble gas) escapes all chemical bonds in the mineral-containing members of the ^{238}U family and reaches the atmosphere. After several steps in the decay chain, the long-lived (22 years) daughter nuclei ^{210}Pb is formed.

While crust minerals containing ^{238}U are a source of ^{222}Rn, the presence of ^{210}Pb and other radon daughters depends on the position on the continent or sea floor, the geological properties of the ground, the climate and the general conditions of the ground surface layer. However, ^{210}Pb is also released from industrial processes such as the sintering of ores containing some amount of ^{238}U, the burning of coal or the production and use of agricultural phosphate fertilizers.

Naturally occurring ^{7}Be and ^{210}Pb, together with ^{137}Cs, after reaching the atmosphere, follow the same transport routes as other man made pollutants. This means that they can be used as tracers in investigations of the atmospheric transport of aerosols and other pollutant particles.

Preliminary results from the first study [12] of the potential use of mosses to determine the spatial distributions of ^{7}Be and ^{137}Cs deposition will be discussed in this chapter. In a second study [13], the activity concentrations of ^{7}Be, ^{210}Pb and ^{137}Cs of moss samples taken at a regular time interval (a one-week sampling frequency) from the very same place across the four seasons of the year were measured to obtain (and interpret) the temporal dynamics of airborne radionuclide deposition.

The results show that the temporal variations of moss radionuclide activity can provide information comparable to that acquired by either precipitation collectors or air sampling techniques. This study helps to increase awareness of the potential for the use of mosses for temporal and spatial analyses of the atmospheric deposition of airborne radionuclides.

2. Mosses as Biomonitors

Biomonitoring includes a broad array of environmental techniques that examine the relationship(s) between biological variables and pollution. Biological variables can range from the damage to living cells to ecosystem-level impacts. Pollutants can be atmospheric, terrestrial or aquatic. The tools of biomonitoring are bioaccumulators, organisms that accumulate pollutants, and bioindicators, organisms that respond to specific pollutants in a consistent, measurable fashion [14].

The relevant information obtained from biomonitoring methods (e.g., using plants or animals) is commonly deduced from either changes in the behavior of the monitored organism or from the concentrations of specific substances in the organism's tissues. With the proper selection of organisms, the general advantage of a biomonitoring approach is primarily related to the permanent and common occurrence of the organism in the field, even in remote areas, the ease of its sampling, and the absence of any necessary expensive technical equipment [15].

Because mosses do not have a root system, they predominantly obtain mineral substances from the air. Water can provide nutrients for mosses when the surface is saturated. Some species are very sensitive to pollution, but others are very tolerant to air pollution and have been used as indicator species. The most recent data in the literature indicate that between 15 000 and 25 000 moss species are present worldwide [16, 17]. Regardless of the differences in the estimates of the number of moss species, their floristic diversity classifies them immediately after the Anthophyta. For most mosses, an individual green thallus is generally small in size (up to 5 cm), but in some mosses it may be more than 50 cm long. Frequently, mosses grow in groups, forming clumps, small "pillows" or carpet-like forms. Green plantlets are usually composed of stems or branches with numerous leaves, while simple moss leaves are composed of one or several cell layers. Their cuticles are very thin and have no effect on water loss [18, 19]. Mosses do not have a true root, but tiny root-like rhizoides anchor the stem to the substrate [20].

Mosses are cosmopolitan species. They grow in almost all habitats, except seas, extreme deserts with chloridic and sulfate salts and eroded terrains. Dense moss colonies may be found in the cool forests of all of the bioclimatic zones as well as the peat bogs of the boreal and arctic zones. Pure communities are found in the arctic region, particularly in high-latitude peat bogs. Mosses occur on all continents and grow on a wide range of substrates, including soil, rocks, bark, rotting wood, dung, animal carcasses and leaf cuticles. The only ecosystems that mosses have failed to colonize are oceans [21].

Mosses need the same macroelements and microelements as higher plants, but at lower concentrations [22]. They are able to take up nutrients from the water flowing over them and from the atmosphere. After nutrients reach a thallus, they may remain on the surface in the form of adsorbed particles (Figure 1), in solution in the apoplast (the intracellular space), on the cell wall bound to ion exchange sites or in the cells themselves. Nutrients occurring either at the surface or in the intracellular spaces are weakly bound and may be easily removed from a thallus by the action of moving water, but nutrients that are bound to the cell wall are transported into the cells through purely physico-chemical mechanisms.

Under favorable environmental conditions, extracellular uptake is very fast and may occur for only a few minutes following precipitation events [23].

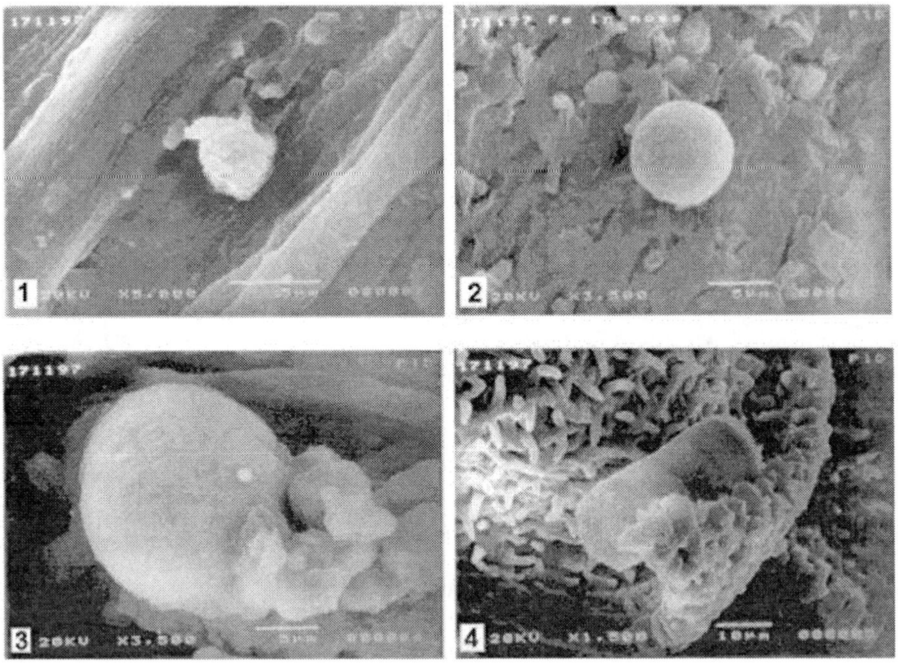

Figure 1. Scanning electron microscope (SEM) images of aerosol particles detected on the surfaces of moss thalli, obtained at JINR in Dubna (1 - Fe particle with Mg impurity; 2 - Spherule of pure iron; 3 - Al-Fe particle cluster with impurities of Zn, Cu and Ti; 4 - Diatomic alga). The elemental contents were determined with an integrated XRF spectrometer.

The rate of chemical uptake depends on numerous factors, such as the chemical nature of the molecule itself, the temperature, the pH and the concentrations of other elements in the local environment. The total amount of ions that can be bound to a moss cell is limited by the number of ion exchange sites, a characteristic that differs between species. The rate of chemical uptake also depends on the age of the thallus. Younger mosses have higher affinities for monovalent cations [24]. The ion exchange process is selective in nature and the affinities for bound bivalent cations are ranked as follows: Pb > Cu > Cd > Co = Fe > Ni > Zn > Mn [25].

In addition, so-called nonphysiological elements may also enter a cell. Unlike essential nutrients, plants may accumulate very high concentrations of heavy metals. Of the entire plant kingdom, only mosses have the ability to tolerate excessively high concentrations of heavy metals.

Mosses respond very quickly to any unfavorable human activity. The species that are more sensitive to high concentrations of heavy metals in their environment are absent from their habitats when adverse living conditions prevail (bioindicator species). In contrast, tolerant mosses accumulate toxic elements such as heavy metals, radionuclides and various organic pollutants (biomonitors) in their thalli. Thus, mosses may be used as indicators of environmental pollution by assessing either their sensitivity to pollution or their ability to accumulate pollutants [26].

Mosses have many advantages over the deposition collectors used to monitor atmospheric pollution. One advantage of mosses is that they are widespread. By measuring the concentration of heavy metals in mosses, a detailed view of the atmospheric deposition across large areas may be obtained. On the other hand, deposition collectors are only located at specific sites. Moss sampling is also very simple and inexpensive.

Heavy metals that occur in the environment may originate from natural and anthropogenic sources. In the atmosphere, the natural sources of these elements include volcanic eruptions, soil deposits, massive fires and marine aerosols. Anthropogenic sources include the chemicals released into the atmosphere from industrial activities, energy plants, traffic emissions, fossil fuel emissions and the uncontrolled pollution that results from wars and accidental releases [4, 26].

There are the two basic ways for elements to reach mosses: (i) wet deposition (rain, snow, dew, fog and other precipitation sources) and (ii) dry deposition (dust). A wide array of elements can be absorbed by moss cells via wet deposition, including Pb, Cd, In, Hg, Ar, Se, Zn and Cu. The mechanism of dry deposition is somewhat different. It includes the transport and direct deposition of diluted pollutants that can enter a moss thallus with or without the aid of water. In addition to industrial pollutants, mosses also accumulate soil elements such as Al, Fe, K, Ca, rare earth elements, U and Th. Assisted by diffusion, these elements are taken up and distributed throughout the whole plant. The changes in the concentrations of certain elements that can occur in moss thallus compared to the concentrations in the surrounding air are remarkable. Due to this feature, mosses may be regarded as natural filters that absorb chemicals, change elemental concentrations and retain particles found in the air of the immediate environment.

The differences in the concentrations of some of the above elements in the absence of a strong anthropogenic source of pollution near the moss sampled are, in the most part, derived from the composition of the surrounding soil. The concentrations of some nutrient elements also affect the concentrations of other heavy metals found in mosses because they occupy the same ion exchange sites on the cell membrane.

For the elements representing characteristic pollutants that are constantly being emitted and transported through the atmosphere, their concentrations in mosses are directly dependent upon their concentrations in the surrounding air as well as certain physical and chemical traits. In this case, the differences in the concentrations of these elements found in mosses collected from two different locations are derived exclusively from the differences in the concentrations of these elements in the air surrounding these two locations. This type of comparison offers the possibility for the use of moss collections in ecological investigations as well as in routine monitoring efforts to identify areas with high pollutant concentrations. After many years of conducting investigations in European countries and worldwide, it has been concluded undoubtedly that the concentrations of heavy metals in mosses are influenced by a great number of factors. In addition to natural variations in macro- and micro-climatic conditions, acidity, temperature, humidity, light, substrate characteristics, altitude, mineral particles, natural biogeochemical cycling processes (e.g., the atmospheric transport of aerosol particles from the marine environment to land), uptake of water with mineral substances and its transport to mosses when washing across thalli, moss species, moss age, growth rates of individual species, seasons, moss sampling characteristics (free-growing, transplanted, mosses from bags), sample preparation and the analytical methods may all affect the concentrations of heavy metals detected in mosses [15, 26]. The effect of these factors is relatively small, however, compared with the human impact reflected through the presence of Pb, Cd and Cu. However, the effects of these variables may be remarkable in the cases of Cr, Ni, V and Fe [27]. Consequently, mosses represent excellent indicators of Pb, Cd, V, and Zn to some extent, but the impacts of the environment on concentrations of elements in mosses may considerably hinder their use in monitoring Ni, Cr, Mn and Fe [28].

On the basis of their morphological and physiological features, terrestrial mosses are good indicators of ecosystem pollution from the atmosphere. We mention here those features that are the most important from the biomonitoring view-point: (i) the absence of a root system prevents ion uptake from the soil or some other substrate, (ii) the undeveloped cuticle enables the uptake of metal ions along the whole cell surface, (iii) the specific composition of the cell wall contributes to rapid ion exchange, (iv) the ability to accumulate and tolerate high metal concentrations, (v) the positive correlations between concentrations in mosses and both dry bulk amounts and wet deposition concentrations [29, 30] and (vi) their cosmopolitan nature enables a broad spatial range for investigations.

Mosses have several advantages when compared to the deposition collectors used to monitor pollution. Mosses are widespread, and their sampling is very simple and inexpensive. Also, moss species are capable of accumulating heavy metals up to higher concentrations than those measured for deposition collectors; thus, the measurements of the analytical methods produce more accurate results because they are farther above the detection limits [27].

The Swedish ecologists Ruhling and Tyler, at the end of the 1960s, were the first to use mosses as biological indicators of the atmospheric pollution caused by heavy metals [1]. Since then, this method has been used very effectively, and the impacts of various factors on the results obtained when using mosses for biological monitoring have been analyzed, and include seasonal fluctuations, sampling procedures, sample preparations and estimates of the similarities and differences between different moss species [31, 32]. The best biomonitors are mosses with branched vegetative bodies that grow in carpet-like formations. Such mosses obtain most of their nutrients either directly from precipitation or dry deposition. To avoid variation in the results obtained, mosses with wide distribution ranges should be selected for a study. Previous investigations conducted in the Scandinavian countries and in other countries of northern Europe have focused on the species *Pleurozium schreberi* (Brid.) and *Hylocomium splendens* (Hedw). Meanwhile, about 30 European countries have been included in moss biomonitoring studies. Mosses have been collected not only from habitats characterized by subarctic climates but also from those with a prevailing warm and dry climate in western Turkey. In this way, certain moss species that are characteristic of the southern regions of Europe have also been included. In the guidelines of the European program of biological monitoring of air pollution, certain other moss species like *Hypnum cupressiforme* (Hedw.) and *Scleropodium purum* (Hedw.) are also recommended. These two species account for 6.9 % of the total moss samples used in moss biomonitoring investigations for the period 1990 - 2005 [3].

In addition to estimating air pollution levels at continental and national scales, mosses have also been used for regional biomonitoring [33, 34]. Such investigations include a dense network of sampling sites and thus produce detailed information on the spatial distribution of air pollution. Also, this type of biomonitoring may be used in the areas neighboring industrial plants and roads [35].

In addition to passive biomonitoring, i.e., using mosses occurring in their natural habitats, air pollution may also be assessed by the so-called active biomonitoring of bags of mosses. Samples are collected from unpolluted areas in bags that are then transferred to the region(s) under consideration. Measurements are carried out prior to and after the exposure, and usually, they are placed in areas where no moss species are found and/or we want to obtain a more accurate quantification of heavy metal deposition as a function of time.

Goodarzi et al. [36] demonstrated the use of moss bags to predict the long-term soil accumulation of As, Cd, Cu, Hg, Pb and Zn near a zinc-lead smelter in Canada. Mercury deposition has been estimated using moss bags in Finland and China [37, 38]. Moss bags have also been successfully used to monitor the pollution of urban regions or mid-range scale pollution [14, 39, 40].

Radioactive elements may accumulate in mosses to very high concentrations. In mosses, an increased concentration of radioactive cesium was recorded after nuclear tests and after the Chernobyl accident [41 - 43]. These modest datasets show that mosses may be very effective biological monitors of radionuclides.

3. Sampling and Measurement

Sampling procedure. A representative sample enables a real estimation of the harmful effects of air pollution. This section presents a critical consideration of the standard sampling procedures applied to mosses when the air pollution over a large region is analyzed. In temperate zones, moss samples are collected between April and October, namely, during the spring and summer. In arid regions, sampling is performed during the period with precipitation.

According to the guidelines of the manual for moss sampling produced by the International Cooperative Programme on Effects of Air Pollution on Natural Vegetation and Crops (ICP Vegetation), it is necessary to take at least one sample per 400 km^2. The number of samples per unit area may be larger, depending upon the biomonitoring program goals, usually for the biomonitoring of air pollution in some smaller realm, namely, a region.

Because it is necessary to avoid the influence of local pollution sources, the sampling strategy should include the following procedure: select a representative, non urban area far from main roads and settlements, at least 300 m and 100 m away from any house or road, respectively; on forests or plantations, sampling should be done in small open spaces to preclude any significant effect of canopy drip; each sample should be composed of 5 to 10 subsamples collected from a 50 $\times$ 50 m area; samples should be collected using plastic talc free gloves; avoid smoking and other activities that may ruin the original concentrations of elements in mosses. For sampling designs for large areas, it is necessary to collect at least 3 samples per species from sites characterized by the highest and the lowest contamination levels to obtain the highest and lowest concentrations of ions. Each sampling location should possess data on its geographical location. Dead material and litter should be removed from the samples and only the segments of the last three years of growth are used for the analyses. Samples should be kept refrigerated, deep frozen or dried at room temperature and stored under those conditions until chemical analysis [44].

Chemical analytical techniques. For the determination of metal concentrations, several analytical techniques have been used in previous moss surveys. The most frequently used techniques are inductively coupled plasma spectrometry (both ICP optical emission spectrometry and ICP mass spectrometry) and atomic absorption spectrometry (flame AAS as well as graphite furnace AAS). Both techniques are characterized by high sensitivity. The preparations of samples for ICP and AAS techniques are quite similar. Moss samples should be dried at 40°C because some of the more volatile metals may be lost from mosses at higher temperatures. Usually, just one gram of each dry moss sample is used for digestions. Special care should be taken that just the last three years of growth of moss is selected. The 1-g samples are usually digested in de-ionized water and concentrated high-purity nitric acid. The samples are heated on a hot plate or in a microwave oven. After cooling and filtering, solutions are ready for analysis.

Neutron activation analysis. Neutron activation analysis (NAA) is an important physical analytical technique used for the multi-element analysis of major, minor, trace and rare elements. This is a reference analytical technique that can determine very low concentrations of the individual elements. Sample preparation for neutron activation analysis is much simpler. After the drying and selection of the last three years of growth segments, no chemical preparation is necessary. Although the usual NAA procedure involves a nuclear reactor and sophisticated detector equipment and this technique is rarely used in ecological studies done with biomonitors, in this Chapter a brief description of this technique will be given because most of the data presented were collected using NAA.

Neutron activation analysis is an isotope specific analytical technique for the qualitative and quantitative determination of elemental contents. The method is based upon the conversion of stable atomic nuclei into radioactive nuclei by irradiation with neutrons and the subsequent detection of the gamma (γ) radiation emitted during the decay of these radioactive nuclei. Activation by neutrons may result in radionuclides from all elements (that have radioactive daughter products) present in the sample, with sometimes strongly different production rates. Radiative capture ($n + {}^{A}Z \rightarrow {}^{A+1}Z^{*} \rightarrow {}^{A+1}Z + \gamma$) is the most often used reaction; after the capture of a neutron, the irradiated isotope ${}^{A}Z$ turns into the excited compound state of the ${}^{A+1}Z^{*}$ isotope. Isotope ${}^{A+1}Z$ can be non-stable and decay emitting gamma radiation, which is used for quantitative and qualitative analyses.

The analytical characteristics of instrumental neutron activation analysis (INAA) are: it is a non-destructive analysis. The tested portion does not have to be converted into a solution, thus dissolution losses do not exist; the method is based upon processes that occur in the atomic nucleus. The chemical form and physical state of the elements do not influence the activation and decay processes; the method can be applied in a self-validating manner. A unique combination of the gamma-ray energies, their intensity ratios and the half-lives of the radionuclides exist, which provides opportunities for consistency checks. Many adjustable experimental parameters, sometimes over several orders of magnitude, allow for the optimization of the experimental design and the identification of potential problems. Elements such as C, H, N, O and Si, which are often found as major components of many matrices commonly analyzed by NAA, do not activate well and contribute little or no gamma-ray activity, and the method is suitable for determinations of masses in the order of 10^{-6} - 10^{-9} g and less, depending on the element to be determined.

Table 1 shows the isotopes used for the detection of 39 elements in mosses, half-lives and the energies of the gamma peaks in the spectrums that are usually used for identification and calculation of the concentration of each element. Environmentally meaningful elements such as Cd and Pb are also determined by atomic absorption spectrometry as a complementary analytical technique.

Table 1. Isotopes produced in neutron activation analysis, their half-lives and the energies of their characteristic gamma rays

element	isotope	Half-life	Energy (keV)
Na	^{24}Na	15.0 h	2753
Mg	^{27}Mg	9.45 min	1014.4
Al	^{28}Al	2.24 min	1778.9
Cl	^{38}Cl	37.3 min	2167.5
K	^{42}K	12.4 h	1524.7
Ca	^{49}Ca	8.7 min	3084.4
Sc	^{46}Sc	83.8 d	889.2
V	^{52}V	3.76 min	1434.1
Cr	^{51}Cr	27.7 d	310.1
Mn	^{56}Mn	2.58 h	1810.7
Fe	^{59}Fe	44.5 d	1291.6
Co	^{60}Co	5.27 y	1332.4
Ni	^{58}Co	71.3 d	810.8
Cu	^{66}Cu	12.7 h	1039.4
Zn	^{65}Zn	243.9 d	1116.0
As	^{76}As	26.3 h	559.1
Se	^{75}Se	119.8 h	264.7
Br	^{82}Br	35.3 h	776.5
Rb	^{85}Rb	18.7 d	1076.6
Sr	^{85}Sr	50.6 d	514.0
Mo	^{99m}Tc	60.0 h	140.5
Ag	^{110m}Ag	249.7 d	657.8
Cd	^{115}Cd	44.6 d	527.9
Sb	^{124}Sb	60.2 d	1691.0
I	^{128}I	25.0 min	442.3
Cs	^{134}Cs	2.06 y	795.8
Ba	^{131}Ba	11.8 d	496.8
Hf	^{181}Hf	42.4 d	482.0
Ta	^{182}Ta	115 d	1221.4
W	^{187}W	23.9 h	685.8
Au	^{198}Au	2.7 d	411.8
La	^{140}La	40.3 h	1596.5
Ce	^{140}Ce	32.5 d	151.2
Sm	^{153}Sm	46.8 h	103.2
Tb	^{160}Tb	72.1 d	879.4
Yb	^{169}Yt	30.7 d	198.0
Lu	^{177}Lu	6.75 d	208.4
Th	^{233}Pa	27.4 d	312.0
U	^{239}Np	2.35 d	228.2

The results presented in this chapter were obtained in the Franck Laboratory of Neutron Physics (FLNP) at the Joint Institute of Nuclear Research in Dubna, Russia, at the pulsed fast reactor IBR-2. The average thermal neutron flux of the IBR-2 reactor is 10^{12} neutrons s^{-1} cm^{-2}, and at the pulse peak, the neutron flux reaches a value of 10^{17} neutrons s^{-1} cm^{-2}. The use of thermal neutrons, due to their large cross sections (up to 100 barns) and the high fluxes produced in IBR-2 reactor, has the advantage of requiring a relatively short activation time, from seconds to hours. However, in some cases, the resulting γ-ray spectra are noisy, which negatively influences the precision. Unwanted noise gamma radiation can be produced from the products of interfering nuclear reactions. For example, ^{28}Al created by the standard neutron capture reaction $^{27}Al(n, \gamma)$ ^{28}Al on natural ^{27}Al can be obscured by the products of a couple more reactions ($^{28}Si(n.p)^{28}Al$ or $^{31}P(n,\alpha)^{28}Al$) that occur in the reactor in parallel with the desired reaction.

The neutron capture cross section is a decreasing function of the neutron energy, which roughly obeys v^{-1} when low, where v is the neutron velocity. However, in the epithermal energy region (1 eV - 1 MeV), the resonance capture of neutrons takes place. Prominent resonance peaks appear on the decreasing curve due to the energy dependence of the neutron capture for some nuclides. The positions of the peaks and their amplitudes depend on the structure of the observed nuclide. This means that epithermal neutrons can cause selective activation and reduce the effects of competing reactions carried out by thermal neutrons. Several materials, such as Cd and B, have extremely large cross sections for thermal neutron capture (up to 10^4 barns) and can be employed to separate epithermal form thermal neutron components. The cross sections for epithermal neutrons are smaller than the cross sections for thermal neutrons. Therefore, selective epithermal neutron activation requires longer activation times (orders of magnitude of days in some cases); however, the use of epithermal neutrons can significantly improve precision and lower detection limits compared with thermal neutrons activation analysis. The pulsed fast reactor IBR-2 (Joint Institute for Nuclear Research (JINR), Frank Laboratory of Neutron Physics, Dubna) is a unique device in that it has a very high epithermal-to-thermal neutron ratio. This allows for the efficient selective activation of selected nuclei in the epithermal energy region and sensitive determinations of small concentrations with high accuracy [45].

For NAA measurements, the samples were cleaned of extraneous plant material and dried to a constant weight. The samples were neither washed nor homogenized. Green-brown parts of the moss were selected because they correspond approximately to the last three years of growth. Previous experience from the use of NAA in moss biomonitoring has shown that samples of 0.3 g are sufficiently large to be used without homogenization. The samples were pelletized before irradiation using simple press-forms. For short irradiation times, unwashed samples of about 0.3 g were heat-sealed in polyethylene bags. The moss material was exposed to a thermal neutron flux of 1.23×10^{12} cm^{-2} s^{-1}. To determine short-lived isotopes, the samples were irradiated for 3 min. After irradiation, two gamma-spectrometric measurements were performed; the first one for 5 min after 2 - 3 min of decay, and the second one for 20 min after 9 - 10 min of decay. For long irradiation times, samples of similar weight (about 0.3 g) were packed into aluminum cups. Long-lived isotopes were determined after irradiation for 100 h in the cadmium-screened channel 1. The flux of fast neutrons in this channel was 4.2×10^{12} cm^{-2} s^{-1}. After irradiation, the samples were repacked into clean containers and measured after 4 - 5 and 20 - 23 d for 45 min and 3 h, respectively. The gamma-spectra were registered using HPGe detectors. Characteristic energy peaks, presented in Table 1, were used for the identifications of nuclides and the calculations of their concentrations in the irradiated samples.

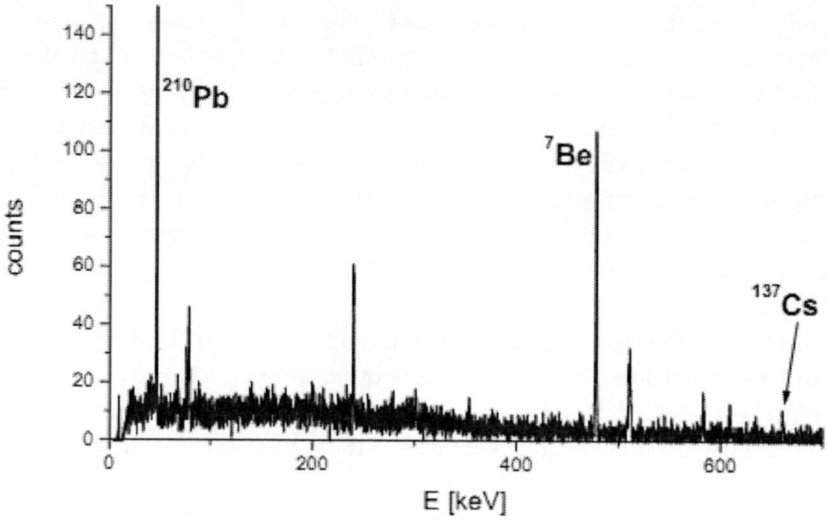

Figure 2. Characteristic spectrum of a moss sample measured for a period of $26 \cdot 10^3$ s.

Measurement of airborne radionuclides by gamma spectrometry. The sampling of mosses for measurements of airborne radionuclides does not require as much care as the sampling of mosses for measurements of heavy metals because contamination during sample handling does not affect the contents of ^{7}Be of ^{210}Pb. Samples can even be taken from urban environments. Collected mosses were dried to a constant mass. One difference is that more sample is necessary for gamma spectrometry (up to 50 g, depending on the counting procedure). The gamma spectrometry technique for the measurement of airborne radionuclides in mosses is a relatively simple procedure. Gamma spectrometry does not require any chemical preparation of samples. Dried moss sample is packed into some calibrated geometric container and gamma spectra are collected. There are several gamma peaks of interest: 46.5 keV is emitted after the decay of ^{210}Pb; 477.6 keV is the ^{7}Be line, and the 661.7 keV gamma line of ^{137}Cs is well known. Two main requirements should be accomplished: a low background shield surrounding the detector to allow for the detection of low activities and good energy efficiency at low energies to get satisfactorily good statistics for the counts of the low gamma energy of ^{210}Pb. Figure 2 shows the low energy portion of a characteristic gamma spectrum of a moss sample. The prominent gamma peaks for ^{210}Pb and ^{7}Be can be seen. The spectrum was measured for 26×10^3 s.

For the standard measurement procedure, dried moss was packed into cylindrical plastic containers (0.7 mm wall thickness) with a diameter of 67 mm and a height of 31 mm. Two containers were filled with from 22 to 48 g of dry material. The third day after sampling, the packed mosses were measured as usual. A low-background HPGe detector, with a 36 % efficiency, shielded by 16 cm of lead and a 1.5 mm inner Cu shield was used. The gamma spectrum of a single sample was collected until a statistical uncertainty of the 477.6 keV ^{7}Be line of less than 5 % was achieved. The longest sample counting time was 48 ks. Next measurements were carried out using an extended range HPGe detector equipped with a Be window to determine ^{210}Pb concentrations. The detector was shielded by 18 cm of lead, 1 mm of tin and 1.5 mm of copper. The detector efficiency was 32 %. The measurement time was from 22×10^3 to 88×10^3 s depending on the detector availability. The statistical uncertainty of the 46.5 keV ^{210}Pb line was also less than 5 %. The accuracy of the efficiency calibration was tested using an International Atomic Energy Agency (IAEA) source made from dry grass.

4. Selected Results

A moss survey in which the concentrations of several heavy metal were measured in a certain selected area offers a rich dataset for further analysis. Depending on the primary goals, the results can be presented a number of different ways. The scale of the investigation can vary, from a small region, e.g., a 40 - 50 km in diameter area around a well known industrial source of air pollution, to an area as large as continent. The research can focus on very different elements, depending on the character of pollutant emitters in investigated area. In a moss survey of almost all European countries, 12 elements were measured: As, Cd, Cr, Cu, Fe, Hg, Ni, Pb, V, Zn, Al and Sb. If NAA is chosen as the analytical technique, more than 40 elements can be determined in one sample. The determination of a large number of heavy element concentrations allows for the analysis of correlations in their spatial patterns of atmospheric deposition. Another very important aspect of the analysis of moss survey results is the examination of temporal trends of selected heavy elements in various areas. The very first moss survey at the European scale was conducted in 1990, and after four surveys at five-year intervals, temporal trends in the accumulation of some heavy elements in moss samples in Europe can be determined.

In this Section, some of the most important aspects of the analysis of the results are presented. Examples are presented from two subsequent moss surveys conducted in Serbia [46, 47]. Serbia participated in the European moss surveys twice, in 2000 and 2005. In the first survey, the northern part of the country was sampled. In the second one, moss samples were collected from areas across almost in certain selected the entire country. In 2001, a detailed study of the copper basin located in east Serbia was conducted. This region is one of the most polluted areas in Europe.

Data related to the presence of airborne radionuclides in moss samples are not very abundant. The results from a couple studies conducted in Serbia will be used to show some possible applications of the moss technique to investigations of airborne radionuclide.

4.1. Air Deposition of Heavy Elements in Serbia

4.1.1. Basic Statistics

The very first set of data gathered after a successful sampling is the concentrations of elements analyzed in moss samples. Even simple statistics for the dataset can offer valuable information. In Table 2, the minimum, maximum and median values of Cu measured in three different countries during the 2005 moss survey are presented [3]. The maximum concentration of Cu in Serbia is significantly higher than the corresponding values detected in Germany and Switzerland. It is evident that a very high deposition of Cu takes place in Serbia because of the mining industry (there is a well known copper basin in the region of the town Bor in east Serbia). The low values of copper in the Swiss moss reflect the fact that there is no copper industry in Switzerland. Considering that the minimal values of Cu deposition in

Germany and Serbia are very similar and that the median value of the Cu concentrations in Serbia is not significantly greater than the German median value, it is evident that concentrations of Cu are significantly higher at just a limited number of sampling locations in Serbia.

The values of arsenic measured in moss samples collected in Serbia, France and Spain during the 2005 moss survey are presented in Table 3. The minimum and median concentrations of As in Serbia are significantly higher than the corresponding values for France and Spain. Considering that there are no significant industrial sources of As in Serbia, the elevated concentrations of As in Serbian mosses could be the result of other processes.

Table 2. Concentrations (in *ppm*) of copper in Serbia, Germany and Switzerland measured in the 2005 moss survey

Country	minimum	maximum	median
Serbia	3.0	451	11.1
Germany	3.54	41.7	7.27
Switzerland	0.33	7.96	1.20

Table 3. Concentrations of arsenic in Serbia, France and Spain (in *ppm*) measured in the 2005 moss survey

Country	minimum	maximum	median
Serbia	0.22	21.6	1.41
France	0.070	13.5	0.37
Spain	0.029	6.84	0.18

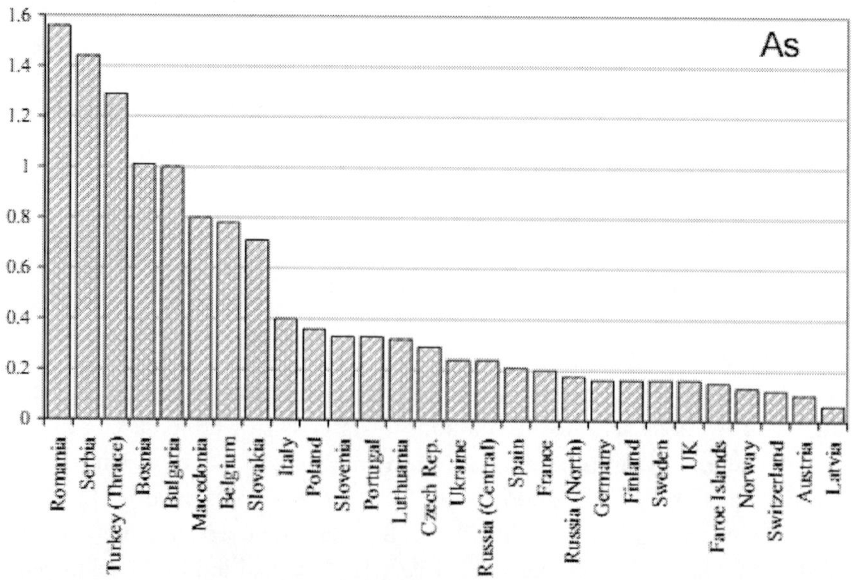

Figure 3. Median values of As concentrations (in ppm) measured in different European countries in the 2005 moss survey.

Most of north Serbia is cultivated land, and the area of forests is not very large. It is highly probable that these mosses are exposed to dust to a greater extent than the Norwegian mosses, for example. This means that soil elements can be more abundant in the Serbian mosses. Arsenic can originate from minerals in the soil. It is well known that in some regions of Serbia underground water contains relatively high concentrations of As.

A comparison of the values of As measured in all of the European countries participating in the 2005 moss survey is given in Figure 3. It is evident that the Balkan countries have greater concentrations of As in their moss samples. This is a consequence of the mineral composition of the soil in this relatively broad area. Subsequently, the simple statistics for the measured concentrations of some heavy elements can provide more information than just the atmospheric deposition of industrial pollutants released into the atmosphere.

4.1.2. GIS Maps of Element Spatial Distributions

The spatial distribution of the heavy elements analyzed is one of the most important results of the moss survey. Maps made using GIS (geographic information system) technology can show the depositional patterns of elements, which can indicate sources of pollution and the most probably paths of atmospheric transport. The depositional patterns of heavy elements are mostly determined by the nature of the technological processes releasing pollutants into the atmosphere, for example, the emission of dust particles into air during the processing of some ores. Dust particles are relatively large and heavy (compared with aerosol particles) and usually cannot be transported large distances by air.

When copper ore is crushed and screened before flotation, large amounts of cold dust particles can be released. It is well known that the copper industry is a strong local source of air pollution, and high concentrations of Cu in mosses are rarely detected at distances greater than 50 km from a copper mine. Figure 4 shows the depositional pattern of Cu in Serbia for the 2001 in survey devoted just to the region of the copper basin in Eastern Serbia near the town named Bor.

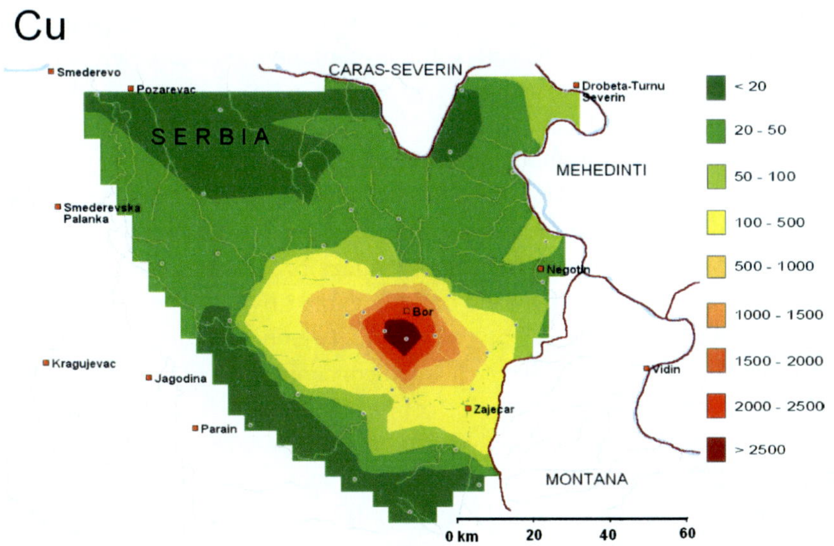

Figure 4. GIS map of the spatial distribution of Cu concentrations (in *ppm*) measured in moss samples collected in Eastern Serbia during the 2001 survey.

The values for several elements measured in moss samples collected along a transect through the Bor copper mine in the northwest - southeast direction (the direction of the prevailing winds in this area) are depicted in Figure 5. It can be seen that the copper industry emits several elements, mostly colored metals. It is interesting that the spatial distribution of the metals presented is very similar. The concentrations of these elements decrease rapidly with distance from the mine. At 40 km from the source, the concentrations of Cu and other elements are two orders of magnitude lower than corresponding values in the vicinity of the mine.

During burning processes, most heavy elements are released into the atmosphere in the form of very fine ash particles. Elements emitted into the atmosphere as the result of the burning of fossil fuels can be transported much greater distances than dusts from metal industries. In these cases, it is not as easy to distinguish pollution sources and explain deposition patterns as in the case of copper. More than 70 % of anthropogenic nickel emissions originate from crude oil refining and electricity and heat production at power stations burning fossil fuels. There are about 70 g of nickel per one ton of oil. Figure 6 shows a GIS map of the spatial distribution of nickel obtained from the 2005 moss survey for the northern part of Serbia. The highest Ni concentrations in moss samples were recorded in the region denoted by number 1. The oil refinery Novi Sad is located there. The elevated level of Ni in mosses in region 3 is due to another one oil refinery located in the vicinity of Belgrade. Some industrial facilities are also located in area 2, southwest from Belgrade. Trans-border transport is probable if some industrial operations are located near the state border. The concentrations in area 5 in Figure 6 could be an example of the trans-border air transport of pollutants. The gradient in the Ni concentrations measured in moss samples near the east Serbian border indicates that a source of nickel emissions could be in industrial region located at the border in nearby Romania. In the three cases mentioned, the source of Ni can be identified. However, there is no industrial infrastructure responsible for Ni emissions in the area denoted by number 4. Long-range transport from the oil refinery in Novi Sad theoretically could cause elevated Ni concentrations in this area, but south winds are unusual in this part of Serbia (less than 10 % of the days per year). Northern Serbia is an agricultural region, part of the "black earth belt". Nickel is a relatively abundant element in black earth (72 g t^{-1}). Mosses in cultivated agricultural regions are exposed to dust, and one of possible explanations of the higher concentrations of Ni in mosses collected in region 4 is the dry deposition of dust from the surrounding agricultural fields.

Similar distributions of elements can indicate a common origin. Figure 7 presents GIS maps of Fe and U concentrations measured in mosses. The only active iron smelter in Serbia is located southeast from Belgrade. Slightly higher values of Fe were detected in this region. A coal burning power station (located in the region denoted by the number 2 in Figure 6) that releases Fe and U found as trace elements in fossil fuel can be seen in Figure 7. Moss samples collected near the Romanian border have elevated concentrations of both Fe and U as a result of the atmospheric deposition from the Bor copper metal industry as well as the Romanian industrial infrastructure located near the Serbian border. However, the spatial distributions of Fe and U in the northern part of Serbia are almost identical.

The distributions of both Fe and U are similar to distribution of Ni in this region. Considering that iron and uranium can be found in soil, one logical explanation is that dust particles are the source of Ni, Fe and U in the moss samples collected in northern Serbia.

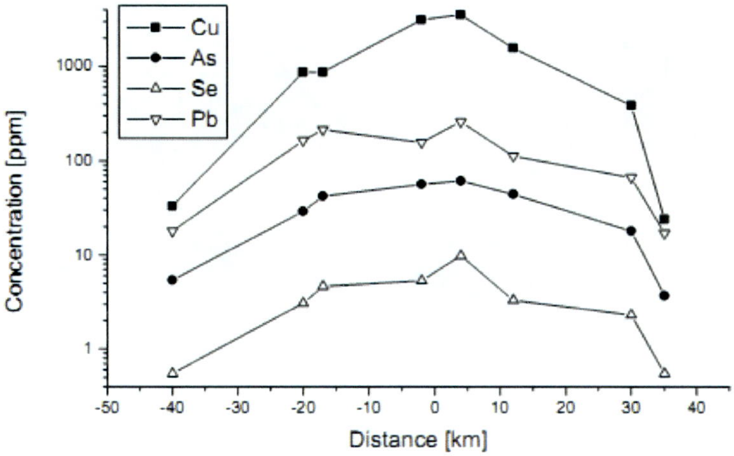

Figure 5. Measured values of concentrations of several elements as a function of the distance from a copper smelter.

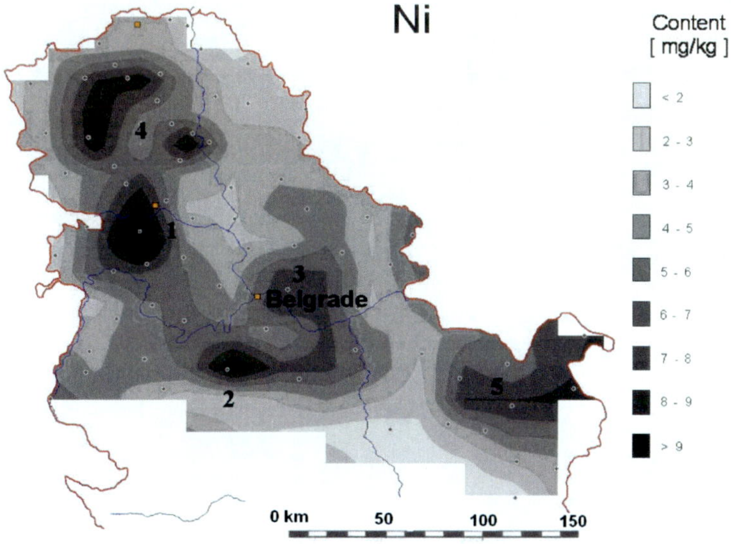

Figure 6. GIS map of the spatial distribution of Ni in moss samples collected in northern Serbia during the 2005 moss survey.

The halogen elements I and Br have low boiling temperatures and high solubility. Both elements are released into the atmosphere by some thermal processes, can be transported long distances. Due to their high mobility, it is sometimes very difficult to identify sources of I or Br pollution because deposition may take place several tens of kilometers far from the place of emission. Figure 8 shows the deposition pattern of Br measured in the 2000 moss survey in northern Serbia. In the region with high bromine deposition, no possible industrial infrastructure that emits Br could be identified. Bromine can be found in oil in relatively high concentrations, up to 10 mg l^{-1} and oil processing and burning can be a source of Br and I. The depositional pattern of bromine depicted in Figure 8 could be the result of the large fires in the Novi Sad oil refinery that were the result of NATO military action during 1999.

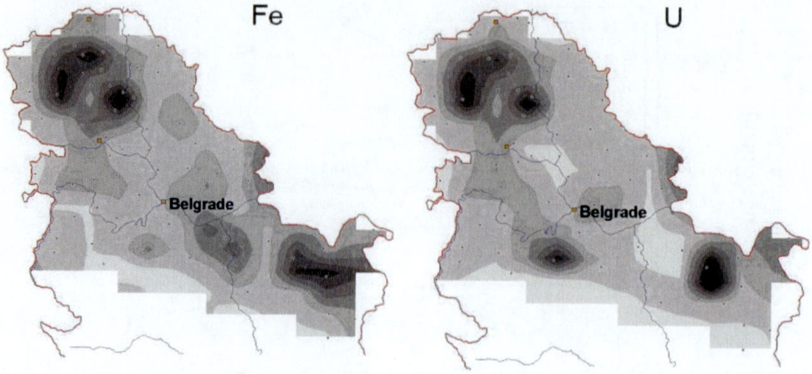

Figure 7. GIS map of the spatial distributions of Fe and U in the part of Serbia sampled for the 2005 moss survey.

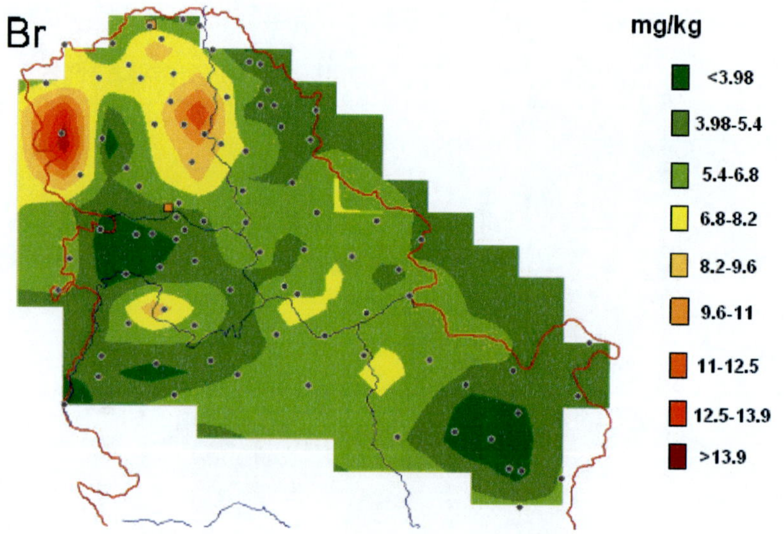

Figure 8. GIS map of the spatial distribution of Br in the part of Serbia sampled for the 2000 moss survey.

More than 40 thousand tons of crude oil was set fire at the beginning of May 1999. For several days of non-controlled fires, the south wind transported the smoke relatively far from the burning refinery. One possible explanation for the presence of Br in the agricultural region in northern Serbia could be applications of pesticides. The possibility that bromine originating from sea aerosols is transported relatively deep into the continent also cannot be excluded. The spatial distribution of iodine from atmospheric deposition is very similar to the one for Br, and the emissions sources cannot be easily identified.

4.1.3. Temporal Trends

After the completion of four moss surveys, between 1990 and 2005, in a number of European countries, some general temporal trends in the deposition of heavy elements were revealed [3]. For example, between 1990 and 2005, the average median arsenic concentration in mosses measured in five European countries declined by 72 %. In the same period, a 57 % decrease in lead deposition was reported. In contrast, the Cr concentration remained almost constant between 1990 and 2005.

A systematic survey of the atmospheric deposition of heavy metals in Serbia was conducted just twice [46, 47], in 2000 and in 2005. This is not enough information to obtain reliable evidence of temporal trends of elemental deposition; however, some changes in deposition patterns can be seen between the two successive surveys. First, the sampling of mosses in northern Serbia was performed in the summer of 2000. This was one year after the NATO military action when a significant part of the local industrial infrastructure was damaged or destroyed. Two large oil refineries, near Novi Sad and Belgrade, were destroyed and thousand of tons of oil were set on fire. During several months of NATO military action, high amounts of pollutants from large oil fires were emitted into the atmosphere. In the decade prior to the NATO bombing campaign, most of the Serbian industrial facilities operated at a significantly lower level due to the crisis in the region. Between 2000 and 2005, some revitalization of Serbian industry took place. However, because of the level of the military action, it never reached the level from before the start of the crisis. Nevertheless, the consequences of a single violent emission of air pollutants and the renewal of industrial activities can be analyzed through the comparison of the data gathered in 2000 and 2005.

The network of moss sampling was set up in the first moss survey in 2000, and samples were collected from the same locations in 2005. The simplest analysis of the temporal trends is the comparison of some statistical parameters. In Table 4, several statistics for V, Cl and Se are presented.

A simple comparison of the data presented in Table 4 indicates some general trends in the deposition of these heavy elements. All the statistical data for V show that the deposition of V was higher in the first survey than in the second one. This indicates a decrease in vanadium emissions and deposition in 2005. The maximum value of Cl in 2000 is significantly higher than that measured in 2005. However, the average and median values are almost the same in the two surveys. Moreover, the median value of Cl in the 2005 survey is higher than the 2000 median. This indicates that there was a high deposition in one relatively small area in 2000. A similar trend can be seen for the deposition of Se. The maximum value of Se measured in the 2000 survey is much higher then the maximum registered in 2005. However, the minimum, average and median values measured in the 2005 survey are higher than the corresponding values obtained in 2000. This indicates that the emission and deposition of Se in Serbia was higher in 2005 than in 2000; however, in first survey there was one spot with an extremely high deposition of Se.

Table 4. Statistical values for concentrations of selected elements (in *ppm*) measured in the 2000 and 2005 moss surveys in Serbia

	V 2000	V 2005	Cl 2000	Cl 2005	Se 2000	Se 2005
Minimum	2.85	2.68	131	104	0.0463	0.0963
Maximum	38.8	31.5	1030	790	9.81	2.40
Average	11.5	8.02	270	269	0.366	0.438
Median	9.42	6.37	236	251	0.164	0.361

This spot is located in the copper basin in eastern Serbia near the town of Bor. It is interesting to note that, in 2000, higher values of Se (up to 50 % greater) were detected at several locations near the damaged oil refinery in Novi Sad.

In the second moss survey conducted in 2005, sampling was repeated at the 75 sites established in the first 2000 survey. This allowed us to calculate the ratio of heavy element concentrations between years measured at the 75 sampling sites. The spatial distributions of the 2000/2005 ratios can be presented easily by using GIS technology. The maps of the 2000/2005 ratios indicate that there are areas with higher concentrations measured in the 2000 moss survey. Figure 9 shows the spatial distribution of the 2000/2005 ratios of V and Al concentrations. It is well known that V and Al are very often strongly correlated, and the comparison of V and Al concentrations serves as one internal measure of the quality of the measurement procedure. In the region of the copper industry, denoted by number 1, the concentrations of V in moss samples were 3.5 times higher during the 2000 survey than at the same places in the 2005 survey. A very similar trend can be seen for Al. In the Bor region, mosses contained about 3.8 times more Al in 2000 than five years later. This could be the consequence of the lower levels of production in the copper industry. However, the differences in the V and Al concentrations measured between the two surveys at location 2, near the oil refinery in Novi Sad, is the result of the deposition of smoke and ash particles after large the oil fires during the 1999 NATO bombing campaign. Al is a major element of the Earth's crust and soil. Oil combustion is the major anthropogenic source of vanadium emissions because V plays an essential role in some marine organisms that form the parent material of the oil. In the vicinity of the Novi Sad and Belgrade oil refineries seriously damaged in 1999, the concentrations of vanadium measured in mosses in 2000 were 3 times higher than the corresponding concentrations detected in 2005. A similar trend can be seen for the comparison of the Al results. The concentrations of Al measured in the samples collected in the vicinity of damaged refineries in 2000 were 3.5 times higher than the concentrations of Al measured in the 2005 survey. Al is characteristic element of concrete and other building materials and violent fires can release these substances in air. A similar analysis can be done with Cl and Br. The ratios of the Cl and Br concentrations measured in 2000 and 2005 were calculated, and GIS maps of these ratios are shown in Figure 10.

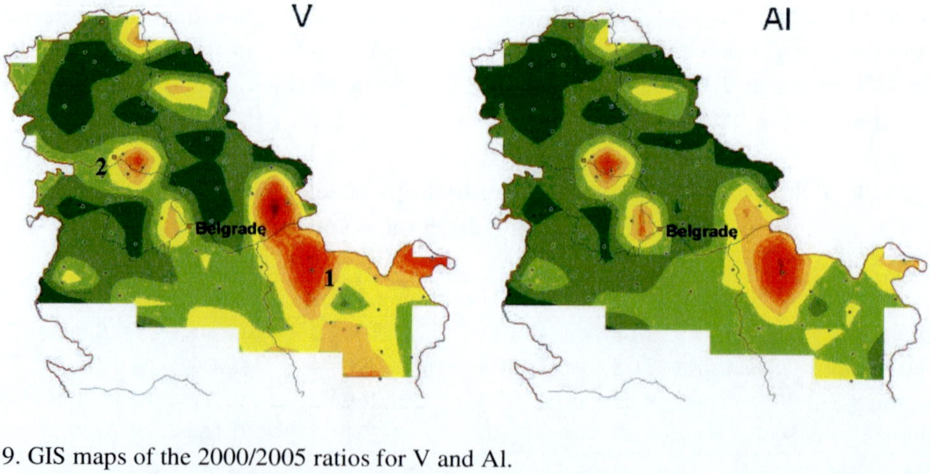

Figure 9. GIS maps of the 2000/2005 ratios for V and Al.

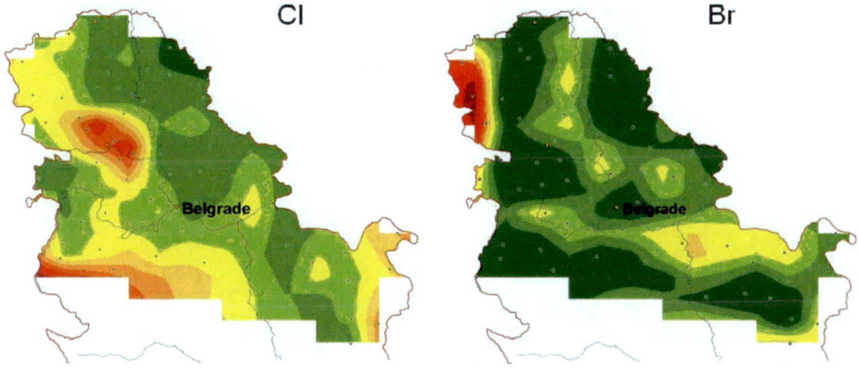

Figure 10. GIS maps of the 2000/2005 ratios for Cl and Br.

The concentrations of Cl measured in the vicinity of Novi Sad (the area denoted by number 2 in Figure 9) in the 2000 moss survey are 2.5 times larger than the concentrations recorded at the same place five years later. The spatial distribution of Br measured in the 2000 moss survey is presented in Figure 8. The maximum deposition of Br was recorded in two areas, northwest and northeast from the Novi Sad. In the area northwest from Novi Sad, the concentrations of Br in moss samples collected in 2000 were about 2.5 times higher than the concentrations measured at very same place in 2005 In the area northeast from Novi Sad, the concentrations of Br measured in 2000 were about 1.5 times higher than in 2005.

4.1.4. Correlations between Elements

There are several pairs of elements that are usually strongly correlated due to their similar chemical behaviors and their appearance together in nature. These elements can pass through some technological processes together and be deposited jointly. Some pairs of strongly correlated elements can be used as internal controls for measurement procedures. If, in some moss sample, the correlation between two such elements is not preserved, it can be an indication of a possible error in one of several phases of the experimental procedure. Figure 11 depicts two pairs of traditionally highly correlated elements. The La-Ce scatter graph displays the concentrations measured in the 2005 moss survey. It is well known that rare earth elements are usually well correlated due to their very similar chemical properties. The measured values of La and Cs concentrations in the 194 moss samples are very strongly correlated. The other scatter graph shows the Fe and Sc concentrations measured in moss samples taken for the 2000 survey. There are several more strongly correlated pairs, such as Al-V, Fe-Cr, Tb-Y and several more combinations of rare earth elements as well as different combinations of colored metals, among other examples.

One of the standard methods used in the analysis of measured data is to check moss samples in which the correlation between the two strongly correlated elements is not conserved.

For example, in Figure 11, two moss samples have significantly higher concentrations of Fe than are expected for the concentration of Sc measured in these samples (the two points circled in Figure 11). Upon further examination, these moss samples, with relatively high concentrations of Fe (compared to their Sc concentrations), were collected in the vicinity of an iron smelter.

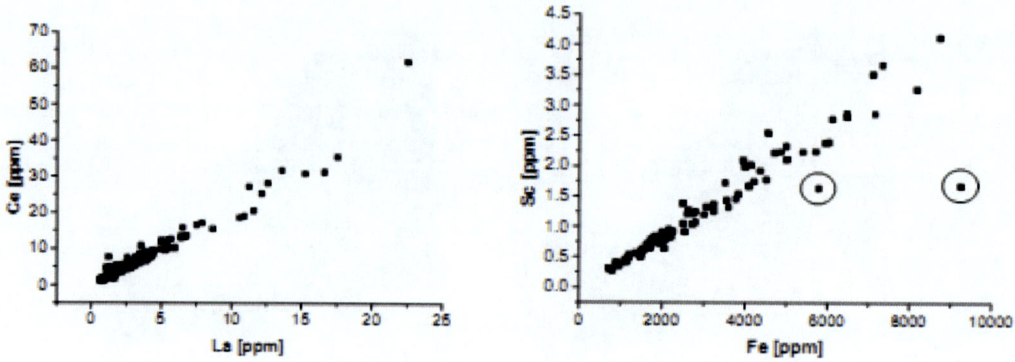

Figure 11. La-Ce and Fe-Sc correlations.

A more sophisticated statistical method that is usually used to analyze the interrelationships among a large number of variables is factor analysis (FA), a frequently used multivariate technique. If the concentrations of *m* elements measured in *n* samples are the final results of some moss survey, the data produce a matrix with an *m x n* order. This matrix can include a large number of data. Some of the elements could be correlated due to similar chemical behaviors or due to the fact that some elements are emitted from the same source and transported and deposited by similar mechanisms Factor analysis is a multivariate technique used to reduce a matrix of data to their lowest order with the minimum loss of information. In this statistical method, the variability among measured values (observed variables) is described by the introduction of unobserved variables, which are called factors. The number of factors is always less than the number of original variables, or measured data. If some factor has high loadings for several elements, those elements may be correlated because they have the same source and similar mechanisms of transport and deposition. The identification of factors in the set of moss survey data can provide clues as to which elements are correlated and, consequently, elucidate their sources and possible mechanisms of transport and deposition.

The moss data collected from both surveys were subjected to factor analysis. Table 5 presents the factor analysis results for the 2005 moss survey. Factor 1 has particularly high loadings for the lanthanides, Hf, Ta, U and Th. In addition, V, Al, Na, Mg and are also mainly associated with this component. These elements are all typical of crustal material, and most probably this component reflects, at least in part, contamination of moss samples with soil particles. High loadings of Fe, Co, Ni, Sc and Cr are characteristic of the ferrous-nickel metal industry. However, these same elements are regular constituents of most types of soil. In the southerly latitudes, where the climate is drier than in Northern Europe, mosses are more likely to pick up mineral soil particles. Almost all of the northern part of Serbia consists of cultivated agricultural regions, and the mobility of dust particles can be a significant factor in air deposition.

Under these circumstances, it becomes difficult to distinguish between industrial and other anthropogenic sources for the elements concerned. Still, when considering the geographical distribution of this component, as exemplified by the U and Fe concentrations (Figure 7) as well as Ni contents (Figure 6.), it is likely that soil dust mostly contributes to Factor 1. Significant emissions of Fe and similar elements from industrial infrastructure are not very likely because there is no exploitation of Fe ore in Serbia. Factor 1 explains 39 % of the total variance in the data set.

Table 5. Factor analysis results for deposition in northern Serbia in 2005

	Factor 1	Factor 2	Factor 3	Factor 4	Factor 5	Factor 6	Factor 7
Na	**0.934**	0.016	0.130	-0.038	0.079	-0.067	-0.087
Mg	**0.615**	-0.059	0.132	0.320	-0.200	0.339	0.156
Al	**0.946**	0.024	0.133	0.077	0.009	0.109	0.070
Cl	-0.128	-0.025	-0.100	0.271	-0.020	0.023	**-0.773**
K	**0.609**	0.051	0.092	-0.151	-0.101	-0.283	-0.207
Ca	0.035	0.049	0.127	0.324	-0.441	**0.629**	-0.002
Sc	**0.960**	0.050	0.152	0.161	0.022	-0.048	0.017
Ti	**0.940**	-0.018	0.100	0.082	0.049	0.040	0.131
V	**0.874**	0.163	0.091	0.207	-0.025	0.011	-0.012
Cr	**0.776**	0.049	0.042	0.358	-0.117	-0.348	0.135
Mn	0.051	0.178	0.206	0.083	**0.522**	-0.097	**0.592**
Fe	**0.874**	0.032	0.107	0.297	-0.109	-0.167	0.073
Co	**0.692**	0.050	0.631	0.035	0.081	0.031	-0.038
Ni	**0.593**	-0.032	0.156	0.532	0.151	-0.291	0.079
Cu	-0.088	**0.903**	-0.054	-0.026	-0.060	0.043	0.115
Zn	0.399	0.249	0.086	0.402	0.047	0.328	-0.119
As	0.205	**0.678**	0.426	0.267	0.052	0.082	0.271
Se	-0.134	**0.704**	**0.510**	0.043	-0.019	0.010	-0.053
Br	0.106	0.203	0.008	**0.746**	0.216	-0.019	-0.277
Rb	**0.507**	0.081	**0.807**	0.014	0.081	0.114	0.066
Sr	0.177	0.122	**0.834**	-0.032	-0.167	0.149	-0.176
Mo	0.051	**0.656**	0.012	0.261	-0.135	-0.293	-0.292
Ag	-0.137	**0.874**	0.095	-0.018	0.004	0.115	0.226
In	-0.116	0.248	0.195	-0.014	**0.677**	0.117	0.043
Sb	0.210	**0.526**	0.063	0.609	0.130	0.216	0.044
I	0.298	0.015	0.138	**0.816**	-0.124	-0.005	-0.043
Cs	0.184	0.059	**0.677**	0.244	0.074	0.039	0.325
Ba	0.362	-0.033	**0.804**	0.110	0.182	-0.179	0.263
La	**0.961**	-0.045	0.144	0.137	0.056	0.023	-0.007
Ce	**0.952**	-0.073	0.158	0.157	0.022	0.028	0.043
Nd	**0.581**	**0.552**	-0.004	0.134	0.253	0.150	0.187
Sm	**0.926**	-0.061	0.209	0.161	-0.014	-0.067	0.029
Eu	0.149	-0.148	-0.111	0.136	**0.740**	0.038	0.042
Tb	**0.889**	-0.026	0.054	-0.051	0.119	0.161	-0.028
Dy	**0.605**	0.027	-0.081	-0.077	-0.140	-0.003	0.206
Hf	**0.944**	-0.068	0.149	0.140	-0.013	0.044	0.039
Ta	**0.962**	-0.029	0.170	0.061	0.096	0.051	0.012
W	0.160	0.035	-0.059	0.133	-0.328	**-0.736**	0.056
Au	-0.010	**0.608**	-0.095	0.052	0.156	-0.125	-0.217
Th	**0.962**	-0.048	0.156	0.134	-0.004	-0.073	0.044
U	**0.918**	0.010	0.168	0.183	0.089	-0.107	-0.012
Expl. Var	0.390	0.103	0.093	0.075	0.049	0.045	0.043

Factor 2 is an exceptionally clear-cut industrial component, with very high loadings for Cu, As, Se, Mo, Ag, Sb and Au. The main reason for the high pollution and deposition of Cu and other elements is the non-ferrous metal industry in the copper basin, in the town Bor, near the Romanian/Bulgarian border (Figure 12). Archeological findings have shown that the exploitation of copper in the Bor region started in the Neolithic Age; hence, Bor is one of the oldest mines in Europe and a long-established source of pollution of copper and several strongly correlated elements with high values for factor 2. This factor, indicative of the copper industry, accounts for 10.3 % of the variance in the data set.

Factor 3 contains high Se, Rb, Sr, Cs and Ba loadings. These elements are likely to be mainly of natural origin, and their presence in the moss may be associated with the so-called "vascular pump", whereby elements are supplied to the moss by the action of adjacently growing higher plants. However, in the east of Serbia some relatively small mines of Cs and Sr and an accompanying metal industry are active, and industrial activity can be a source of some elements having high loadings for factor 3.

Factor 4, characterized by Br and I, can typically be associated with the atmospheric deposition of aerosols influenced by processes in the marine environment. In this case, however, the highest concentrations of Br and I are typically observed in the samples collected in the very north of Serbia, about 500 km far from Adriatic sea (Figure 8). It is probable that some industrial source is responsible for these emissions of Br and I. The oil industries located in the vicinity of Novi Sad and Belgrade could be sources of both elements. This factor explains 7.5 % of the variance of the data set.

Factor 5 contains Mn and In. Both elements are usually elements emitted by thermal processes and are thus probably the result of industrial activities. Factor 7, with higher loadings for Cl and Mn, is probably indicative of anthropogenic activities.

There is another possible use of factor analysis; the numerical values of a factor can be calculated for all of the sampling sites. The, GIS maps showing the locations with high loadings for a particular factor can be created.

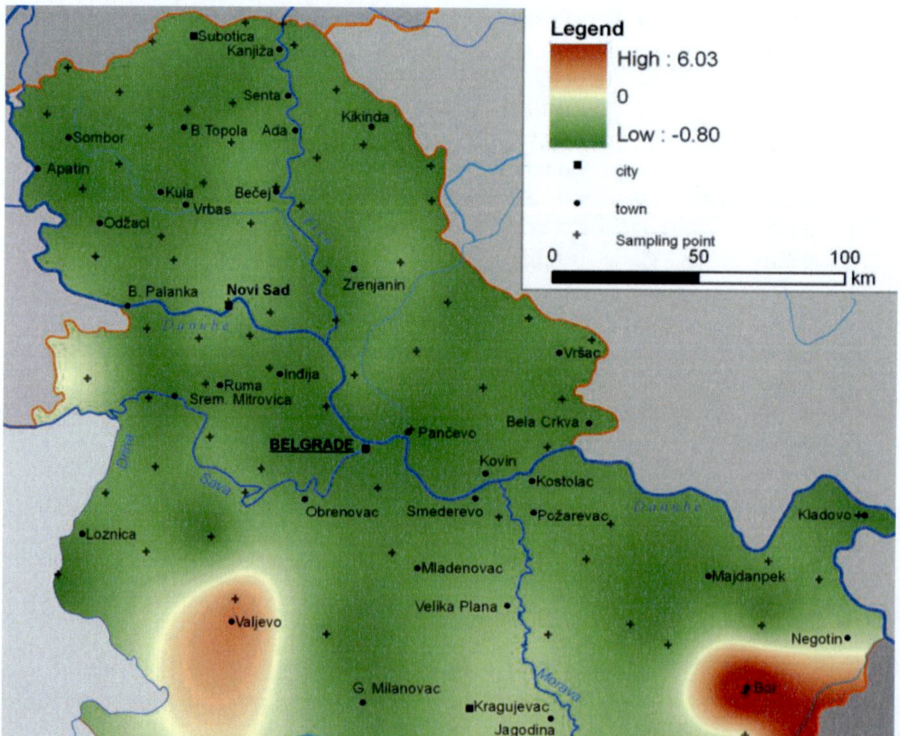

Figure 12. Spatial distribution of the factor 2 loadings calculated for the 2005 moss survey dataset.

Figure 12 presents the spatial distribution of factor 2 over the northern part of Serbia. Factor 2 is characterized by Cu and other nonferrous metals. The high loadings of factor 2 in the region of the Bor copper industry are expected because Cu, As, Se, Mo, Ag, Sb, and Au all occur in high concentrations at a limited number of sites surrounding the town of Bor and relatively low levels almost everywhere else. However, Figure 12 shows that near the town of Valjevo, in western Serbia, there is one area with elevated values of factor 2. Although a metal industry is located in the town of Valjevo and elevated values of factor 2 were obtained here, exceptionally high concentrations of copper and the other accompanying elements were not recorded in this area. This could be explained by the fact that there are some industrial sources of Cu and other nonferrous metals operating in the area of Valjevo. Although the measured values of Cu, As, Se, Mo, Ag, Sb, and Au in the moss samples collected in this area were not the highest ones in Serbia, the concentrations of all of the elements mentioned were strongly correlated and they were all emitted from the same source.

4.2. Some Results for the Deposition of Airborne Radionuclides Measured Using Biomonitors

In order to use ^{7}Be, ^{210}Pb or ^{137}Cs effectively as a tracer to follow different atmospheric processes, their depositional characteristics should be known, and a comprehensive understanding of their variability in the atmosphere should be achieved. Concentrations of some radionuclide and the deposition of ^{7}Be, ^{210}Pb and ^{137}Cs are frequently measured in the lower atmosphere. However, biomonitors are not frequently used in these studies. Atmospheric radionuclides are usually collected by using vacuum pumps or precipitation collectors. In this section, some possibilities for the potential use of biomonitors to study the atmospheric transport and deposition of airborne radionuclides will be presented.

4.2.1. Spatial Distribution of Airborne Radionuclide Deposition

Mosses have been used as monitors of airborne radionuclides, mostly of some fission products, including ^{137}Cs [48, 49]. After the accident in Chernobyl, several studies concerning the spatial pattern of ^{137}Cs deposition were conducted using natural growing mosses as a sampling medium [11, 48]. Papers proposing the potential use of mosses to measure ^{7}Be were published just occasionally [50], and there were no studies regarding the spatial distribution of ^{7}Be. In this section, we present some results obtained from a pilot study of the measurement of ^{7}Be in moss samples collected from distant places.

Samples of carpet-forming mosses were collected from 19 sampling sites distributed along one direction, starting in northern Serbia and continuing to the southern border of the Republic of Serbia. The sampling sites followed the route of the International Freeway E75, and the maximum distance between sampling sites was 400 km. The distance between two adjacent sampling locations was between 20 and 30 km. The altitude gradually increases from north to south from 80 m to 350 m. Prevailing winds carrying wet air masses from the west are characteristic of this area. All samples of fresh plant material were collected within a short time interval, two days, to avoid differences in exposure to additional precipitation and radionuclide deposition from the atmosphere. In Table 6, the minimum, maximum and median values of ^{7}Be and ^{137}Cs measured in this survey are presented.

Table 6. Concentrations of ^{7}Be and ^{137}Cs measured in moss samples [12]

	Minimum [Bq kg^{-1}]	Maximum [Bq kg^{-1}]	Median [Bq kg^{-1}]
^{7}Be	195(24)	560(40)	380
^{137}Cs	7(3)	115(5)	23

The data presented in Table 6 show that measurable activities of ^{7}Be are observed in mosses collected in Serbia. It is very interesting that the distribution of the activity concentrations of ^{7}Be is not uniform. The minimum and maximum measured values differ for by a factor of 2.8. Although the production of ^{7}Be in the stratosphere and upper troposphere does not differ significantly within the 400 km long region, there are considerable differences in ^{7}Be deposition, probably caused by local factors. Thus, the atmospheric transport of ^{7}Be and the factors affecting the variation in deposition over some relatively large area can be studied by the use of biomonitors.

In most moss samples, the concentrations of ^{137}Cs were between 8 (3) and 49 (2) Bq kg^{-1}. In just three samples, much higher concentrations of ^{137}Cs were determined: 115 (5), 105 (6) and 82 (4) Bq kg^{-1}. These values are much lower than the values reported after the Chernobyl accident [48] or the values measured in regions that were strongly affected by the Chernobyl accident [51]. Higher values were measured the vicinity of nuclear power stations [9]. There are two possible ways by which ^{137}Cs can reach mosses: in dust particles from soil and via atmospheric deposition in the same manner as ^{7}Be. Thirty years after the Chernobyl accident, it is estimated that the annual atmospheric deposition should result in an average ^{137}Cs concentration in mosses on the order of 1 Bq kg^{-1}. Concentrations of 100 Bq kg^{-1} cannot accumulate after three or four years of atmospheric deposition of ^{137}Cs to the green parts of moss. The non-uniform spatial distribution and the higher concentrations of ^{137}Cs measured in our samples are probably the result of random re-suspension and re-deposition of cesium emitted from the Chernobyl accident or a prior aboveground weapons test. As Cs can be mobilized in solution, during periods of water saturation it can be washed off the surface of mosses.

4.2.2. Temporal Variations in the Deposition of Airborne Radionuclides

Temporal variations of ^{7}Be, ^{210}Pb and ^{137}Cs concentrations in the air or precipitation have been considered in numerous studies. It is well known that the production of ^{7}Be depends on the intensity of cosmic rays hitting the upper layers of the atmosphere. The long-term 11-year periodicity of ^{7}Be concentrations in the atmosphere is a consequence of the solar modulation of the arriving cosmic radiation [50]. The seasonal variations in ^{7}Be concentrations in the lower atmosphere are influenced by the periodical thinning of the tropopause, during which time the exchange between the stratosphere and the troposphere is more effective [52].

The concentration of ^{210}Pb in the lower atmosphere depends on the rate of ^{222}Rn emanation. While crustal minerals containing ^{238}U are source of ^{222}Rn, the presence of ^{210}Pb and other radon daughters depends on the geological properties of the ground, the distribution of the continents and sea areas and other factors. Temporal variations in ^{210}Pb concentrations in the air depend on the ability of radon to leak from the crust and enter the atmosphere. The general conditions of the ground surface layer, e.g., the soil moisture content or the presence and thickness of snow cover or frozen soil, can significantly affect radon permeability. This means that seasonal variations in ^{210}Pb atmospheric concentrations are expected [52]. Four samples of mosses were collected per month as usual. The mean value of the activity concentrations of ^{7}Be and ^{210}Pb were calculated for all 14 months; the simple arithmetic mean was determined. The results obtained for ^{7}Be [13] are presented in Figure 13. It can be seen that, after almost constant values during the first four months, a prominent increase in the activity concentration of ^{7}Be occurs at the beginning of the summer. There are no systematic measurements of temporal variations of airborne radionuclides in biomonitors, so existing results of air concentrations or deposition measurements of ^{7}Be can be used. Air sampling identified strong seasonal patterns in ^{7}Be concentrations in the lower atmosphere. Measurements of deposition of ^{7}Be also showed prominent temporal variation. The maximum concentrations of ^{7}Be in the air were usually observed in the summer and the lowest in the winter [53].

Almost the same seasonal pattern was obtained from atmospheric deposition measurements [54]. The explanation for this seasonal pattern could be the periodical thinning of the tropopause. During this period, the exchange of air masses between the stratosphere and troposphere is more effective [52]. Both techniques indicate that the minimum measured value of ^{7}Be is about two times lower than the maximum one [53]. The data presented in Figure 14 show that the relative amplitude of the seasonal changes in the measured ^{7}Be activity concentration of collected mosses is very similar to the relative amplitude of the seasonal pattern of ^{7}Be in the air measured by the use of other techniques.

The measured activity concentration of ^{210}Pb does not fluctuate significantly. The maximum deviation from the average monthly value of 530 Bq kg^{-1} is about 11 %. Due to the relatively long half-life of ^{210}Pb (22 years), the activity accumulated in mosses can be much higher than the weekly deposition rate. The simplified approach developed in our previous paper [13] shows that the activity of ^{210}Pb in the green annual segments of mosses is more than 130 times higher than the amount of the average activity of ^{210}Pb deposited during one week. This means that even large variations in the depositional fluxes of ^{210}Pb cannot significantly influence the level of ^{210}Pb activity accumulated during three years (the life time of the green parts of mosses).

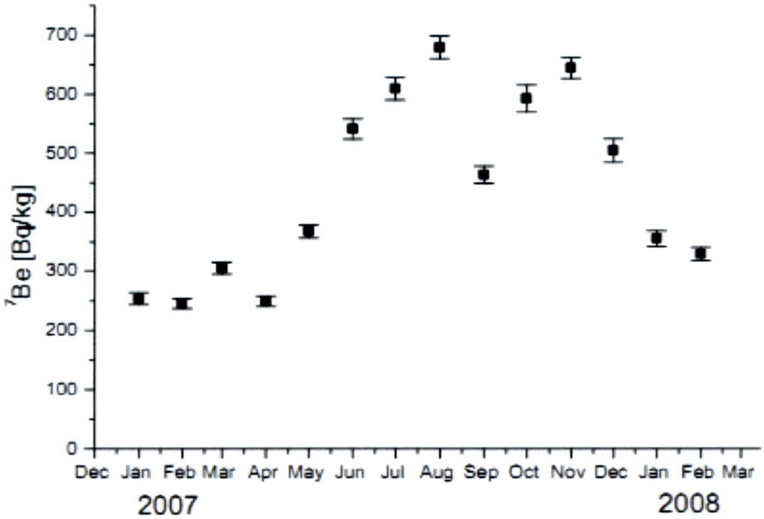

Figure 13. Average values of ^{7}Be activity concentrations measured in moss samples.

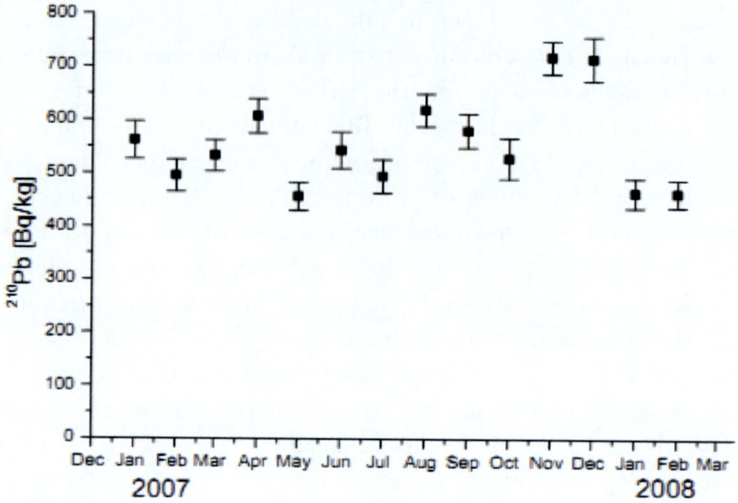

Figure 14. Average values of ^{210}Pb activity concentrations measured in moss samples.

It is evident that the variations in the leakage of radon gas due to the differences in the permeability of the surface soil are lower than the saturation level of ^{210}Pb in mosses. A nearly constant value of ^{210}Pb activity concentration (without any seasonal variation), presented on Figure 14, was obtained in all measurements.

The average monthly activities of ^{137}Cs measured in moss samples show some variations that cannot be treated as seasonal. The maximum measured value was 14 Bq kg^{-1}, and the minimum was 4 Bq kg^{-1}. The depositional flux of ^{137}Cs that remained in the atmosphere after the Chernobyl accident is only on the order of magnitude of 1 Bq m^{-2} per year [55].

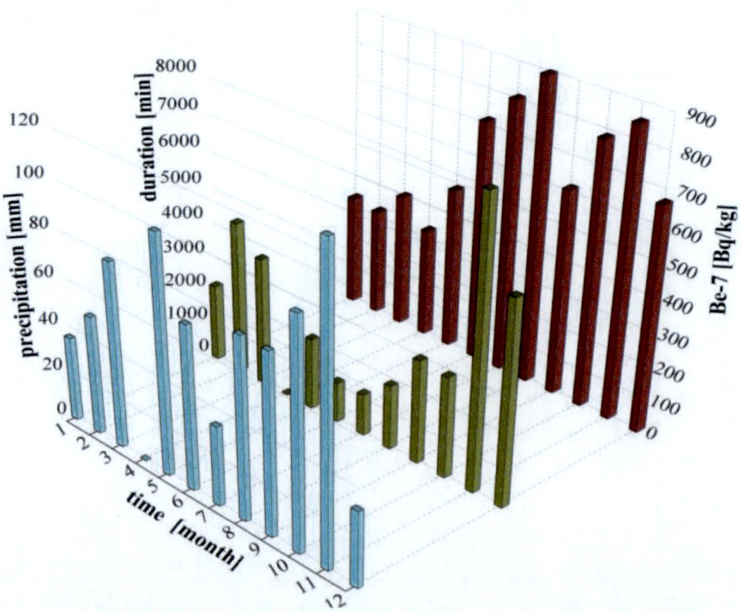

Figure 15. Average monthly values of ^{7}Be activity concentrations in mosses and the monthly amounts of precipitation, as well as the duration of precipitation, during the 12 months of measurements.

Therefore, it is three orders of magnitude lower than average annual deposition of ^{7}Be [54]. The much lower depositional flux is one reason why the activity concentrations of ^{137}Cs in mosses are much lower than the activity concentrations of ^{210}Pb and ^{7}Be. The contributions of ^{137}Cs that has been redistributed through atmospheric dust transport can cause significant perturbations in the cesium contents measured in mosses.

Considering that precipitation can wash out aerosols containing airborne radionuclides from the lower atmosphere, an analysis of the possibility of correlations between ^{7}Be and ^{210}Pb deposition levels and the amount and duration precipitation was conducted. Some moderately strong correlations between precipitation amounts and total depositional fluxes of airborne radionuclides have been observed in previous studies [5, 6, 52]. The activities of ^{7}Be integrated over the sampling period are correlated with rainfall slightly more strongly (r is around 0.8) than the activities of ^{210}Pb and ^{137}Cs [55]. The concentrations of ^{7}Be and ^{210}Pb measured in surface air usually have weak inversely proportional relationships with precipitation [56]. Different types of precipitation occur, and after rather rain or snow, lower concentrations of airborne activity can be expected in the air before sedimentation from higher layers of the atmosphere provides new inputs of radionuclides. Figure 15 presents the ^{7}Be activities measured as well as the values for the monthly amount and duration of precipitation. There is no clear correlation between the measured ^{7}Be activity of moss samples and either the amount or duration of precipitation. Although there was no precipitation at all in the fourth month of measurements (April 2007), a small increase in ^{7}Be activity was observed between April and May, most probably due to dry deposition. It is evident that the deposition of ^{7}Be and its concentration in moss samples depends on the beryllium production in the upper troposphere and it does not correlate with the amount or duration of precipitation.

The data showed that the ^{7}Be activity measured in mosses is not correlated with the duration of precipitation. Figure 15 presents the values of ^{7}Be activity in moss samples and the average monthly duration of precipitation. It is interesting that, in the summer months when the duration of precipitation was the shortest, the highest concentrations of ^{7}Be were recorded.

Using a simplified approach that assumes a constant monthly yield of radionuclides $q(Be)$ in Bq kg^{-1} per week, the activity of ^{7}Be after the time of accumulation t can be expressed as

$$A(t) = \frac{q(Be)\, T_{1/2}(Be)}{\ln 2}(1 - e^{-\lambda \cdot t}) \tag{1}$$

where $T_{1/2}(Be)$ and λ are the half-life and decay constant of ^{7}Be, respectively. If $q_n(Be)$ is the mean yield of ^{7}Be over the n-th time interval of moss observation and A_{n-1} is the activity of the moss at the beginning of the n-th period or the mean activity of the previous $(n-1)$ period, then the time dependence of the ^{7}Be activity in the n-th time interval is

$$A_n(t) = \frac{q_n(Be)\, T_{1/2}(Be)}{\ln 2}(1 - e^{-\lambda \cdot t}) + A_{n-1}\, e^{-\lambda \cdot t} \tag{2}$$

The mean value of the function $A_n(t)$ for the time interval between t_1 and t_2 is

$$\overline{A}_n = \frac{1}{t_2 - t_1}\left[q_n(Be)\frac{T_{1/2}(Be)}{\ln 2}((t_2 - t_1) - \frac{1}{\lambda}(e^{-\lambda t_1} - e^{-\lambda t_2})) + \frac{\overline{A}_{n-1}}{\lambda}(e^{-\lambda t_1} - e^{-\lambda t_2}) \right] \quad (3)$$

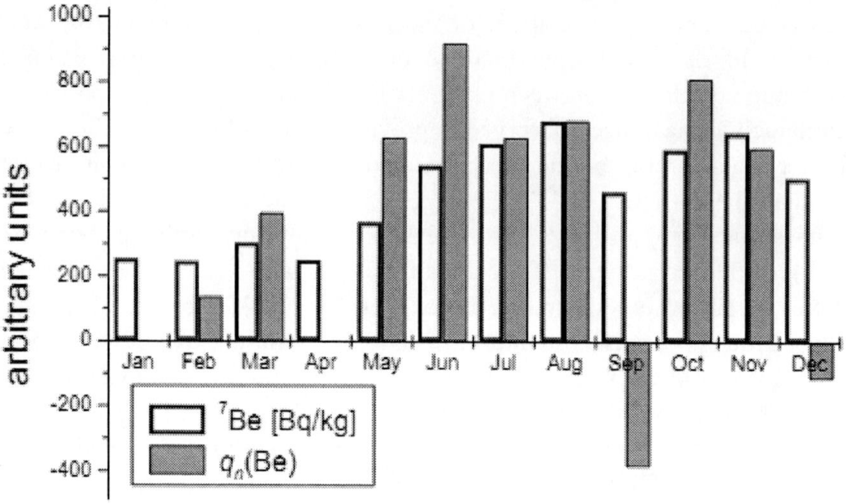

Figure 16. Average monthly ^{7}Be activity concentrations and the estimated values of the monthly yields of ^{7}Be activity $q(Be)$.

The mean yield of ^{7}Be averaged over a one-month period can be calculated using the equation above. Figure 16 shows the calculated values of $q_n(Be)$ presented in arbitrary units to match the values of the average weekly ^{7}Be activity. The value of $q_n(Be)$ calculated using Equation (3) can even be negative, as seen in Figure 16. Negative values of $q_n(Be)$ indicate that, on some occasions, the removal was greater than the uptake of ^{7}Be. The average weekly activity of ^{7}Be in April was lower than in March. In April 2007, there was no precipitation at all, as can be seen from Figure 15; therefore, dry deposition was the only mechanism of airborne radionuclide deposition.

The April value of $q_n(Be)$ is very close to zero. This means that the lower April ^{7}Be activity concentration is the result of decay without the input of new radionuclides, which could be a consequence of a lack of deposition or equal rates of uptake and removal. In contrast, the difference between the ^{7}Be activities in August and September cannot be explained by decay only. The negative value of $q_n(Be)$ in September must be the result of more intense ^{7}Be removal than uptake. A similar result occurs for the period from November to December, too. The most likely mechanism of radionuclide removal from mosses is transport in atmospheric water. Some of the airborne radionuclides attached to aerosol particles can be washed away before the mosses capture them. Under circumstances of high precipitation during a short duration of time, the removal of aerosol particles by precipitation is possible.

CONCLUSIONS

The objective of this work was to present some applications of the moss biomonitoring technique to study the deposition of trace elements and radionuclides from the atmosphere in Serbia. The dataset acquired from a standard moss survey contains the heavy metal concentrations measured in moss samples collected across some area. The results obtained from two moss surveys were used to demonstrate what kind of conclusions related to the atmospheric deposition of heavy metals can be extracted from this information. It was shown that even simple descriptive statistics of the results can be used to help to find out the locations of strong pollution sources. For example, the examination of simple statistics indicates that there is a place of very high deposition of Cu and accompanying elements located in eastern Serbia. The atmospheric deposition of heavy metals is a considerable problem in Serbia in the Bor region, where the maximum deposition levels for elements associated with the extraction of sulfide ores were observed. A comparison of the values of several heavy elements obtained for Serbia with the results collected for other European countries shows that high amounts of As are present in the soils in the Balkan region. Maps of the concentrations of heavy elements made using GIS technology can depict sources of pollution and the most likely atmospheric paths of heavy element transport. Even the simple analysis of GIS maps can show that some elements (Cu and other nonferrous metals) are strong but locally transported pollutants. However, other elements, such as Br and I, can be transported hundreds of kilometers form their sources. Comparisons of the results obtained from successive moss surveys can show temporal trends in heavy metal deposition. Introductions of new technologies or changes in the levels of industrial activities can be seen in simple comparisons of data obtained in two or more surveys. Maps of the ratios of measured concentrations of some elements can elucidate the spatial distribution of regions with elevated deposition levels. The consequences of single events that cause high emissions of certain elements can be also seen in direct comparisons of the results of surveys conducted following this event. In addition, factor analysis can be used to determine possible interrelationships among large numbers of variables.

The moss technique is already well established and widely used in examinations of the atmospheric transport and deposition of heavy metals and trace elements. The preliminary results from our studies also show that the use of mosses has potential for conducting examinations of atmospheric processes by using airborne radionuclides as tracers. The mapping of the ^{7}Be activity distribution in moss samples can provide information about patterns of atmospheric deposition over large areas. Our preliminary results show a noticeable non-uniformity in the ^{7}Be content in moss samples collected across a relatively small region. It is highly possible that certain local factors can significantly influence the atmospheric deposition of airborne radionuclides and heavy elements.

The maximum activity of ^{7}Be in mosses was measured in the summer, and the lowest activity occurred in the winter season. The seasonal pattern and relative amplitude of the measured ^{7}Be activity was very similar for ^{7}Be depositional fluxes and the ^{7}Be air activity measured by using the standard precipitation collector technique as well as air sampling. This indicates that mosses can offer comparable results to air sampling or depositional measurements. There was no seasonal variation of accumulated ^{210}Pb activity concentrations in mosses; thus, any potential non-uniformity in ^{210}Pb distributions should be a consequence of diverse depositional patterns only.

A simplified model using the monthly average values of ^{7}Be activity concentrations measured in mosses collected from same place during four seasons was developed to estimate the mean yield of ^{7}Be. In some cases, the decrease in the average monthly activity of ^{7}Be cannot be explained by absence of deposition and physical decay alone. It is highly possible that high precipitation events can wash out some amount of accumulated ^{7}Be.

Acknowledgements

This chapter was realized as a part of the project “Studying climate change and its influence on the environment: impacts, adaptation and mitigation” (No. 43007), which is financed by the Ministry of Education and Science of the Republic of Serbia within the framework of integrated and interdisciplinary research over the period 2011 - 2014.

Also, The authors are thankful for financial support from the Provincial Secretariat for Science and Technological Development of the Autonomous Province of Vojvodina, Republic of Serbia, in the Frame of the project “Molecular and phenotypic diversity of taxa of economic and epidemiological importance, and endangered and endemic species in Europe.”

References

[1] Å. Rühling, G. Tyler, *Bot. Notiser.*, 122, 321-342 (1968).

[2] Å. Rühling, G. Tyler, *Water Air Soil Poll.*, 2, 445–455 (1973).

[3] European Atlas: Spatial and Temporal Trends in Heavy Metal Accumulation in Mosses in Europe (1990-2005), UNECE ICP Vegetation; H. Harmens, D. Norris, Eds.; Centre for Ecology and Hydrology, University of Wales Bangor: Bangor, GB, 2008.

[4] M.V. Frontasyeva, T.Ye. Galinskaya, M. Krmar, M. Matavuly, S.S. Pavlov, E.A. Povtoreyko, D. Radnovic, E. Steinnes, *J. Radioanal. Nucl. Chem.*, 259, 141-147 (2004).

[5] S.R. Daish, A.A. Dale, C.J. Dale, R. May, J.E. Rowe, *J. Environ. Radioact.*, 84, 457-467 (2005).

[6] F. Cannizzaro, G. Greco, M. Raneli, M.C. Spitale, E. Tomarchio, *J. Environ. Radioact.*, 72, 259-271 (2004).

[7] M.G. Nifontova, *Radioecol. Restor. Radioactive Contam. Sites,* 155-162 (1996).

[8] M. G. Niftonova, *Ecology,* 3, 196-200 (1998).

[9] T.J. Sumerling, *Sci. Total. Environ.*, 35, 251-265 (1984).

[10] R. Delfanti, C. Papucci, C. Benco, *Sci. Total. Environ.*, 227, 49-56 (1999).

[11] E. Steinnes, O. Njåstad, *J. Environ. Radioact.*, 21, 65-63 (1993).

[12] M. Krmar, D. Radnović, S. Rakić, M. Matavuly, *J. Environ. Radioact.,* 93, 53-61 (2007).

[13] M. Krmar, D. Radnović, D.T. Mihailović, B. Lalić. J. Slivka, I. Bikit, *Appl. Radiat. Isot.*, 67, 1139-1147 (2009).

[14] M. Makholm, D. Mladenoff, *Environ. Monit. Assess.,* 104, 1-18 (2005).

[15] B. Wolterbeek, *Environ. Pollut.,* 120, 11-21 (2002).

[16] R. Gradstein, P. Churchill, N. Salazar, *Mem N. Y. Bot. Gard.* 86, 1-577 (2001)
[17] R. Bell, A. Hemsley, *Green Plants Their Origin and Diversity*; 2nd Edition; Cambridge University Press: Cambrige, GB, 2000.
[18] M.C.F. Proctor, *J. Bryol.,* 10, 531-538 (1979).
[19] M.C. F. Proctor, In *The Experimental Biology of Bryophytes;* A.F. Dyer, J.G. Duckett, Eds.; Academic Press: London, GB, 1984, pp. 9-37.
[20] K. Magdefrau, F. Ehrendorfer, *Školska knjiga*, Zagreb, HR (1978) (In Croatian).
[21] Vanderpoorten, B. Goffinet, *Introduction to Bryophytes*; Cambridge University Press: Cambrige, GB, 2009.
[22] J. M. Glime (2007). Michigan Technological University and the International Association of Bryologists [http://www.bryoecol.mtu.edu/].
[23] E. Gjengedal, E. Steinnes, *Environ. Monitor. Assess*, 14, 77-87 (1990).
[24] Rühling, G. Tyler, Oikos 21, 92-97(1970).
[25] E. Nieboer, D.H.S. Richardson, *Ann. Arbor MI,* 339-388 (1981).
[26] H. Zechmeister, K. Grodzinska, G. Szarek-Łukaszewska, In *Bioindicators and Biomonitors Principles, Concepts and Applications;* Elsevier Science, 326-375 (2003).
[27] K. Szczepaniak, M. Biziuk, *Environ. Res.,* 93, 221-230 (2003).
[28] D. Čeburnis, E. Steinnes, K. Kvietkus, *Chemosphere,* 38, 445-455 (1999).
[29] T. Berg, E. Steinnes, *Environ. Pollut.,* 98, 61-71 (1997).
[30] L. Thöni, N. Norbert Schnyder, F. Krieg, *Fresen J. Anal. Chem.,* 354 (5-6), 703-708 (1996).
[31] E. Steinnes, J.P. Rambaek, J.E. Hanssen, *Chemosphere,* 25, 735-752 (1992).
[32] E. Steinnes, J.E. Hanssen, J.P. Rambaek, N.B. Vogt, *Water Air Soil Poll.,* 69, 121-138 (1993).
[33] M. Frontasyeva, I. Smirnov, E. Steinnes, S. Lyapunov, V. Cherchintsev, *J. Radioanal. Nucl. Ch.,* 259 (1), 19-26 (2004).
[34] E. Ermakova, M. Frontasyeva, E. Steinnes, *J. Radioanal. Nucl. Ch.*, 259 (1), 51-58 (2004).
[35] Suchara, J. Sucharová, M. Havlíček, P. Slavík, M. Holá , Š. Maříková, *Abstract Book of 24th Task Force Meeting*, 31 (2011).
[36] F. Goodarzi, H. Sanei, R. Garrett,W. Duncan, *Environ. Geol.,* 43(1-2), 29-38 (2002).
[37] M. Lodenius, *Sci. Total. Environ.,* 213 (1-3), 53-56 (1998).
[38] T. Cao, L. Ana, M. Wang, Y. Lou, Y. Yu, J. Wu, Z. Zhu, Y. Qing, J. Glime, *Atmos. Environ.,* 42, 5390-5402 (2008).
[39] P. Adamo, S. Giordano, S. Vingiani, R. Castaldo Cobianchib, P. Violante, *Environ. Pollut.*, 122, 91-103 (2003).
[40] M. Anicic, M. Tomasevic, M. Tasic, S. Rajsic, A. Popovic, M. Frontasyeva, *J. Hazard. Mater.*, 171, 182-188 (2009).
[41] R. Clymo, In *Production Ecology of British Moors and Montane Grasslands;* O.W. Heal, D.F. Perkins Eds.; Springer Verlag: Berlin, DE, 1978, pp. 187-223.
[42] K. Grodzińska, B. Godzik, G. Szarek, *Prądnik, Prace Muzeum Szafera,* 7-8, 153-158 (1993).
[43] R. Gerdol, S. Degetto, D. Mazzotta, G. Vecchiati, *Water Air Soil Poll.,* 75, 93-106 (1994).
[44] H. Harmens (2010) Heavy metals in european mosses: 2010 survey. Monitoring manual [http://icpvegetation.ceh.ac.uk/].

[45] M.V. Frontasyeva, *Phys. Part. Nuclei*, 42(2), 332-378 (2011).

[46] M.V. Frontasyeva, T.Ye. Galinskaya, M. Krmar, M. Matavuly, S.S. Pavlov, E.A. Povtoreyko, D. Radnovic, E. Steinnes, *J. Radioanal. Nucl. Chem.*, 259, 141-147 (2004).

[47] M. Krmar, D. Radnovic, M.V. Frontasyeva, S.S. Pavlov, Yu.S. Pankratova, In *Advances in Environmental Modeling and Measurement,* D. Mihailovic, B. Lalic Eds.; Nova Science Publishers Inc: New York, US, 2010, pp. 147-158.

[48] C. Papastefanou, M. Manolopoulou, T. Sawidis, *J. Environ. Radioact.*, 9, 199-207 (1989).

[49] D. Adlienė, C. Rääf, Å Magnusson, J. Behring, M. Zakaria, G. Adlys, G. Skog, K. Stenström, S. Mattsson, *J. Environ. Radioact.*, 90, 68-77 (2006).

[50] Kulan, A. Aldahan, G. Possnert, I. Vintersved, *Atmos. Environ.*, 40, 3855-3868 (2006).

[51] Varinlioğlu, S. Topcuoğlu, A. Köse, A.I. Kopya, O. Uyun, A. Azar, H. Karal, *J. Radioanal. Nucl. Chem. Letters*, 187, 435-440 (1994).

[52] Aldahan, G. Possenert, I. Vintersved, *Appl. Rad. Isotp.,* 54, 345-353 (2001).

[53] E. Gerasopoulos, C.S. Zerefos, C. Papastefanou, P. Zanis, K. O'Brien, *Atmos. Environ.,* 37, 1745-1756 (2003).

[54] S. Caillet, P. Arpagaus, F. Monna, J. Dominik, *J. Environ. Radioact.*, 53, 241-256 (2001).

[55] Ioannidou, C. Papastefanou, *J. Environ. Radioact.*, 85, 121-136 (2006).

[56] A.S. Likuku, *J. Environ. Radioact.,* 87, 289-304 (2006).

In: Essays on Fundamental and Applied Environmental Topics ISBN 978-1-61942-522-4
Editor: Dragutin Mihailovic

Chapter 10

SURVEILLANCE OF MOSQUITO POPULATIONS: A KEY ELEMENT TO UNDERSTANDING THE SPREAD OF INVASIVE VECTOR SPECIES AND VECTOR-BORNE DISEASES IN EUROPE

D. Petrić[1,*], M. Zgomba[1], R. Bellini[2] and N. Becker[3]

[1]Laboratory for medical and veterinary entomology, Faculty of Agriculture, University of Novi Sad, Novi Sad, Serbia
[2]Centro Agricoltura Ambiente "G. Nicoli", Crevalcore (BO), Italy
[3]Department of Zoology, University of Heidelberg, Heidelberg, Germany

ABSTRACT

People's increased mobility and international trade play important roles in the dissemination of vectors and the pathogens/parasites that they could transmit. Climate change is likely to become another important consideration in the near future. Since the beginning of the millennium, a number of pathogen introductions into Europe have been recorded. The latest (Ravenna, Italy, 2007) was caused by the tropical Chikungunya virus, which is transmitted by the "Asian tiger mosquito", a species introduced into Italy in 1990. Previously identified phenomenona exhibit complex relationships with climate change, which does not simply comprise global warming but also includes severe weather changes. With regards to the animal kingdom, projected increases in air temperature will have an elevated impact on poikilotherm species, including insects that pose a threat to human health. The responses of insects to these changes (in addition to physiological changes such as the potential for increased vector capacity) could allow for a broadening of their colonized areas and the invasion of new sites. The spread of the sheep disease "blue tongue" and the insects that transmit it from Africa to Europe are widely accepted as consequences of climate change; however, the influence of the high mobility of people and goods as a consequence of globalization should not be underestimated. It is likely that similar scenarios could result in new geographic

* Corresponding author: E-mail address: dusanp@polj.uns.ac.rs; Phone: +381 21 485 3417. Fax: +381 21 450 809.

redistributions of other transmissible diseases and their insect vectors, which will be shaped by the ability of the insects to adapt to environmental changes caused by various factors. Deciphering the true cause of changes in the distribution and behavior of mosquitoes is difficult and complex and depends, to a great extent, on the availability of data obtained by monitoring. In order to assist in vector-borne disease preparedness, the most important invasive vector species, and the reliability and sustainability of different monitoring techniques and surveillance programs will be discussed.

1. Introduction

Drivers for the emergence of infectious diseases include human demographics (e.g., the growth of megacities), international movement of people (travelers and refugees), the smuggling of wildlife, the trade of animals, used tires and ornamental plants and various other aspects of globalization. The drivers of meteorological and climate change are also of growing international and European-focused interest [1, 2]. Warmer cities could favor mosquito breeding and, along with higher air temperatures, shorten extrinsic incubation periods, e.g., for the urban vector *Stegomyia aegypti* [*Aedes aegypti*] [3]. Data on *Culex* spp. mosquitoes, vectors of West Nile virus (WNV), and meteorological factors indicate that higher mosquito populations in a given month can be associated with higher air temperatures and precipitation in the preceding month. Similarly, a study that examined the emergence of WNV in British Colombia, Canada after a spread westward across the continent [4] suggests that higher than mean air temperatures, low snow cover and consequently reduced stream flows may have caused the observed increase in *Culex tarsalis* populations, which facilitated viral amplification and spillover into human and equine populations. The overall pattern of the current studies on mosquito-borne diseases suggests expanded ranges for disease incidence. Meteorological and climate change factors were identified as drivers for some of these patterns, but it is clear that many other factors are involved and may be more important. Nevertheless, other risk factors relating to human activities may be more important for many of these diseases, and uncertainties are often substantial, especially for rarer diseases where there have been improvements in diagnostic tests and surveillance methods [3].

Growing international concerns regarding climate change, which have been expressed in the national communication reports of most European countries within the United Nations Framework Convention on Climate Change (UNFCCC), emphasize a need for the development of climate change mitigation and adaptation strategies. In the area of infectious disease, a key adaptation strategy will be the improved surveillance of vector-borne diseases. However, improvements in monitoring/surveillance and research on whether and how various vector-borne diseases are influenced by meteorological patterns and climate change are also needed, especially interdisciplinary research that considers interactions with other risk factors.

Human activities have initiated the spread of invasive mosquito species and vector-borne diseases (a disease that is transmitted to humans or other animals by an insect or other arthropod), and ongoing globalization and increases in air temperature are greatly accelerating the process. As a result, many vector introductions into Europe have been reported since the beginning of the new millennium. Among the introduced mosquito species, some, such as *Stegomyia albopicta* [*Aedes albopictus*] and *Hulecoeteomyia japonica* [*Aedes japonicus*] are already well established in large areas, whereas others, such as *St. aegypti*, *Georgecraigius atropalpus* [*Aedes atropalpus*] and *Hlecoeteomyia koreica* [*Aedes koreicus*] are still confined to their introduction sites or surroundings; others, such as *Ochlerotatus triseriatus* [*Aedes triseriatus*], have thus far only been intercepted during surveillance programs [5]. While the presence of a competent vector does not by itself result in the transmission of an associated vector-borne disease, it is essential for transmission to occur. Several invasive mosquitoes, now present in Europe, are notorious vectors of diseases around the world, and *St. albopicta* has proven its ability to act as a vector in a European context, transmitting Chikungunya virus in Italy in 2007 [6, 7] and Dengue virus in France and Croatia in 2010 [8]. Consequently, efficient risk assessment and management requires knowledge of the distribution and abundance of such mosquito vector species. For this purpose, field data collection can be implemented through a monitoring program, where the key objective is to provide adequate information for risk-assessments to decision makers, or by a surveillance program, where the key objective is to provide information to guide and evaluate interventions [9, 10]. Anthropogenic activities, such as international trade and tourism, enhance the risk of introducing new vectors and pathogens to previously uninhabited areas where the climate and environmental conditions may also become optimal for them, allowing a rapid increase of vector populations and viral amplification. Thus, there is a clear need to be better prepared at the EU level regarding this threat. The capacity of European countries to obtain data on the presence and abundance of invasive species and to develop efficient control programs and tools for their evaluation needs to be rapidly and consistently improved in order to (i) increase the chances of swiftly detecting and eliminating intruders at the beginning of the colonization process and (ii) support timely risk assessments of arbovirus transmission. Medical surveillance accompanied by entomological surveillance is essential to prevent the spread of arboviruses and to evaluate the risk of viral disease outbreaks. The development of an efficient monitoring network is also critical for verifying the efficacy/effectiveness of control measures. In recent years, the use of the Geographic Information System (GIS) has provided important practical contributions to the investigation of the spatial component of the epidemiology of infectious diseases [11], including vector-borne diseases such as malaria, trypanosomiasis, rickettsiasis and a range of arboviral diseases [12 - 14]. Moreover, the collection of georeferenced epidemiological data can also be useful for cluster identification and geostatistical analyses. The investigation of possible disease and vector-borne disease clustering is fundamental to epidemiology and medical entomology, and one of the aims was to determine whether the clustering is statistically significant and worthy of further investigation, or whether it is likely to be a chance occurrence. Global and local indicators of spatial association like Moran's I [15] or Getis-Ord statistics [16] are often used to measure the data clustering level. Geostatistical techniques are used to produce prediction surfaces and also an error of uncertainty for these surfaces, which provides an indication of how good the predictions are. The characterization of large geographic areas with a high or low abundance of *St. albopicta* may provide information both on the environmental variables that promote species dispersion and development, and on the epidemic diseases risk, which are essential to developing effective disease surveillance programs, particularly for Chikungunya and Dengue.

Currently, many countries are developing surveillance/monitoring programs but no standards are yet defined, and information on procedures and strategies are rarely available to non-specialists. Therefore, the European centre for disease prevention and control (ECDC) launched an initiative in 2011 to produce guidelines for implementing surveillance/monitoring of invasive mosquitoes [17] in order to assist EU Member States and EEA/EFTA countries in implementing invasive mosquito surveillance programs and to promote the harmonization of data collection within continental Europe. Such guidelines may contribute to the standardization of data collection procedures on a pan-European scale, which is urgently needed in order to allow comparisons to be drawn, and will provide precious information and tools for teams willing to begin surveying/monitoring invasive mosquitoes and vector-borne diseases.

2. Invasive Vector Species and Vector-Borne Diseases in Europe

The Asian tiger mosquito is on a rampage. Entomologists are impressed, public health officials are nervous, and many of the rest of us are swatting furiously [18].

Present-day human activities enable the transportation of mosquitoes from one continent to another within a matter of hours to a few days. International trade and the increased transcontinental mobility of humans facilitate the dispersal and, in some cases, the establishment of exotic mosquito species in other countries with favorable climatic conditions. Exotic species are thus shuttled from their native geographic ranges to recipient biotopes where they have never been present before. If some of these exotic species possess mechanisms that allow them to adapt to the new conditions and reproduce in the recipient ecosystem, they are termed "invasive". Within the mosquito family (Diptera: Culicidae), three species are notable for their dispersal potential and their significance as vectors of human diseases: *St. aegypti*, *St. albopicta* and *Hl. japonica*. Their desiccation-resistant eggs, wide host preference range, ability to exploit a wide range of natural and artificial breeding places (container-breeding species) and adaptation to temperate climates including winter diapause (except *St. aegypti*) enable the permanent establishment of viable populations in temperate regions [19, 20]. Invasive species pose a threat to biodiversity by homogenizing biota with cosmopolitan species that usually endanger and replace native counterparts. Once misbalanced, the restoration of native diversity becomes impossible. Invasive mosquito species also pose a threat to human and/or animal health as a biting nuisance and as vectors of transmittable mosquito-borne diseases.

All three species are characterized by their high vector competency for arboviruses. *St. aegypti* and *St. albopicta* are the primary and secondary vectors, respectively, for Dengue fever (DF) and Dengue hemorrhagic fever (DHF), which affect more than 40 % of the human population worldwide, especially in mega-cities of the tropics [21 - 25]. *St. albopicta* is the most important vector for the Chikungunya virus [26]. Recently, this species was involved in the transmission of Chikungunya virus to humans in Italy in 2007 and was also most likely involved in the first confirmed autochthonous dengue cases in France and Croatia in 2010 [8, 27 - 29]. In addition to Dengue and Chikungunya, other viruses such as Batai, Inkoo, Lednice, Sindbis, Tahyna, Usutu and West Nile have shown some activity, and the Rift Valley and Japanese encephalitis viruses are likewise threatening human health in Europe [28].

The "Asian tiger mosquito", *St. albopicta*, originating from Southeast Asia, has undergone a noteworthy expansion of its range in the last few decades [19]. Due to its immense invasive capacity, it is listed in the inventory of "100 of the World's Worst Invasive

Alien Species" (http://www.issg.org). With the increase in the international trade of used tires, this species has spread across very large distances and between continents [30]. In Europe, it was first reported in Albania in 1979 [31] and later in Italy in 1990, where it was probably introduced through the import of used tires from the USA [32, 33]. Over the next few years, the species rapidly dispersed to other regions of Italy [34], and it has now been reported in France [35], FR Yugoslavia [36], Belgium [37], Switzerland [38], Greece [39], Croatia [40], Spain [41], Slovenia, Bosnia and Herzegovina [42], the Netherlands [43], Germany and Serbia (Petrić unpublished) (Figure1).

A distribution predicted by an MCDA model, a short-term, minimal impact scenario, using 3 variables (annual precipitation and January and summer air temperatures) [44] suggests that areas in Central Europe up to the southern fringes of Sweden and in the Balkans have become significantly more suitable for the development of tiger mosquitoes (Figure 2). *St. albopicta* is an efficient vector of Chikungunya and Dengue viruses and filariasis.

The "Asian rock pool" or "Asian bush" mosquito, *Hl. japonica,* is an Asian species native to Japan, Korea, South China, Taiwan and the Russian Federation. In 1998, it occurred for the first time in the USA (New Jersey and New York) and is now distributed over at least 22 other states [45]. In Europe, this species was established in Belgium and has successively been detected in Switzerland and Germany, where it is rapidly spreading [46, 47]. *Hl. japonica* is a competent vector of several arboviruses, including WNV in Europe and Japanese encephalitis virus (JEV) worldwide, and this species is considered a significant public health risk [48, 49].

The "African tiger mosquito" or "Yellow fever mosquito", *St. aegypti*, has spread across almost all tropical and subtropical countries over the past four centuries.

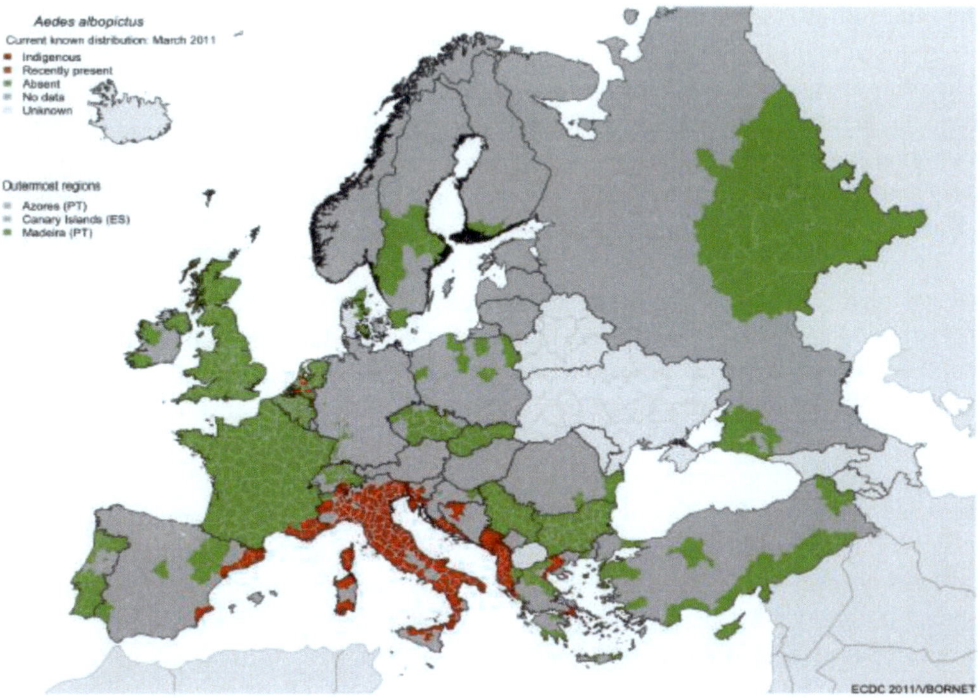

Figure 1. Distribution of *St. albopicta* in Europe (VBORNET vector maps: http://ecdc.europa.eu).

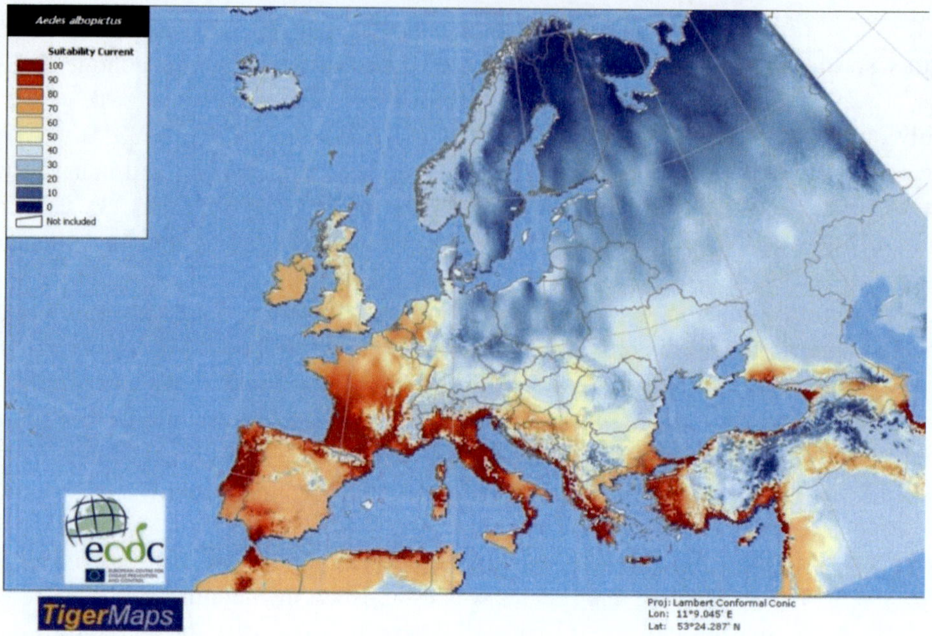

Figure 2. Predicted risk of establishment of *St. albopicta* in Europe, MCDA model, 3 variables - annual precipitation and January and summer air temperatures [45].

Populations have increased especially in areas where household water storage in containers is common and where waste disposal services are inadequate. *St. aegypti* disappeared from Southern Europe at the beginning of the last century but recently, in 2004, was introduced in Madeira and has since started to spread around the Black Sea. It has also been introduced to the Netherlands through the used tire trade [50].

The primary dispersal mode of these three invasive mosquito species by human activity has been through the transport of desiccation-resistant eggs in cargo. The most important types of goods responsible for this passive transport are used tires, which are generally stored outdoors and thus collect and store rain water that is indispensable for mosquito development [51]. Businesses that process and/or trade used tires should be given a high priority for the monitoring of exotic fauna and flora. Another documented source of introduction is through ornamental plants, e.g., "Lucky Bamboo" (Dracaena spp.) from Southeast Asia, which is transported in containers with standing water, making it an ideal insectaria in transit. "Lucky Bamboo" was the primary reason for the introduction of *St. albopicta* from Southeast Asia to California [52]. Similarly, multiple introductions of the Asian tiger mosquito to the Netherlands in commercial horticultural greenhouses have been linked to the intensive trade of this plant [43, 50]. The pathway through which *Hl. japonica* was introduced to Switzerland and Germany is not yet clear. One hypothesis is that this species was introduced via used tires or by airfreight through Zürich. However, it seems that *Hl. japonica* is most abundant in flower vases in cemeteries, indicating that used tires may not be the only reason for the widespread occurrence of this species. Another possibility is that *Hl. japonica* was, and may still be, introduced with ornamental plants (e.g., the box tree *Buxus* spp.) in transoceanic containers originating from Asia. *Buxus* spp. are common plants in cemeteries and are frequently imported from eastern Asia.

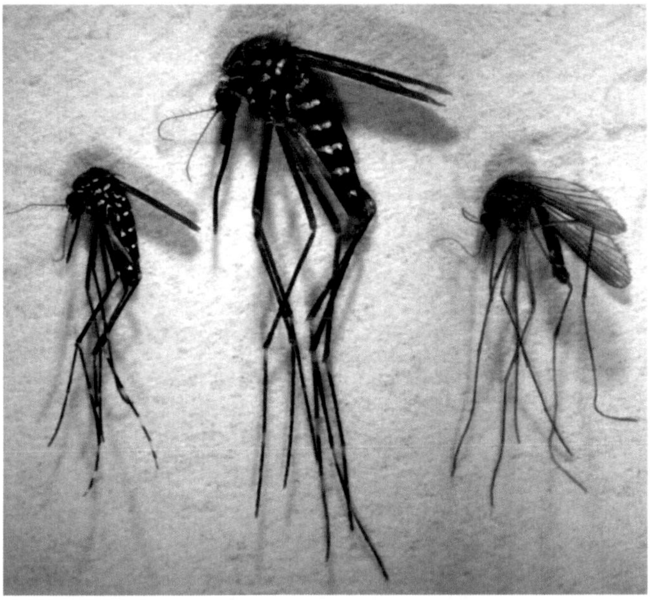

Figure 3. Invasive and indigenous European vector species: a) *St. albopicta*; b) *Hl. japonica*; c) *Culex pipiens* [55].

A strong indication that *Hl. japonica* is imported together with plants from eastern Asia is the fact that another invasive insect, the box tree pyralid moth, *Glyphodes perspectalis* (Walker) (Lepidoptera: Crambidae), occurs in the vicinity of Basel, Lörrach, Rheinfelden, Aargau and other parts of northern Switzerland. Both insects, the moth and the mosquito, occur in China, Japan and Korea. It may therefore be surmised that both species were introduced at the same time to parts of Europe via the same trade routes in association with ornamental plants such as box trees. The moth was first recorded in Germany in 2007 and has now been observed in six European countries. Due to their high humidity and cool air temperatures, refrigerated transoceanic containers offer ideal conditions for the transport of living insects [53]. Therefore, harbors, ports and inland air or road terminals that receive transoceanic containers from infested countries should be routinely monitored. Rest areas and parking lots along highways originating in areas infested with exotic species can also serve as sites of introduction [38, 54].

3. Monitoring Invasions - Tools and Formulas

Thorough assessments are necessary to prevent and/or control the introduction and infestation of new mosquito species. Nuisance alerts by the public are stressed by many experts to be a valuable early warning sign of invasion. A roadmap has been established [17, 56] to develop assessment guidelines for different scenarios that may be applicable to countries/regions where invasive mosquito species are: (i) not present; (ii) expanding their range; (iii) established in most suitable biotopes; and (iv) present but not monitored. Guidelines have also been proposed for monitoring population densities, especially during disease outbreaks. One of the key issues connected to inevitable site inspections is the selection of sampling sites (hot spots) and the creation of a priority check list. This issue considers areas with a high risk of importation from remote places (harbors, used tire facilities, airports, plant import companies, freight containers, container terminals and graveyards close to risky sites - particularly for *Hl. japonica*) and from adjacent areas (rest areas and petrol stations along traffic paths closest to the infested areas, and facilities for local transport and trade). Other key issues are how to (i) choose collection/trapping methods according to analyses of the available resources and adapt these strategy to local conditions; (ii) define the grid size and the number of traps per grid; (iii) engage the municipality/local people in monitoring activities; and (iv) communicate information to the public.

For the monitoring of population densities in the case of disease outbreaks, it is important to emphasize the need to standardize trapping techniques, data storage and processing procedures and the indices to be used. It is also important to tune entomological monitoring to the data obtained through virus surveillance programs (e.g., selection of sampling sites according to preexisting patterns of human case distributions).

It has been documented in many countries that dry ice–baited light traps, which are mostly used for autochthonous species monitoring, present a low attractiveness to *Stegomyia* females [57]. Despite the fact that sentinel traps with or without dry ice are providing quite a good sample size of adult *St. albopicta* populations [58, 59], the surveillance programs in Europe and around the world are currently using ovitraps, simple tools developed to attract ovipositing *Stegomyia* females and sample the eggs that they lay [60]. Historically, monitoring species belonging to the genus *Stegomyia* (*St. aegypti* and *St. albopicta*) has been achieved through the use of ovitraps [61]. This method provides several advantages over other methods, including high sensitivity (it can detect the presence of the insect even at low densities), ease of field management, achievability even by unskilled staff, and low material costs [62]. Nevertheless, ovitrap reliability, in terms of quantitative estimation of adult population densities, is controversial and questionable [19, 63, 64]. Of the disadvantages, we mention the restricted capacity to attract only certain species, the unavailability of adults that may be screened for virus, and the difficulty in determining species at the egg state.

During the "Meeting on vector-related risk of introduction of Chikungunya and Dengue fever and spread of *Ae. albopictus* and *Ae. japonicus* within Europe", organized in May 2011 in Speyer, Germany by the World Health Organization (WHO) and European Mosquito Control Association (EMCA), outlines for "Strategy and need assessment of monitoring of invasive mosquitoes" were specified by a working group of 14 European experts (conveners Mauro Tanola and Dušan Petrić) [56]. The most appropriate tools for the monitoring of eggs (ovitraps - mass positioning of ovitraps has been used in France to prevent the spread of tiger mosquitoes), larvae and pupae (standard dipper, aspiration tube), and adults (sticky ovitraps, gravid traps, dry ice baited trap, sentinel/sentinel + dry ice, suction traps, sweeping nets, human bait sampling) were listed. The positive and negative aspects of the various tools were discussed by experienced users, and the relative advantages of traditional construction [28, 65, 66] and improvements tailor-made for the sampling of invasive mosquito species were argued.

In this chapter, the authors will pay considerable attention to the monitoring of invasive/vector mosquito species by ovitrap, in consideration of simplicity, cost effectiveness and the reassuring results obtained by this method in Italy and Southern France, where Asian tiger mosquitoes are most widespread and are considered a real public health threat. It may be stressed that in situations where the target species are not attracted or poorly attracted by ovitrap, it may be more appropriate to consider other more convenient tools. Therefore, (i) the surveillance system for *St. albopicta* in Italy was the first to be established in Europe and is well developed and covers quite a large area and (ii) one of the authors is involved in its creation and development, and the advantages and disadvantages of this system will be discussed in detail later in this chapter.

The indices traditionally used to evaluate *Stegomyia* population densities (which could potentially also be applied to other species with similar oviposition habits, such as *Dahliana geniculata* [*Aedes geniculatus*], *Gc. atropalpus*, *Hl. japonica*, *Hl. koreica,* and *Oc. triseriatus*) and the efficacy of control campaigns, such as the House Index (HI: percentage of houses with at least one active breeding site), the Container Index (CI: percentage of containers with larvae), and the Breteau Index (BI: number of active breeding sites per 100 premises), are widely used as empirical standard parameters in developing countries [68]. However, results obtained using these indices are of limited value in European countries because of the differences in socio-economic and structural conditions that characterize human dwellings and the differences in the availability of breeding sites in public areas. Other indices that are more appropriate for European urban areas are the PPI (number of pupae/premise) and PHI (number of pupae/hectare), which defines the mosquito density per unit area, considering both public and private domains. Moreover, the traditional indices show some disadvantages when implemented in epidemiological studies. The CI only considers the percent of positive containers and not their absolute number (either per unit area, per premise, or per person). The HI is more accurate than the CI, because it refers to the number of houses, but it is again limited because it does not account for the number of positive containers. The BI is the only index that combines the data on the positive containers with their density per premise. In addition, the main limitation of the three indices is the lack of information referring to the real productivity of the containers, their relation to the adult population size and their applicability in the large European cities. More recently [69, 70], the Adult Productivity Index (API) was proposed as a new tool, based on the sum of positive containers of different typologies considering their specific relative densities. Another sampling method is the Pupal Demographic Survey (PDS), which seems to be more appropriate for epidemiological studies focusing on the estimation of the vector density transmission threshold [71]. The PDS exploits the strong correlation between the number of pupae and the number of adults in a defined area, based on the low incidence of natural mortality usually affecting the pupal state. Studies on the correlation between traditional indices and adult population densities show controversial results: while some [70] evidenced a good correlation between BI and both the larval and the adult densities, others [70] found no correlation between traditional indices and the PHI or number of pupae/person.

In a recent study conducted in Italy by one of the authors, a statistically significant correlation between CI and the mean number of eggs/ovitrap was found [72]. In the 2007 Chikungunya outbreak area in Italy, *Stegomyia* population indices (HI, CI, BI) appeared to be well correlated amongst each other but not with the PHI [72]. A good correlation was obtained between the PHI and the weekly mean number of eggs/ovitrap collected from 7 to 14 days after the premises' inspection (Table 1). Similarly, the number of females/ha, estimated on the basis of the number of sampled pupae, was correlated with the number of eggs collected in the week after the sampling. This result indicates that the number of eggs estimated by ovitraps can be used to determine the mean number of biting females per unit area.

Table 1. Pearson product moment correlations (R) between mosquito population indices and the mean number of eggs/ovitraps/week collected the week before, the week of and the week after the inspection [72]

Population Indices	Mean number of eggs/week/ovitrap		
	Previous week	Inspection week	Week after inspection
HI - House Index	0.0867	- 0.1117	- 0.3778
CI - Container Index	0.3194	0.0482	- 0.4175
BI - Breteau Index	0.0623	- 0.1465	- 0.4313
PPI - Pupae/premise	- 0.0289	- 0.2553	- 0.5118
PHI - Pupae/ha	0.1703	0.3396	0.8622**

* $P < 0.05$; ** $P < 0.01$. HI: percent of houses with at least one positive container. CI: percent of infested containers. BI: Number of positive containers/100 houses. PPI: number of pupae per premise. PHI: number of pupae per hectare.

The number of eggs that can be found within each ovitrap is affected by the skip-oviposition behavior of females, by local environmental conditions and by the concurrent availability of other oviposition sites [19]. Therefore, it is possible that the mean number of eggs/ovitrap may increase for few weeks during source removal campaigns, or that rainfall during a dry period may cause a temporary reduction of eggs/ovitrap.

The number of ovitraps to be placed and the choice of their locations are two important issues that must be addressed for a reliable estimation of the population densities in an urban area, and it is crucial to obtain comparable information for vector surveillance [73]. The optimal number of ovitraps varies according to (i) the phase of colonization of the region and (ii) the ability of the species to disperse. Mosquitoes, like most insects, actively disperse either up- or down-wind (depending of the wind speed and the flight speed of the mosquito species) over a distance that strongly depends on the species in question [74]. It has been determined that *St. albopicta* females disperse to a maximum distance of 600 - 800 m from their breeding sites [75]. In general, the distribution of a species at the beginning of the colonization phase is patchy and aggregated, and its pattern of dispersal depends on habitat features and weather conditions. In order to achieve consistent reliability levels, monitoring protocols require more ovitraps in areas at the initial step of colonization (low mosquito density and high data aggregation) relative to the areas at a mature step of colonization (high mosquito density and more uniform spatial dispersion). The ratio between the variance and the mean number of eggs laid in the ovitraps, indicated as VMR (Variance Mean Ratio), is often used as a density distribution-related parameter that provides information on the dispersal pattern of the species: VMR is equal to 1 when the species is randomly dispersed; lower than 1 when the species is uniformly dispersed; and higher than 1 when the species distribution is aggregated or clustered [65]. The variance (s^2) is also related to the mean density (m) by Taylor's power law [76, 77], which has largely been used to quantify the aggregation degree and the statistically significant sample size for insect monitoring

$$s^2 = a \times m\, b, \quad (1)$$

where a is a constant depending on environmental conditions; m is the mean egg density value and b is a constant for the species and measures data aggregation similar to VMR, i.e., when b is greater than 1 it indicates that the data are clustered.

The equation used [78] to define the minimum sample unit size for an urban area is

$$N = [Z_{\alpha/2} / D]^2 \times a \times m^{b-2}, \quad (2)$$

where Z is the standard normal distribution value for a given probability α [79]; D is the monitoring precision level required (according to the literature, D = 0.1 is considered a sufficient value [80], while 0.2 < D < 0.3 has been considered optimal for the binomial sampling of *St. aegypti* [81]); m is the mean egg density value; and a and b are the Taylor's coefficients.

The adequacy and reliability of the monitoring system can be evaluated by measuring the Relative Variation (RV) [65], i.e., the ratio

$$\frac{Standarderror\ of\ the\ mean\ number\ of\ eggs\ /\ ovitrap\ /\ week}{Meannumber\ of\ eggs\ /\ ovitrap\ /\ week}$$

The value of RV = 0.25 is usually adequate for most extensive sampling surveys, although in certain intensive programs an RV = 0.1 may be required [80]. Highly aggregated mosquito populations will likely produce a high RV.

In Northern Italy, in the period from 1994 - 2008, standard ovitraps consisted of shiny black plastic cups (400 ml capacity), filled 2/3 with water, with a masonite strip fastened to the inner edge to provide a suitable surface for oviposition [62, 82]. The size, shape, and material of the traps may affect their attractiveness to ovipositing females and the number of eggs laid on the strips [19, 62, 64]. A one-week check interval is usually adopted in order to prevent the ovitrap from becoming a mosquito reproductive site [83, 84], unless insecticides (preferably *Bacillus thuringiensis israelensis* - *B.t.i.*) are added to prevent larval development in the cups [85]. An important but not conclusive study [86] found that adding *B.t.i.* to the ovitrap water enhanced ovitrap attractiveness to *St. albopicta* females in laboratory trials but not in the field. During preoviposition behavior, visual cues drive gravid females to locate a suitable oviposition site [87]. Afterward, female behavior is mediated by a complex variety of chemical cues from the water source, including (i) volatile and nonvolatile molecules produced from the fermentation of plant organ infusions [88] and (ii) molecules related to water chemical composition, through which females can detect nutrient quality and the presence of conspecific and competitor larvae, pathogens, and predators [89]. In some cases, specific pheromones have been identified [90].

In a field study in the Emilia-Romagna Region of Italy, the oviposition activity index (OAI) [91] was evaluated according to the formula

$$OAI = (N_t - N_c) / (N_t + N_c), \quad (3)$$

where N_t is the mean number of eggs laid in each treatment cup while N_c is the mean number of eggs laid in each control cup. Index values can vary from + 1 to - 1, with positive values indicating that treatment cups are more attractive.

OAI results calculated for *B.t.i.*-treated ovitraps checked for the presence of larvae every 14 days and documented increases in the number of eggs collected by about 17.4 % over the control ovitrap that only contained water. This finding could be particularly useful in situations where *St. albopicta* densities are low.

The study supported by the Emilia-Romagna Region Public Health Department aims to develop a large scale *St. albopicta* monitoring network based on mean egg density data collection [92] (Figure 4).

This monitoring method can be achieved at a low cost but needs to be well designed in order to provide reliable information for the estimation of population densities in large urban areas where the health risks of mosquito-borne diseases are highest. Standard ovitrap monitoring methodologies, in combination with the GIS, geostatistical analysis and computer-based mapping techniques have proven useful as practical tools for entomological and epidemiological studies and operational use.

The regional monitoring system was sufficiently reliable to determine spatial variations within the *St. albopicta* data at the municipality level. In fact, the results indicate that the *St. albopicta* mean egg density data aggregated for municipality are spatially correlated and significant at a distance of less than 30 km, particularly between 0 and 7.5 km. Cross-validation results indicate that the estimated egg densities at unsampled locations are reasonably acceptable with some limits due to the non-uniform distribution of the data. Mean egg density data aggregated for municipality were sufficient to produce a spatial interpolation at the municipality level.

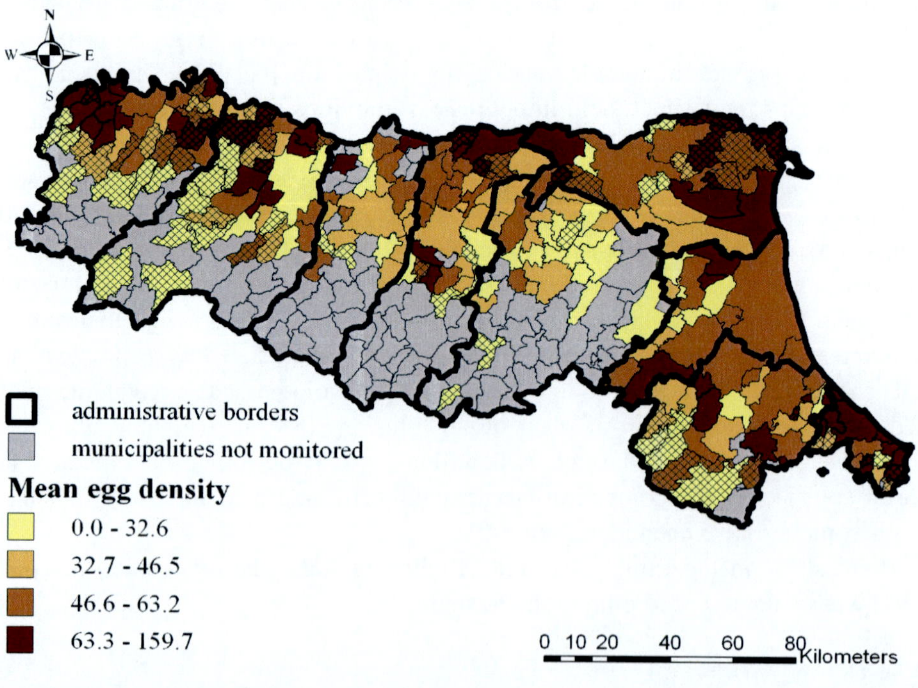

Figure 4. Choropleth map of mean egg density (number of eggs/ovitrap/week) during the field study in the Emilia-Romagna Region (Italy) calculated for 22 monitoring weeks. Legend values are subdivided into quartiles; wired polygons represent municipalities with sampling designs that were not statistically efficient for measuring true population densities for RV > 0.3 [92].

a)

b)

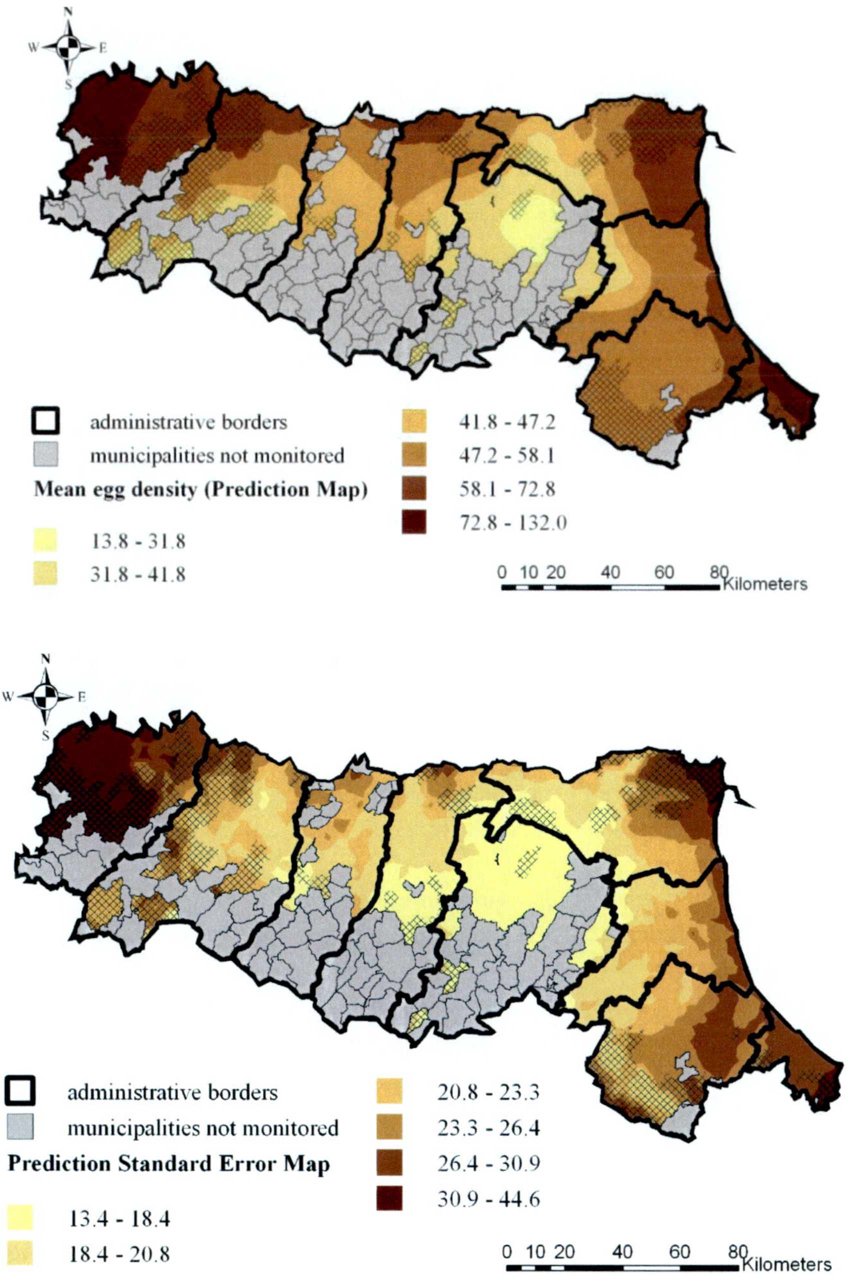

Figure 5. Universal kriging interpolation map (a) and standard error map (b) of mean egg density in the Emilia-Romagna Region in Italy (municipalities with RV > 0.3 were not considered in the interpolation calculation and are indicated on the map by wired areas) [92].

Model parameters obtained by variogram analysis were used in an ArcGIS Geostatistical Analysis to obtain the prediction map (Figure 5a), the quality of which has been examined by creating a prediction of standard error (Figure 5b). The predicted standard errors quantify the degree of uncertainty for each location on the surface.

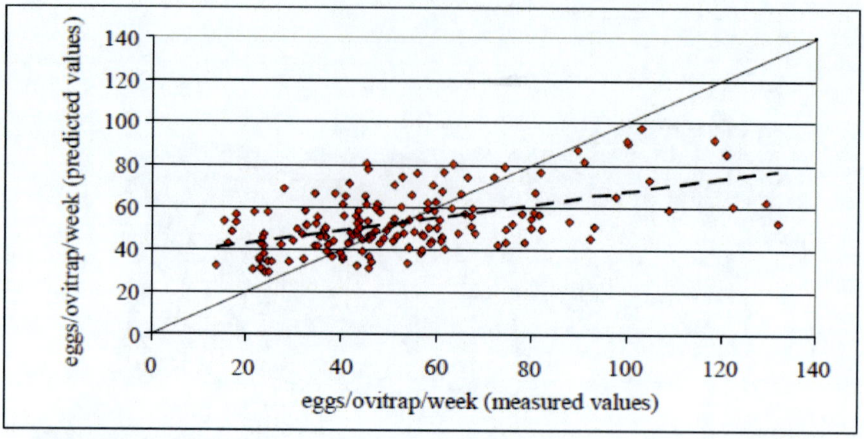

Figure 6. Cross validation results for the Emilia-Romagna Region (Italy). Scatter plot of the predicted versus measured values (the slope is lower than one; the kriging interpolation tends to underpredict large values and overpredict small values) [92].

The standard error map shows low errors in five out of the nine provinces (Bologna, Modena, Reggio-Emilia, Parma and Ravenna) and high and medium errors in the remaining four (Ferrara, Forlì-Cesena, Rimini, and in particular Piacenza on the west).

Cross-validation shows low errors near municipalities with about 53 eggs/ovitrap/week (intercept between the 1:1 correlation line and the best fit line; Figure 6) and large errors at higher egg densities.

A high correlation was calculated between mean egg density and elevation classes. The shaded elevation map, acquired from satellite images, is overlaid on the interpolated egg density map (Figure 7). Both layers show similar spatial trends, indicating a positive relationship between the ovitrap data and altitude. This positive example opens the opportunity for likely comparisons with other informative layers, such as Normalized Difference Vegetation Index (NDVI), air temperature and rainfall distributions, and land use/land cover maps.

The extrapolation and interpolation of data need to be conducted with caution, and the production of computer-generated maps that appear to be more informative than the data upon which they are based should be avoided. Bearing this in mind, contour smoothed maps obtained from geostatistical analyses and cluster maps obtained from cluster detection can be overlaid on other smoothed informative layers to identify environmental variables such as elevation, rainfall distribution, mean air temperature, relative humidity that could influence seasonal mosquito population densities in the region. These maps can also be overlaid on epidemiological data to identify health risks. Another field of application for the spatial analyses of *St. albopicta* egg density data could be the evaluation of the efficacy of the control programs performed in different municipalities, the quality of which significantly affects the mosquito population density and the attribution of a municipality to either a high or low egg density cluster. Critical points in adopting geostatistical analyses of entomological data for creating large-scale interpolation maps can be found in the difficulty of assessing a standard procedure to find the best variogram model, and in finding the appropriate dataset (mean eggs, total eggs, rank, etc.) to satisfy the prerequisite of data stationarity, which is necessary to obtain the best interpolation.

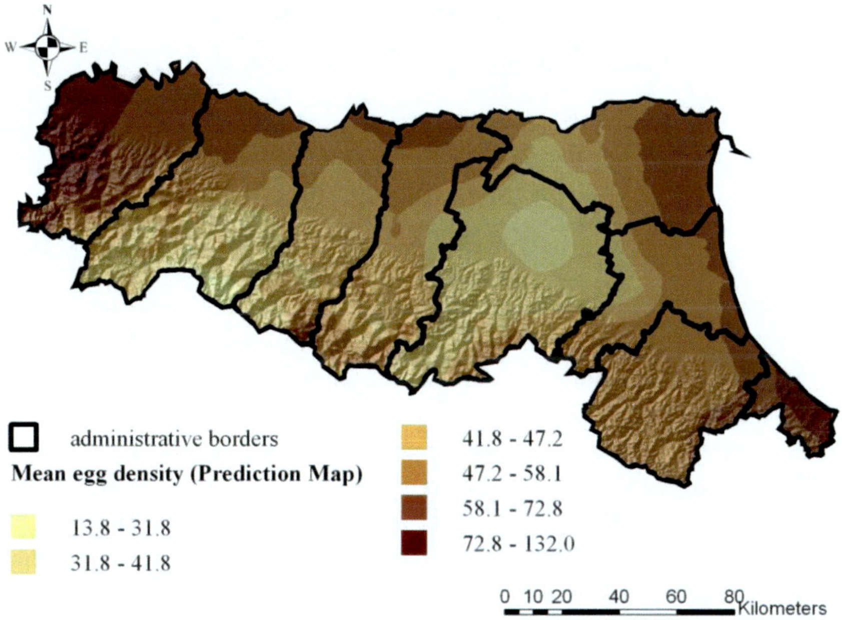

Figure 7. Example of an environmental informative layers overlay: shaded elevation map (Void-filled seamless SRTM data V2, 2006, International Centre for Tropical Agriculture (CIAT), available from the CGIARCSI SRTM 90m , Database: http://srtm.csi.cgiar.org) overlaid to interpolated egg density map [92].

The ability of European countries to obtain data on the presence and abundance of invasive species and to develop efficient control programs and tools for their evaluation needs to be rapidly and consistently improved in order to increase the chances for early detection and elimination of invaders at the beginning of the colonization process. There is a higher chance of suppressing/eliminating the invaders if convenient control methods are applied when the colonized area is still limited, as has been demonstrated on several occasions in Italy, France and Belgium [17].

St. albopicta control programs are currently being applied in Northern Italy, Catalonia (Spain), Switzerland, Croatia, Greece and The Netherlands, all of which are based on the use of ovitraps as a tool for mosquito population density estimation. In Emilia-Romagna, during the breeding season of May - October 2008, 2741 ovitraps were activated in the urban areas of 242 municipalities according to standard criteria and were checked weekly. The universal kriging interpolation was used to estimate the seasonal abundance of the species at unsampled locations, and spatial cluster analysis was used to identify particular areas that had statistically significant high or low mosquito densities. The overall data pattern was highly clustered and autocorrelated, and the choropleth and LISA cluster maps showed high egg densities in the north, northeastern and southwestern areas of the region as described earlier in this chapter.

The German Mosquito Control Association (KABS) has been conducting indigenous mosquito surveillance as a component of a comprehensive mosquito control program since 1991. Approximately half a million mosquitoes have been collected with CO_2-baited traps, ovitraps and larval dippers every year, primarily in Southwest Germany, and the species were identified in order to promptly detect possible invaders and apply the most convenient control methods.

4. Monitoring Changes (Vector-Pathogen Behavioral Shifts)

The European Commission identifies research on vectors of human diseases, together with malaria, HIV/AIDS and tuberculosis, as a priority in the struggle towards poverty reduction. Currently, it has been acknowledged that the distribution, density and ecology of vectors are highly sensitive to environmental changes caused, in part, by changing climatic conditions that allow vectors to spread to areas beyond their native tropical habitats (including Europe). These conclusions are strongly supported by the ECDC and Member States' public health institutes. In terms of disease transmission and the invasion of new areas, the threat posed to the community by anopheline mosquitoes, which transmit malaria, the Asian tiger mosquito and other invasive vector species contrasts with a considerable lack of vector and vector-borne disease surveillance activities in many European countries.

Environmental conditions have significant effects on the development of mosquito vectors and on the pathogens/parasites themselves. Aside from providing/withholding appropriate breeding sites for the mosquitoes, precipitation (rain), air temperatures and relative humidities higher than 60 % additionally influence the transmission of the parasites. Therefore, as air temperatures in Europe rise, so do concerns regarding the resurgence of malaria. These concerns were also expressed in the recent Assessment Reports of the Intergovernmental Panel on Climate Change (IPCC 2007) [93]. Models predict an increase of global air temperatures in the interval of 1.8 °C to 4 °C by the year 2100, which could be associated with a considerable augmentation in the number of vector-borne diseases.

Because mosquitoes are poikilothermic, the development of the pathogen/parasite in the mosquito depends on the outside air temperature. The development of the protozoan parasites causing malaria can only be completed if minimal summer isotherm (June - July - August, JJA) air temperature values are above 16.5 °C for *Plasmodium falciparum* and 14.5 °C for *P. vivax*. The higher the air temperature, the faster the sporozoites are formed, and the faster their life cycles are completed. Therefore, the risk of being infected with malaria increases with increasing air temperatures. However, there is also a correlation between mosquito longevity and air temperature; mosquitoes live longer at lower air temperatures than at higher air temperatures. If air temperatures are too high, both the mosquitoes and the associated parasites can die. Based on this relationship, there is an optimal air temperature range for the transmission of *Plasmodium* spp. (22 - 28 °C for *P. vivax* and its main potential vector in Europe, *Anopheles messeae*, and 26 - 32 °C for *P. falciparum* and its African vector *An. gambiae*). One of the foreseen scenarios is that indigenous, European *Anopheles* mosquitoes can, under favorable conditions, transmit imported malaria parasites. As a general rule, co-adaptation between the vector and the parasite must have occurred over the course of evolution to enable them to synergize. Recent research in the United Kingdom and Germany has demonstrated that *P. falciparum* can multiply [94], and that both oocysts and sporozoites (which accumulate in the salivary glands) can develop in *An. plumbeus* females that have ingested blood infected with *P. falciparum* [95]. Native *An. plumbeus* is therefore first on the list of European anophelines that could potentially transmit the causative agent of the deadly malaria tropica. The *An. plumbeus* female normally lays its eggs in tree holes, where the larvae and pupae develop. The behavior of these mosquitoes has changed dramatically in rural areas of Germany, especially in the south. They have started using abandoned cesspools (previously used to collect liquid manure and now collecting rain water) as breeding sites in areas were cattle are no longer reared. While only a few mosquitoes can breed in natural tree holes, the cesspools represent huge artificial "tree holes" in which millions of mosquito larvae can develop. Thus, major changes in agricultural practices have supported changes in oviposition behavior. In the past, due to their high levels of organic waste, cesspools were not appropriate breeding sites for *An. plumbeus*. However, the abandonment of the use of cesspools, and their replacement with cisterns in which only lightly contaminated rainwater is collected, has produced mass breeding areas for *An. plumbeus*. Disturbances caused by these mosquitoes in residential areas have been reported more frequently in the last 10 years. Female *An. plumbeus* are extremely anthropophilic blood feeders. Their oviposition behavior has been altered to exploit breeding sites in the immediate surroundings of human dwellings, which when considered together with their vector-parasite relationships, promotes them to the position of a dangerous new vector. In order to evaluate the feasibility of the above-stated predictions, the results of air temperature changes that have taken place over the past 60 years in southwestern Germany (air temperatures, Mannheim's weather station, 1947 - 2006) were analyzed. Air temperatures have increased continually over the last few years to approximately the same degree in both the summer and winter months. The warmest years since air temperatures have been recorded have occurred within the last 10 years. On average, the increase in air temperature has been approximately 1.2 °C (1977 - 2006). During the same period, the increase in the summer months was 1.4 °C (Figure 8). An air temperature increase of 1.6 °C would significantly accelerate the development of the malaria vector *An. messeae*, making it 2.3 days shorter. Therefore, the gonotrophic cycle would be shortened and the number of *An. messeae* generations would increase and, thus, the overall threat of vector-borne diseases in Europe will most likely increase. Even though malaria parasites were eradicated in Central Europe, they are now being imported into Europe from the tropics more frequently. The increase in tourism-related travel, above all other changes, is responsible for the transport of the parasites. Approximately 10 000 imported malaria cases per year are registered in Europe, of which more than 1000 cases per year occur in Germany [96]. Most of the cases represent infections with the hazardous *P. falciparum,* but because they are generally treated quickly, further transmission of the parasites by European mosquitoes is prevented to a great extent. Under the present conditions, infections in Europe are rarely lethal. In addition to affecting tourists, infected mosquitoes can also travel from the tropics in airplanes, causing so-called airport malaria in people who have never visited the tropics. The question remains as to whether these imported malaria cases could become a threat to the European human population. Co-adaptation between a vector and parasite similar to the scenario described above has occurred recently between the Chikungunya virus and Asian tiger mosquito. Since 2004, several million indigenous cases of Chikungunya virus disease have occurred in Africa, the Indian Ocean, India, Asia and, recently, in Europe. The virus, usually transmitted by *St. aegypti* mosquitoes, has now repeatedly been associated with a new vector, *St. albopicta*. Analysis of full-length viral sequences reveals three independent events of viral exposure to *St. albopicta*, each followed by the acquisition of a single adaptive mutation providing a selective advantage for transmission by this mosquito [97]. This alarming and recent, unique example of "evolutionary convergence" (a phenomenon that is rarely observed in nature) occurring in nature illustrates the dangerously rapid ability of pathogens to adapt to ecological changes, driven directly by human activities (anthropogenic spreading of *St. albopicta* around the world, e.g., by the used tire trade).

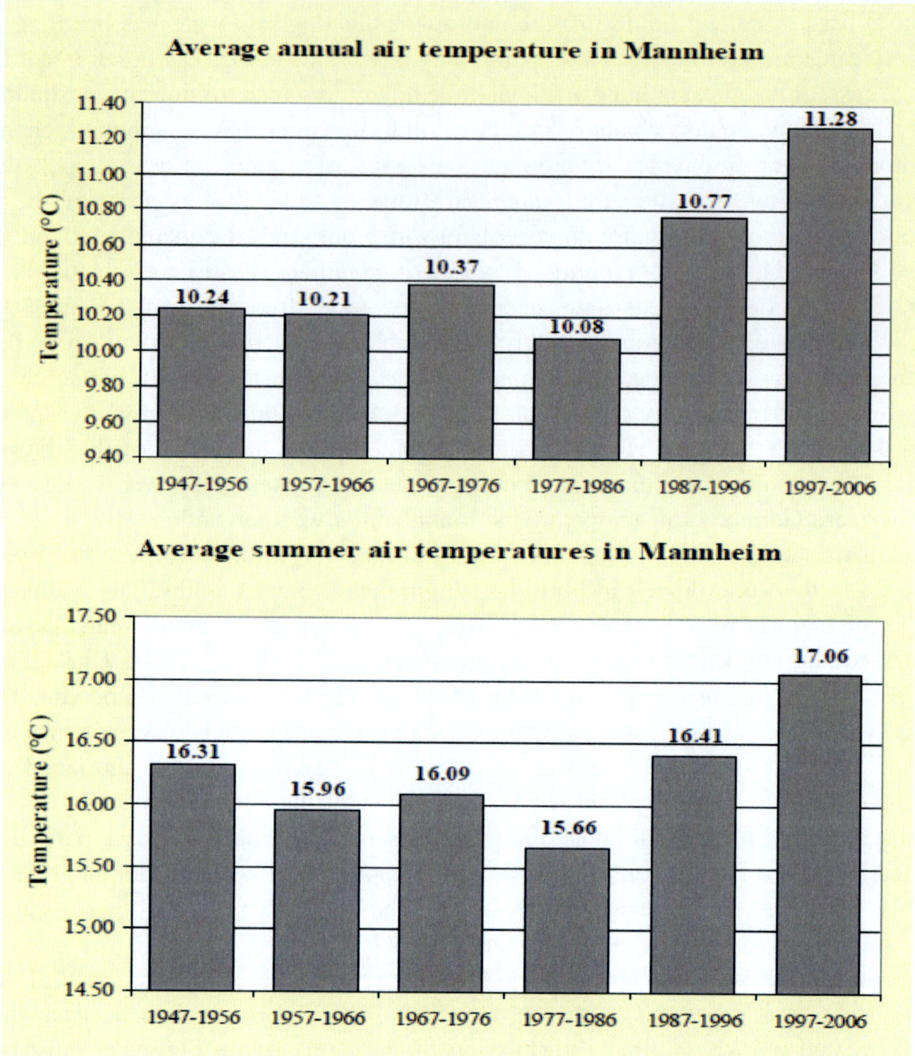

Figure 8. Average annual and summer air temperatures in Mannheim (Germany) presented in ten-year time intervals.

Another change that is occurring is in the biting behavior of *Culex pipiens*, the most important vector of WNV in Europe. Over the last 10 years *Cx. pipiens pipiens* biotype *molestus* females have been frequently attacking humans outdoors, a feeding habit typical in Southern Europe but never observed in Northern Serbia during a period of intensive monitoring from 1980 to 1990. Highly ornithophilic biotype *pipiens* of the same species have been observed to feed on humans when artificially stimulated with CO_2 (Petrić, unpublished observations).

The Asian tiger mosquito, *St. albopicta*, originates from Southeast Asian forests where it mainly develops in water-filled coconut shells or bamboo stumps. It only later become synanthropic, adapting to artificial containers. They can now be found in containers such as water barrels, car tires and other places where small pools of water may collect. Due to the change in its oviposition preferences, this exotic mosquito species has undergone an astonishing expansion of its range within the last decades, driven directly as a consequence of human activities, mainly the international used tire trade. Since 1979, *St. albopicta* has been found in Africa, the Americas and Europe, and, more recently, also in the Pacific region. It is expected to spread worldwide to tropical and subtropical regions and occasionally to regions with sub-temperate climates. Once established, national trade and traffic lead to rapid distribution within the geographic zone. When modeling the potential for this mosquito to spread in Germany based on climatic predictions, the following predictors are used: 18 °C + summer isotherms (JJA), - 3 °C isotherms for the coldest month (January), and an annual precipitation of 500 mm. To determine the reproduction period, the number of days of frost during the overwintering period and, interestingly, the seasonal emergence of apple tree blossoms (which differs by region) are taken into account. While mean air temperatures reveal macroclimate conditions, phenological data, such as the commencement of apple tree flowering, take all relevant micro-, meso-, and macroclimatic factors into account [95].

In the Upper Rhine Valley, due to elevated air temperatures and local, often heavy precipitation, the time period during which mosquito control measures are taken has widened significantly (Figure 9). In the 1980s, control measures began in mid-April and are now already beginning in March.

The campaigns used to end in mid-September but have continued into October in the past few years. Multivoltine species are not the only species to have widened their time window for development and increased their number of generations per year; the habitually monovoltine snow melt mosquito species *Ochlerotatus cantans* has started to produce a second generation per year (Becker, unpublished data). Similar changes have been observed within the European population of the codling moth (*Cydia pomonella* Linnaeus), which is producing an additional third generation per year in many places due to, as hypothesized by authors, global warming [98]. Presumably, climate change will result in higher air temperatures and fluctuating rainfall that will significantly increase the sequence of control campaigns, with an accompanying rise in the cost of the campaigns even when no vector-borne parasites are present.

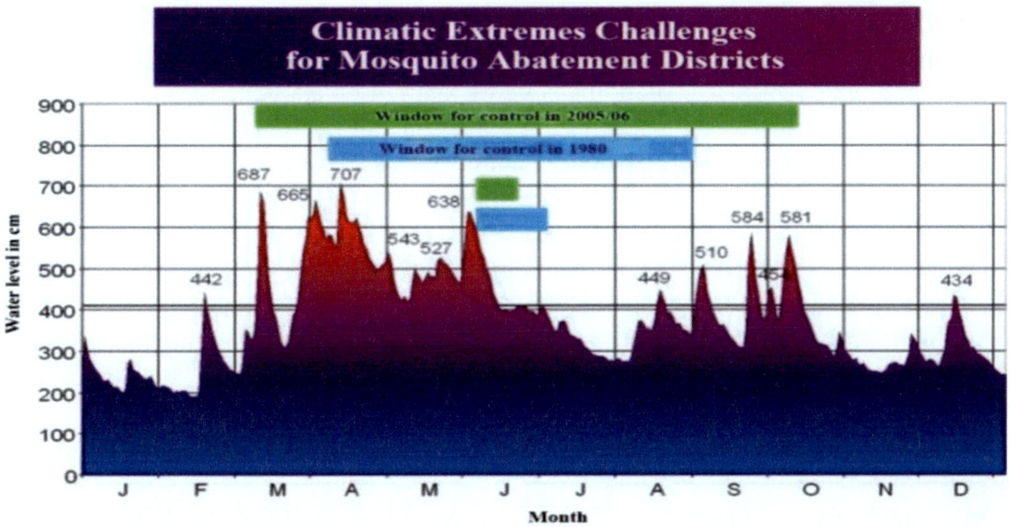

Figure 9. Climate extremes and influence on mosquito development and control [95].

There have already been some changes in mosquito occurrence and behavior that can be attributed to local climate warming in the Czech Republic: (i) a south-Palaearctic species *An. hyrcanus* that was not previously present was found in CO_2 trap catches; (ii) occasional exophagy and anthropophily of the anautogenous biotype of *Cx. pipiens* has been observed in lower parts of the country; (iii) late spring "breeders" *Aedimorphus vexans* [*Aedes vexans*], and *Oc. sticticus* now breed together with snow melt mosquitoes (*Oc. cantans* or *Oc. catahylla*) and (iv) *Cs. annulata* overwinters in the larval stage which was previously exclusively the case in Southern Europe (Frantisek Rettich, personal communication).

5. Monitoring Vector-Borne Diseases

The West Nile virus, an RNA arbovirus (*Flaviviridae*, *Flavivirus*), is the most extensively distributed flavivirus of the Japanese Encephalitis serocomplex group worldwide. The virus was first isolated in 1937 from the blood of a woman with neurologic disorders in the West Nile district of Uganda [99]. A large number of wild and domestic bird species may serve as reservoirs of WNV, from which it is transmitted by mosquito vector species. Mosquitoes of the genus *Culex* are generally considered to be the principal vectors of WNV worldwide [28]. A broad spectrum of mammalian species, including humans, horses, cats and rabbits, can be infected naturally or experimentally with WNV [100]. Humans are terminal hosts that are unsuitable for further transmission of the virus by mosquitoes because of low viral titers and a short duration of viremia. Most human infections are asymptomatic. Clinical manifestations can range from uncomplicated febrile illness to fatal meningitis or encephalitis [101]. Severe neurologic cases have been reported in about 1 % of infected patients [102].

The presence of WNV was registered in the western Mediterranean and southern Russia in the early 1960s [103]. The significance of WNV in these areas is not only the fact that there has been expansion into new areas but also the possibility of changes in viral virulence, which has been registered in some of the recent outbreaks. Mild cases of fever, which were apparently more frequent in previously described epidemics, have been replaced by outbreaks of cases with severe neurological manifestations and deaths. In the USA, WNV was first registered in 1999 when it led to an epidemic of fatal encephalitis in 12 % of the infected patients [104]. Since then, as of end of 2010, 30 600 cases of clinically manifested human infections, 12 668 cases of meningitis/encephalitis and 1206 fatalities have been reported to the CDC (www.cdc.gov).

Among the outbreaks registered in recent decades in Europe, an encephalitis epidemic in southeastern Romania in 1996 was the first large urban outbreak [105]. The virus continued to circulate in Romania after the epidemic. In the period from 1997 - 2000, 39 cases of human infection were registered, resulting in 5 (13 %) deaths [106]. In 2010, a total of 57 cases of WNV infection were identified in Romania with a case fatality rate of 8.8 % [107]. In the same year, an outbreak of human WNV infections occurred in Central Macedonia (Greece) [108] with 32 fatalities, which serves as another timely reminder that WN fever is an emerging vector-borne disease in Europe.

An example of a good monitoring practice can be taken from the Emilia-Romagna Region (Italy) where the WNV surveillance plan 2009 locally adopted the surveillance measures indicated by the National plan [109]. In particular, among the surveillance activities, the choice was made to monitor wild, non-migratory birds, such as corvids (the crow family), which are considered to be the most sensitive indicators among birds (for WNV lineage 1), and can be captured easily (it should be mentioned that raptors, rather than corvids, are most sensitive to WNV lineage 2). Regarding equine surveillance, the Regional plan emphasized the education of veterinary practitioners, focusing on the inclusion of WNV in differential diagnoses and the achievement of rapid reporting. A major feature of this plan was to establish an extremely sensitive system of passive surveillance. In addition to passive surveillance, active monitoring of horses was implemented in the area involved in the 2008 outbreak, including Ferrara and the neighboring provinces [110]. Evidence of WNV circulation in 2008 was found in animals [111, 112], humans [113], and mosquitoes. This highlighted the need to implement an integrated surveillance system that would describe the phenomenon comprehensively. Such a system should facilitate the collection of data to evaluate spatial distributions and time trends of viral circulation and allow sharing of information. For this reason, the 2009 Emilia-Romagna Regional Surveillance Plan implemented activities beyond those of the National Plan (Italy), revised the surveillance system of human cases, activated intensive entomological monitoring, and enlarged the surveillance area to involve all the provinces along the Po River.

The main aim of a human surveillance system should be the early detection of infection in humans and the estimation of its diffusion through the systematic analysis of newly emerging clinical cases, in order to manage specific interventions (e.g., blood transfusion and organs transplantation). In countries with undeveloped surveillance programs, historical records can also indicate the areas where limited surveillance resources should focus. Human surveillance is performed by serology or viral genome detection on blood and cerebrospinal fluid for all suspected cases suffering from acute meningoencephalitis in the at-risk area. Active surveillance of people who live or work in areas of documented viral circulation should also be performed. In addition, blood and cerebrospinal fluid samples from subjects living or residing for at least one night in the surveyed area in 2009 were sent to the Regional Reference Centre for Microbiological Emergencies.

The Emilia-Romagna Region veterinary WNV surveillance system is active from May to October, performing passive and active surveillance on horses and non-migratory wild birds. In Italy, all suspected signs of WN disease in horses must be reported to the veterinary authority. Suspected cases are confirmed as positive by a reverse transcription – polymerase chain reaction (RT-PCR) performed on the central nervous system [114] or a WNV virus neutralization (VN) test (cut-off titre 1:10) in microtitre plates or an IgM enzyme linked immunosorbent assay (ELISA) [115, 116]. In the provinces of Ferrara, Bologna, Modena, Ravenna, and Reggio Emilia, for every 1600 km^2, 28 seronegative unvaccinated equine sentinels, which is sufficient to detect an incidence above 10 % (CI 95 %), were selected in the spring of 2009. They were serologically tested twice after the selection, at the beginning of August and the beginning of September. Samples collected were screened by a homemade competitive ELISA [117].

Bird surveillance was carried out in all the provinces along the Po River, in the plains area of the Emilia-Romagna Region. For every 1 600 km^2, a monthly sample of about 40 wild birds caught or shot within specific wildlife population control programs was collected. Samples of organs (brain, heart, and kidney) from each bird were pooled and examined by RT-PCR [114].

The entomological (mosquito) surveillance system was based on the weekly to monthly (frequency depends on local resources) collection of mosquitoes from fixed stations and from sites where birds, humans, or horses signaled WNV activity. Mosquito collections for WNV screening were conducted using 92 CO_2 baited traps positioned in fixed stations. Moreover, mosquito collections were performed promptly using CO_2 and gravid traps in sites where positive horse and human cases had been detected. In Serbia, CO_2 traps alone are used. In Germany, of the 643 mosquito pools assayed (cell culture and RT-PCR) for the presence of Sindbis virus (SINV), ten were SINV RNA positive pools, all of which originated from samples from gravid traps comprising *Cx. torrentium*, *Cx. pipiens* and *An. maculipennis* s.l. [67]. This confirms the usefulness of this type of trap in virus surveillance programs.

The surveillance system was active in the period between April and October. Collected mosquitoes were pooled (maximum 200) by species, date, and site of collection and examined by RT-PCR [114]. In addition, overwintered mosquito females were collected during March and the beginning of April by manual aspirator in rural buildings in the area where WNV was active in 2008.

The veterinary and entomological surveillance actions described above detected WNV activity from the end of July 2009, about 2 - 3 weeks before the onset of the first human neurological case. Mosquitoes and birds were the first indicators of circulating WNV. Human cases occurred later in the season. Passive surveillance of horses also seems to be suitable as an early tool for the detection of WNV activity, but it will be less sensitive in the future because an intensive program of horse vaccination was started in June 2009.

The results of the entomological surveillance confirm that the CO_2 trap is a reliable and valuable tool for early detection of WNV. *Culex pipiens*, the most abundant mosquito species in the region, is the only vector species incriminated because no other species collected in the field were infected.

The quick and intensive spread of WNV over the past two years suggests that the whole Po plain may be affected in the future. In forthcoming years, the surveillance of wild birds and insects will be used to forecast the extension and spread of WNV. The information gathered will be used to direct or optimize actions intended to prevent virus transmission, such as vector monitoring and control, information campaigns to improve personal protection, and deploying screening tests on blood, tissues, and organs for transplant.

Another arbovirus that deserves attention is the Usutu virus (USUV) (*Flaviviridae*, genus *Flavivirus*, belonging to the Japanese encephalitis serocomplex). Isolated for the first time in South Africa in 1959 [118], USUV was first detected in Central Europe in 2001, where it caused high mortality in blackbirds around Vienna, Austria [119].

USUV seems to have similar life cycles to WNV, mainly exploiting birds as reservoirs and *Cx.pipiens* as a vector, but some differences have been noticed in the Emilia-Romagna arboviruses surveillance program, which have led to hypotheses of a possible involvement of different reservoirs (other bird species and/or mammals) in the USUV life cycle [120]. The detection of USUV genomes in *St. albopicta* also stresses the need for special attention and further research.

A universally applicable arbovirus surveillance system does not exist; thus, local mosquito surveillance systems should be tailored according to (i) the probability of arbovirus activity and (ii) the resources available for surveillance.

The importance of mosquito surveillance was underscored by Reeves [121] in his statement "...each epidemic... that was evolved in recent years could have been prevented or abated early in the course of its development by means of surveillance and vector abatement".

6. MONITORING SYSTEM – ORGANIZATION

In the summer 2007, an epidemic of Chikungunya occurred in the villages of Castiglione di Cervia and Castiglione di Ravenna (Italy), involving about 250 cases with secondary focuses in other Emilia-Romagna urban areas [7]. This epidemic strongly impacted the organization of *St. albopicta* monitoring and control activities and required that Public Administrations put more effort into the management of the vector. The main objective was then to develop a control strategy target to achieve the containment of the vector below the critical density population threshold for preventing new epidemic events, considering both Chikungunya and Dengue.

In order to develop a homogeneous approach to the health problems caused by *St. albopicta* over the whole region, the General Directorate for Health and Social Policies of Emilia-Romagna has promoted a Regional Group for the surveillance and control of *St. albopicta* (Figure 10) [78]. Its aim was to share the knowledge of all the participants in the monitoring network (local health units and municipalities), particularly among the members of the scientific group that provided expertise in entomology, epidemiology, meteorology, and informatics.

The microhabitat characteristics of the trap stations are fundamental to the effectiveness of the traps [73]. Therefore, to maximize the standardization of the environmental parameters and to avoid differences in efficacy among traps (depending on relevant environmental characteristics of the station, e.g., shading degree, vegetation type, and humidity level), the choice of the location and the positioning of the traps must be the responsibility of skilled technicians.

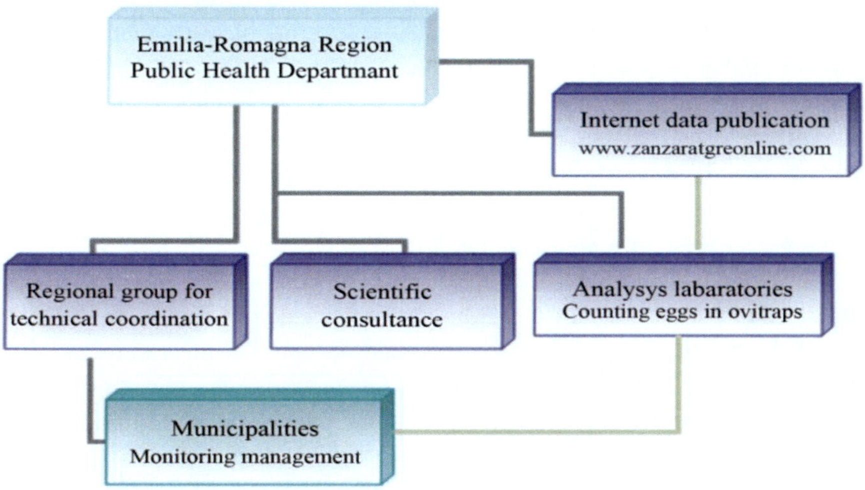

Figure 10. *Stegomyia albopicta* monitoring system organization in the Emilia-Romagna Region (Italy) [78].

Table 2. Costs of the 2008 monitoring program of the Emilia-Romagna Region (Italy)

Activities	Costs
Egg counting (ARPA, Ferrara and Parma Universities)	€ 80 000.00
Ovitrap positioning and consultants	€ 33 960.00
Routine ovitrap management (Municipalities)*	€ 409 864.00
Total	€ 523 824.00

* The overall cost has been estimated considering 7 € per ovitrap and 26 collection turns [78].

In the Emilia-Romagna *St. albopicta* monitoring network, the routine field management of ovitraps is conducted by municipal technicians. The Provincial Laboratories Network of the Regional Agency for Environmental Protection (ARPA), the Universities of Parma and Ferrara, and the Museum of Natural History of Parma are involved in the classification and counting of the collected eggs.

Ovitrap monitoring currently covers the period between weeks 21 - 41 with 11 collections (every second week). The data are published bi-weekly on a dedicated website (www.zanzaratigreonline.it) that allows the generation of reports according to three levels of access: all registered users can get basic statistics on the mean *St. albopicta* population density and on the trends of the infestation at the provincial scale (first level of access); municipality operators can also get data statistics related to their territory (second level of access); and regional group members have access to data at the provincial and municipal scale and also to the data provided by each single ovitrap (third level of access). The costs of the monitoring system in the year 2008 are shown in Table 2. The 2011 costs are estimated to be about half of the 2008 costs, with the same level of precision.

A homogenous survey of each inhabited area was obtained by using GIS (Geographic Information Systems - ESRI ArcView) to divide each area into a number of quadrants equal to the number of ovitraps that needed to be activated. Each ovitrap was georeferenced using GPS-equipped palmtops and labeled by an identification code. The stations were maintained for the whole season and as desired in the following years. Within each quadrant, the ovitrap was placed in a green, shaded, and easily accessible area. It was positioned on the ground, with a free space of at least 1 m above it. In the last protocol, each ovitrap consisted of a cylindrical black plastic pot (capacity 1000 ml, diameter 11 cm), filled approximately 2/3 with about 800 ml of dechlorinated water with 1 ml/liter *B.t.i.* One 14.5 × 2.5 cm strip of masonite was fixed to the ovitrap with a metal clip as an egg deposition substrate [85]. At each biweekly check, the deposition substrate and the solution were replaced after performing a careful cleaning in order to remove any eggs. The masonite strips were then delivered to the Laboratory Network for classification and counting.

Many environmental variables, such as the dimensions of the town, the incidence of small premises or quarters with large buildings, the presence of green areas, the degree of maintenance of the premises and the courtyards, and the number of private and public catch basins, affect the density and dispersion of *St. albopicta* in urban environments. In addition, the control programs adopted by the different municipalities may differentially affect the reliability of the ovitrap monitoring.

In the central part of the reproductive season, the ovitrap monitoring network provided data with low Relative Variation (RV < 0.2), thus indicating that the minimum ovitrap numbers in the urban areas > 600 ha could be further reduced, thereby lowering the maintenance costs [92]. In urban areas in Italy, a correlation exists between the mean number of eggs laid by *St. albopicta* females in the ovitraps and the mean adult population density calculated by entomological surveys, such as the Pupal Demographic Survey (PDS) [72]. PDS is a survey method based on the estimation of female populations by pupal density and is used in most epidemiological studies on *St. aegypti* [60]. In Italy, the most productive breeding containers that *St. albopicta* colonizes are the catch basins in both public and private areas, which are estimated to account for more than 90 % of adult production [72]. Nevertheless, catch basin sampling is highly time consuming, and the development of a monitoring network based on PDS would require unsustainable costs for a routine surveillance program.

The low cost of this monitoring program based on the use of ovitraps makes the system more sustainable even during non-epidemic periods. In addition, the availability of continuous data on mosquito populations provides information on the real-time development of infestations and can therefore provide knowledge that is readily available in case of an epidemic emergency. The network has provided useful information to implement systems to prevent and control outbreaks of arboviruses such as Chikungunya and Dengue and has created a WEB-GIS application that facilitates the analyses of the monitoring data by the regional coordinating group [92].

The organizational scheme coordinated by the Emilia Romagna Public Health Department has helped to manage the system without major problems. In particular, data were rapidly and directly transferred from the field to the institutions charged with activating the control programs and also to citizens. The system has proved to be highly efficient and prepared to face any accidental arrival of viral infection reservoirs that could cause outbreaks if the disease vector insect is present at high-density levels. The occasional collection of *Da. geniculata* eggs showed that the ovitraps are likely suitable breeding sites for other mosquito species, such as *St. aegypti* and *Hl. japonica*, even though the ability to detect possible new species through egg observation has not been proved. In specific cases, eggs could be sampled randomly, submitted for hatching in laboratories and reared to larvae and adults to achieve specific identification.

Medical surveillance accompanied by entomological surveillance is essential to prevent the spread of arboviruses and to evaluate the risk of viral disease outbreaks. The development of an efficient monitoring network is also an essential tool for verifying the effectiveness of control measures. Currently, a pilot phase of a new monitoring system based on fortnightly ovitrap checking [85] is being evaluated in order to further reduce the costs of the monitoring network.

The capacity of many other European countries to detect the early presence of invasive species and define their abundance and colonized area needs to be rapidly improved in order to increase the chances of early detection and elimination of invaders at the beginning of the colonization process and/or to develop efficient control programs. Moreover, in areas where the invading species is established, monitoring of further spread and abundance is needed for timely risk assessment of arbovirus transmission.

It has been demonstrated on several occasions within different countries and environmental conditions that it is possible, and perhaps highly convenient in term of cost-benefit balance, to eliminate an invading mosquito species by promptly applying intensive suppression methods if the colonized area is still well delimited.

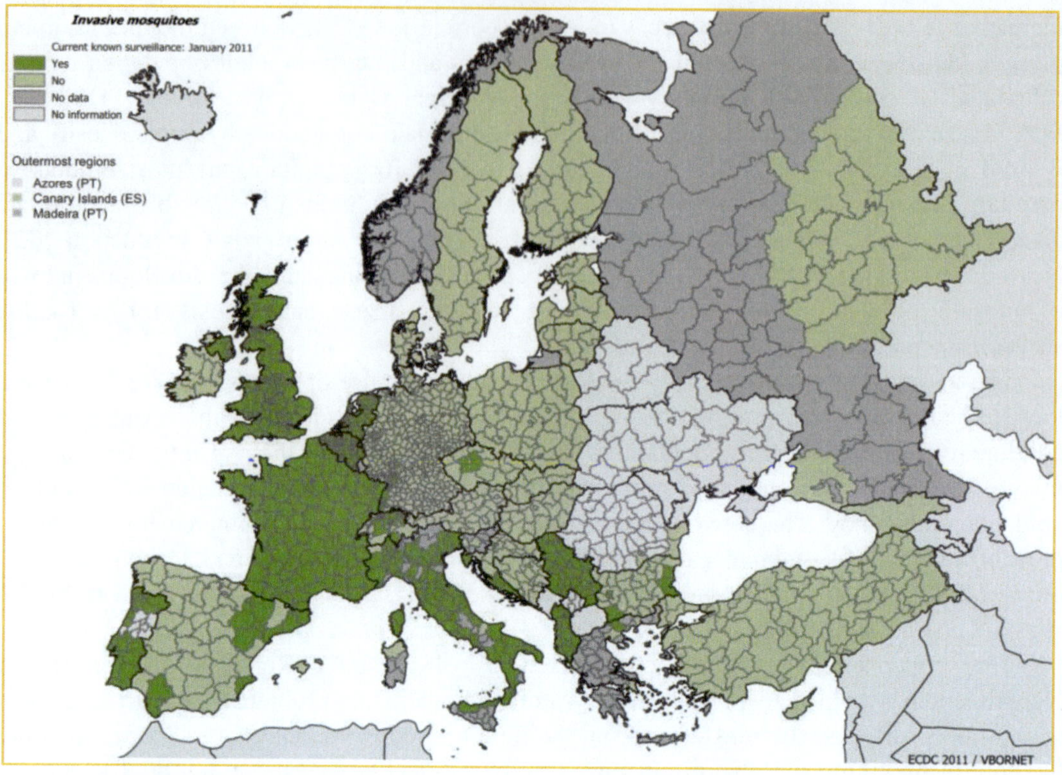

Figure 11. Current known surveillance activities in Europe (VBORNET: http://ecdc. europa.eu).

At least some kinds of surveillance and monitoring networks (research and/or control based) are already organized in many European countries, including Albania, Belgium, Bulgaria, Croatia, the Czech Republic, France, Germany, Greece, Italy, Montenegro, the Netherlands, Portugal, Serbia, Slovenia, Spain, Switzerland and the United Kingdom (Figure 11). ECDC, Stockholm, Sweden has been supporting several projects that aim to increase the capacities of European countries for surveillance and control of invasive mosquito species and vector-borne diseases: (i) the TigerMaps project, which included a multi-model approach to model and predict the spread of *St. albopicta* in Europe taking into consideration the current presence/absence data, expert knowledge and a variety of IPCC-derived climate scenarios; (ii) VBORNET, the European Network for Arthropod Vector Surveillance for Human Public Health and (iii) Vi-Map, which aims to map European health vulnerabilities to climate change related to communicable disease.

The European Spatial Agency (ESA) is also on the ground by promoting better exploitation of remote sensing satellite capacities in the field of vector surveillance and by supporting the VECMAP initiative inside the Integrated Applications Promotion (IAP) ESA ESTEC, an integrated spatial tool and service for modeling the distribution of mosquito vectors of disease.

Conclusions

Vector monitoring and vector-borne disease surveillance programs are of primary importance for the following purposes: (i) early detection of invasive exotic mosquito species; (ii) observation of their spreading; (iii) examination of alterations in vector and pathogen populations; (iv) detection of climate moderated changes, adaptation and mitigation; (v) development, implementation and evaluation of control measures; and (vi) epidemiological studies. Data on the trapped mosquitoes should be maintained to create a historical record of mosquito species found in association with different habitats and pathogens to allow early detection of adaptations.

To achieve the best results, entomological and medical surveillance should be bound together and integrated with meteorology, geographic spatial techniques (spatial statistics, geostatistics, GIS) and informatics/statistics studies.

When dealing with mosquito and pathogens, a thorough knowledge of the past and present is essential but not yet established. Both groups of organisms have a tremendous capacity to adapt and change, sometimes with incredible speeds that are not easy to conceive. Modeling is one of the ways to learn about the future, but monitoring/surveillance are of paramount importance to observe changes and to correct and improve our vision.

Acknowledgements

This chapter was realized as a part of the projects "Studying climate change and its influence on the environment: impacts, adaptation and mitigation" (No. 43007), "Surveillance of game health and introduction of novel biotechnology detection methods of infectious and zoonotic agents - risk analysis to humans, domestic and wild animals and environmental contamination" (TR 31084) which are financed by the Ministry of Education and Science of the Republic of Serbia, within the framework of integrated and interdisciplinary research over the period from 2011 - 2014 and "Survey of West Nile virus in vector and seroprevalence in human population" (114-451-2142/2011-01), which is financed by Provincial Secretariat for Science and Technological Development, AP Vojvodina.

References

[1] J.C. Semenza, B. Menne, *Lancet Infect. Dis.*, 9(6), 365-75 (2009).

[2] European Centre for Disease Prevention and Control (ECDC) (2009). Meeting Report: First meeting of ECDC Expert Group on Climate Change. Stockholm: ECDC [http://www.ecdc.europa.eu/en/publications/Publications/0910_MER_First_Meeting_of_ECDC_Expert_Group_on_Climate_Change.pdf].

[3] N. Wilson, D. Lush, M.G. Baker, *Euro Surveill.*, 15(30), pii=19627 (2010).

[4] D. Roth, B. Henry, S. Mak, M. Fraser, M. Taylor, M. Li (2010). *Emerg. Infect. Dis.* [http://www.cdc.gov/EID/content/16/8/1251.htm].

[5] F. Schaffner, W. Van Bortel, ECDC, *VBORNET Newsletter,* 2, 6-8 (2010).

[6] R. Angelini, A.C. Finarelli, P. Angelini, C. Po, K. Petropulacos, P. Macini, C. Fiorentini, C. Fortuna, G. Venturi, R. Romi, G. Majori, L. Nicoletti, G. Rezza, A. Cassone, *Euro Surveill.*, 12(36), pii-3260 (2007).

[7] P. Bonilauri, R. Bellini, M. Calzolari, R. Angelini, L. Venturi, F. Fallacara, P. Cordioli, P. Angelini, C.Venturelli, G. Merialdi, M. Dottori, *Emerg. Infect. Dis.*, 14, 852-853, (2008).

[8] I. Gjenero-Margan, B. Aleraj, D. Krajcar, V. Lesnikar, A. Klobučar, I. Pem-Novosel, S. Kurečić-Filipović, S. Komparak, R. Martić, S. Đuričić, L. Betica-Radić, J. Okmadžić, T. Vilibić-Čavlek, A. Babić-Erceg, B. Turković, T. Avšić-Županc, I. Radić, M. Ljubić, K. Šarac, N, Benić, G, Mlinarić-Galinović, *Euro Surveill.*, 16(9), pii=19805 (2011).

[9] M. Braks, *et al.*, on behalf of WP4, ECDC, *VBORNET 2011. Towards an Integrated Approach in Monitoring and Surveillance of Vector-Borne Diseases in Europe* (In prep.).

[10] P. Nsubuga, M.E. White, S.B. Thacker, M.A. Anderson, S.B. Blount, C.V. Broome, T.M. Chiller, V. Espitia, R. Imtiaz, D. Sosin, D.F. Stroup, R.V. Tauxe, M. Vijayaraghavan, M. Trostle, In *Disease Control Priorities in Developing Countries;* 2nd Edition; Oxford University Press: New York, US, Chapter 53: Public Health Surveillance: A Tool for Targeting and Monitoring Intervention, 2006; pp. 997-1018.

[11] L.A. O'Dwyer, D. L. Burton, *Aust. N.Z. J. Publ. Heal.*, 22, 819-823 (1998).

[12] A.M. Liebhold R.E. Rossi, W.P. Kemp, *Annu. Rev. Entomol.*, 38, 303-327 (1993).

[13] U. Kitron, *J. Med. Ent.*, 35, 435-445 (1998).

[14] S. Brooker, S.I. Hay, D.A. Bundy, *Trends in Parasitol.*, 18, 70-74 (2002).

[15] A.D. Cliff, J. K. Hord, *Spatial Processes: Models and Applications*; Pion: London, GB, 1981; p. 266.

[16] A. Getis, J.K. Ord, *Geogr. Anal.*, 24, 189-206 (1992).

[17] F. Schaffner, R. Bellini, D. Petrić, E.-J. Scholte, *Invasive Mosquitoes: Guidelines for Implementing Monitoring.* ECDC Direct Service Contract (In prep).

[18] M. Enserink, *Science*, 320(5878), 864-866 (2008).

[19] W.A. Hawley, *J. Am. Mosq. Contr. Assoc.*, (Suppl.) 4, 1-39 (1988).

[20] C.G. Moore, C.J. Mitchell, *Emerg. Infect. Dis.*, 3, 329-334 (1997).

[21] S.B. Halstead, *Bull. WHO*, 58, 1-21 (1980).

[22] S.B. Halstead, *WHO Chronicle*, 38, 65-67 (1982).

[23] S.B. Halstead, *World Health Stat. Q.*, 45, 292-298 (1992).

[24] N. Becker, S. Djakaria, A. Kaiser, O. Zulhasril, H.W. Ludwig, *Bull. Soc. Vector. Ecol.*, 16, 176- 182 (1991).

[25] N.G. Gratz, *Annu. Rev. Entomol.*, 44, 51-75 (1999).

[26] P. Reiter, D. Fontenille, C. Paupy, *Lancet Infect. Dis.*, 6, 463-464 (2006).

[27] A. Beltrame, A. Angheben, Z. Bisoffi, G. Monteiro, S. Marocco, G. Caller (2007). *Emerg. Infect. Dis.* [http://www.cdc.gov/EID/content/ 13/8/1264.htm].

[28] N. Becker, D. Petrić, M. Zgomba, C. Boase, M. Madon, C. Dahl, A. Kaiser, *Mosquitoes and Their Control.* Springer: Heidelberg, Dordrecht, New York, 2010; p. 577.

[29] J. Schmidt-Chanasit, M. Haditsch, I. Schöneberg, S. Günther, K. Stark, C. Frank, *Euro Surveill.,* 15(40), pii=19677 (2010).

[30] P. Reiter, *J. Am. Mosq. Contr. Assoc.*, 14, 83-94 (1998).

[31] J. Adhami, N. Murati, *Revist. Mjekesore.,* 1, 13-16 (1987).

[32] A. Sabatini, V. Raineri, G. Trovato, M. Coluzzi, *Parassitologia*, 32, 301-304 (1990).

[33] G. Dalla Pozza, G. Majori, *J. Am. Mosq. Contr. Assoc.,* 8, 318-320 (1992).
[34] R. Romi, *Notiziario I.S.S.*, 7, 7-11 (1994).
[35] F. Schaffner, S. Karch. *C. R. Acad. Sci.,* III 323, 373-375 (2000).
[36] D. Petrić, I. Pajović, A. Ignjatović Ćupina, M. Zgomba, *Abstract of Symposia of the Entomologists of Serbia*, Goč, RS, 26-27 (2001).
[37] F. Schaffner, W. Van Bortel, M. Coosemans, *J. Am. Mosq. Contr. Assoc.*, 20, 201-203 (2004).
[38] E. Flacio, P. Lüthy, N. Patocchi, F. Guidotti, M. Tonolla, R. Peduzzi, *Boll. Soc. Ticinese Sc. Natl.*, 92, 141-142 (2004).
[39] A. Samanidou-Voyadjoglou, E. Patsoula, G. Spanakos, N.C. Vakalis, *Eur. Mosq. Bull.,* 19, 10-12 (2005).
[40] A. Klobučar, E. Merdić, N. Benić, Ž. Baklaić, S. Krćmar,. *J. Am. Mosq. Contr. Assoc.*, 22, 147-148 (2006).
[41] C. Aranda, R. Eritja, D. Roiz, *Med. Vet. Entomol.*, 2, 150-152 (2006).
[42] E.J. Scholte, F. Schaffner. In *Emerging Pests and Vector-Borne Diseases in Europe*; W. Takken, B.G.J. Knols, Eds.; Book series: Ecology and Control of Vector-Borne Diseases; Wageningen Academic Publishers: Wageningen, NL, 2007; Vol. 1, pp. 241-260.
[43] E.J. Scholte, F. Jacobs, Y.M. Linton, E. Dijkstra, J. Fransen, W. Takken, *Eur. Mosquito Bull.*, 22, 5-9 (2007).
[44] F. Schaffner, G. Hendrickx, E.-J. Scholte, J. Medlock, E. Ducheyne, and P. Angelini R. Bellini, R. Eritja, C. Jeannin, E. Merdić, D. Petrić, R. Romi, A. Talbalaghi, C.Venturelli, R. Zamburlini and other Colleagues from Related European Network of Scientists. *2009 ECDC TigerMaps Technical Report* (2009).
[45] V.L. Saenz, L.H. Townsend, R.M. Vanderpool, M.J. Schardein, R.T. Trout, G.C. Brown, *J. Am. Mosq. Contr. Assoc.*, 22, 754-755 (2006).
[46] F. Schaffner, C. Kaufmann, A. Mathis, *Med. Vet. Entomol.*, 23, 448-451 (2009).
[47] N. Becker, K. Huber, B. Pluskota, A. Kaiser, *Eur. Mosquito Bull.*, 29, 88-102 (2011).
[48] M.R. Sardelis, M.J. Turell, *J. Am. Mosq. Contr. Assoc.*, 17, 137-141 (2001).
[49] M.R. Sardelis, MJ. Turell, R.G. Andre, *J. Am. Mosq. Contr. Assoc.*, 19, 159-162 (2003).
[50] E.J. Scholte, M. Dik, B. Schoelitsz, M. Brooks, F. Schaffner, F.R. Foussadier, M. Brak, J. Beeukes, *Euro Surveill.*, 15(45), pii=19710 (2010).
[51] A.B. Knudsen, *Parassitologia*, 37, 91-97 (1995).
[52] M.B. Madon, J.E. Hazelrigg, M.W. Shaw, S. Kluh, M.S. Mulla, *J. Am. Mosq. Contr. Assoc.*, 19, 298 (2004).
[53] P. Reiter, R.F. Jr. Darsie, *J. Am. Mosq. Contr. Assoc.,* 44, 396-399 (1984).
[54] B. Pluskota, V. Storch, T. Braunbec, M. Beck, N. Becker, *Eur. Mosquito Bull.*, 26, 1-5 (2008).
[55] F. Schaffner, G. Hendrickx, *Proceedings of the WHO - EMCA Meeting on Vector-Related Risk of Introduction of Chikungunya and Dengue Fever and Spread of Ae. albopictus and Ae. japonicus within Europe",* Speyer, DE (2011) (In prep.).
[56] K.L. Chan, *Mosq. Borne Dis. Bull.*, 1, 79-88 (1985).
[57] A. Farajollahi, B. Kesavaraju, D.C. Price, G.M. Williams, S.P. Healy, R. Gaugler, M.P. Nelder, *J. Med. Ento.*, 46(4), 919-925 (2009).
[58] P.J. Obenauer, P.E. Kaufman, S.A. Allan, D.L. Kline, *J. Med. Ento.*, 46(4), 900-908 (2009).

[59] C.W. Thaggard, D.A. Eliason, *Mosq. News*, 29(4), 608-612 (1969).
[60] D.A. Focks, *A Review of Entomological Sampling Methods and Indicators for Dengue Vectors*, WHO: Geneva, CH, 2003; p. 40.
[61] R. Bellini, M. Carrieri, G. Burgio, M. Bacchi, *J. Am. Mosq. Control Assoc.*, 12, 632-636 (1996).
[62] A. R. Holck, C. L. Meek, J. Holck, *J. Am. Mosq. Control Assoc.*, 4, 97-98 (1988).
[63] L.-Y. Zhang, C.-L. Lei, *Ann. Trop. Med. Parasitol.*, 102(5), 399-407 (2008).
[64] D. Petrić, M. Tanola, A. Plenge-Bönig, G. Hendrickx, A. Ibáñez-Justicia, H. Kampen, A. Rose, C. Kuhn, F. Lidron, K. Huber, A. Jöst, W. P. Pfitzner, S. Sualp Caglar, R. Lühken, *Proceedings of the WHO - EMCA Meeting on Vector-Related Risk of Introduction of Chikungunya and Dengue Fever and Spread of Ae. albopictus and Ae. japonicus within Europe,* Speyer, DE (2011) (In prep.).
[65] M.W. Service, *Mosquito Ecology: Field Sampling Methods*; 2nd Edition; Elsevier: London, GB, 1993; p. 1007.
[66] N. Becker, D. Petric, M. Zgomba, C. Boase, C. Dahl, J. Lane, A. Kaiser, *Mosquitoes and Their Control*; Kluwer Academic/Plenum Publisher: New York, US, 2003; p. 498.
[67] H. Jöst, A. Bialonski, V. Storch, S. Günther, N. Becker, J. Schmidt-Chanasit, *J. Clin. Microbiol.*, doi:10.1128/JCM.00037-10.
[68] WHO, *Weekly Epidemiological Record,* World Health Organization, 47, 73-84 (1972).
[69] W. Tun-Lin, H. Kay, A. Barnes, *Am. J. Trop. Med. Hyg.*, 53(6), 594-601 (1995).
[70] W. Tun-Lin, B.H. Kay, A. Barnes, S. Forsyth, *Am. J. Trop. Med. Hyg.*, 54(5), 543-547 (1996).
[71] D.A. Focks, D.D. Chadee, *Am. J. Trop. Med. Hyg.*, 56(2), 159-167 (1997).
[72] M. Carrieri, P. Angelini, C. Venturelli, B. Maccagnani, R. Bellini, *J. Med. Ento.* (2011) (In press).
[73] R.W. Craig, S.A. Long, R.C. Russell, S.A. Ritchie, *J. Am. Mosq. Contr. Assoc.*, 22, 635-640 (2006).
[74] D. Petrić, M. Zgomba, R. Bellini, R. Veronesi, A. Kaiser, N. Becker, *Proceedings of the 3rd International Conference of Insect Pests in the Urban Environment*, Prague, CZ, 437-445 (1999).
[75] C. Liew, C.F. Curtis, *Med. Vet. Entomol.*, 18, 351-360 (2004).
[76] L. R. Taylor, *Nature*, 189, 732-735 (1961).
[77] L.R. Taylor, *Annu. Rev. Entomol.*, 29, 321-357 (1984).
[78] M. Carrieri, A. Albieri, P. Angelini, F. Baldacchini, C. Venturelli, S. Mascali Zeo, R. Bellini, *J. Vector Ecol.*, 36(1), 108-116 (2011).
[79] G.D. Buntin,. In *Handbook of Sampling Methods for Arthropods in Agriculture;* L.P. Pedigo, G.D. Buntin, Eds.; CRC Press, Inc.: Boca Raton, FL, US, 1994; pp. 99-115.
[80] T.R.E. Southwood, P.A. Henderson, *Ecological Methods*; Blackwell Science Ltd.: Oxford, GB, 2000; pp. 7-62.
[81] M. Mogi, W. Choochote, C. Khamboonruang, P. Suwanpanit, *J. Med. Ento.*, 27, 509-514 (1990).
[82] G. Celli, R. Bellini, M. Carrieri, *Boll. Ist. ''G. Grandi'' Univ. Bologna*, 48, 211-217 (1994).
[83] S.A. Ritchie, *Mosq. News*, 44(1), 77-79 (1984).
[84] R. Pilani, A. Caprioglio, R. Bellini. *Abstract of 3rd Workshop European Mosquito Control Association*, Osijek, HR, 13-14 (2004).

[85] M. Carrieri, A. Masetti, A. Albieri, B. Maccagnani, R. Bellini, *J. Am. Mosq. Contr. Assoc.*, 25, 149-155 (2009).
[86] A.C. Stoops, *J. Vector Ecol.*, 30(1), 41-44 (2005).
[87] C.R. Williams, S.A. Long, R.C. Russell, S.A. Ritchie, *J. Am. Mosq. Control Assoc.*, 22, 635-640 (2006).
[88] A.L. Sant'Ana, R.A. Roque, A.E. Eiras, *J. Med. Entomol.*, 43(2), 214-220 (2006).
[89] L. Blaustein, M. Kiflawi, A. Eitam, M. Mangel, J.E. Cohen, *Oecologia*, 138(2), 300-305 (2004).
[90] M.A. Braks, W.S. Leal, R.T. Carde, *J. Chem. Ecol.*, 33(3), 567-578 (2007).
[91] W.L. Kramer, M.S. Mulla, *Environ. Entomol.*, 8(6), 111-1117 (1979).
[92] A. Albieri, M. Carrieri, P. Angelini, F. Baldacchini, C. Venturelli, S. Mascali Zeo, R. Bellini, *Bull. Insectology*, 63 (2), 209-216 (2010).
[93] IPCC Climate Change 2007. Fourth IPCC Assessment Report (2007).
[94] P. Marchant, W. Eling, G.-J. van Gemert, C.J. Leake, C.F. Curtis, *Parsitol. Today,* 14(9), 344-345 (1998).
[95] N. Becker, *Parasitol. Res.*, 103, 19-28 (2009).
[96] T. Löscher, *Dtsch. Arztebl.*, 95(43), 1913-1915 (1998).
[97] X. de Lamballerie, E. Leroy, R.N. Charrell, K. Ttsetsarkin, S. Higgs, E.A. Gould (2008) *Virol. J.*, 5, 33 [http://www.virologyj.com/content/ 5/1/33].
[98] I. Pajač, B. Barić, S. Šimon, K.M. Mikac, I. Pejić, *J. Food Agric. Environ.*, (In press).
[99] K.C. Smithbum, T.P. Hughes, A.W. Burke, J.H. Paul, *Am. J. Trop. Med. Hyg.*, 20, 47-192 (1940).
[100] N. Özer, *Turk. J. Biol.*, 29, 125-135 (2005).
[101] G.L. Campbell, A.A. Marfin, R.S. Lanciotti, D.J. Gubler, *Lancet Infect. Dis.*, 2(9), 519-529 (2002).
[102] F. Mostashari, M.L. Bunning, P.T. Kitsutani, D.A. Singer, *Lancet*, 358(9278), 261-264 (2001).
[103] Z. Hubalek, J. Halouzka, *Emerg. Infect. Dis.*, 5, 643-650 (1999).
[104] D. Nash, F. Mostashari, A. Fine, J. Miller, D.O.`Leary, K. Murray, *New Engl. J. Med.*, 344, 1807-1814 (2001).
[105] T.F. Tsai, F. Popovici, C. Cernescu, G. Campbell, N. Nedelcu, *Lancet*, 352(9130), 767-771 (1998).
[106] G.L. Campbell, C.S. Ceianu, H.M. Savage, *Ann. N. Y. Acad. Sci.*, 951, 94-101 (2001).
[107] A. Sirbu, C.S. Ceianu, R.I. Panculescu-Gatej, A. Vázquez, A. Tenorio, R. Rebreanu, M. Niedrig, G. Nicolescu, A. Pistol, *Euro. Surveill.*, 16(2), pii=19762 (2011).
[108] A. Papa, K. Danis, A. Baka, A. Bakas, G. Dougas, T. Lytras, G. Theocharopoulos, D. Chrysagis, E. Vassiliadou, F. Kamaria, A. Liona, K. Mellou, G. Saroglou, T. Panagiotopoulos, *Euro Surveill.*, 15(34), pii=19644 (2010).
[109] P. Angelini, M. Tamba, A.C. Finarelli, R. Bellini, A. Albieri, P. Bonilauri, F. Cavrini, M. Dottori, P. Gaibani, E. Martini, A. Mattivi, A.M. Pierro, G. Rugna, V. Sambri, G. Squintani, P. Macini, *Euro Surveill.*, 15(16), pii=19547 (2010).
[110] P. Macini, G. Squintani, A.C. Finarelli, P. Angelini, E. Martini, M. Tamba, et al. *Euro Surveill.*, 13(39), pii=18990 (2008).
[111] P. Calistri, A. Giovannini, G. Savini, F. Monaco, L. Bonfanti, C. Ceolin, et al. *Zoonoses Public Hlth.*, 57(3), 211-9 (2010).
[112] Emilia-Romagna West Nile Regional Surveillance Plan (2009) [http://www.saluter.it].

[113] G. Rossini, F. Cavrini, A. Pierro, P. Macini, A.C. Finarelli, C. Po, et al. *Euro Surveill.*, 13(41), pii=19002 (2008).

[114] Y. Tang, C.A. Hapip, B. Liu, C.T. Fang, *J. Clin. Virol.*, 36, 177-82 (2006).

[115] OIE (2008). Manual of Diagnostic Tests and Vaccines for Terrestrial Animals. Cap. 2.1.20 West Nile fever, [http://www.oie.int].

[116] F. Monaco, R. Lelli, L. Teodori, C. Pinoni, A. Di Gennaro, A. Polci, et al., *Zoonoses Public Hlth.*, doi: 10.1111/j.1863-2378.2009 (2009).

[117] D. Lelli, A. Moreno, E. Brocchi, E. Sozzi, E. Canelli, G.L. Autorino, et al. *Proceedings of 3rd National Workshop of Veterinary Virology*, Valenzano (Bari), IT, 63 (2009).

[118] M.C. Williams, D.I. Simpson, A.J. Haddow, E.M. Knight, *Ann. Trop. Med. Parasitol.*, 58, 367-374 (1964).

[119] H. Weissenbök, J. Kolodziejek, A. Url, H. Lussy, et al., *Emerg. Infect. Dis.*, 8, 652-656 (2002).

[120] M. Tamba, P. Bonilauri, R. Bellini, M. Calzolari, A. Albieri, V. Sambri, M. Dottori, P. Angelini, *Vector-Borne Zoonot.*, DOI: 10.1089/vbz.2010.0055 (2010).

[121] W.C. Reeves, *Foreword* In *St Louis Encephalitis*; T.P. Monath, Ed.; American Public Health Association: Washington, DC, US, 1980; pp. xi–xviii.

In: Essays on Fundamental and Applied Environmental Topics ISBN 978-1-61942-522-4
Editor: Dragutin Mihailovic

Chapter 11

PREDOMINANT ENTEROVIRUS SEROTYPES IN THE VOJVODINA REGION (SERBIA)

I. Hrnjaković Cvjetković[1,2,*], J. Radovanov[1] and V. Milošević[1,2]
[1]Centre of Virology, Institute of Public Health of Vojvodina, Novi Sad, Serbia
[2]Faculty of Medicine, University of Novi Sad, Novi Sad, Serbia

ABSTRACT

Enteroviruses are small, non-enveloped, single-stranded RNA viruses that belong to the family *Picornaviridae*. At the moment, there are more than 100 known human serotypes classified in 4 species, and new types are being discovered every year. Enteroviruses infect millions of people worldwide. The clinical manifestations of enteroviral infections range from mild respiratory illness to severe diseases such as myocarditis, neonatal sepsis, type I diabetes and aseptic meningitis. Enteroviruses are excreted in large quantities in the feces of infected individuals and remain viable for long periods of time in the environment. They spread mainly by the fecal-oral route, through direct contact with secretions from an infected person and contaminated water and food. Enteroviruses demonstrate significant serotype variation by time, geographic location, and disease. Serotyping enables the identification of endemic circulating virus types, types responsible for outbreaks, and the monitoring of changes in the pathogenicity, virulence, and epidemic potential of different enteroviral types. To identify the predominant types of enteroviruses included in the etiology of aseptic meningitis in the Vojvodina region (Serbia), a total of 70 samples (29 cerebrospinal fluid specimens, 27 stool specimens, 9 rectal swabs, and 5 throat swabs) were obtained from pediatric patients suspected of having aseptic meningitis. The presence of enteroviruses was detected by isolation in cell cultures and confirmed through the use of an indirect immunofluorescence assay. Isolate serotyping was performed using a neutralization assay and revealed 11 different viruses: coxsackie virus A16, coxsackie virus B3, 4 and 5, and echovirus 2, 4, 6, 11, 13, 16, and 30. The most prevalent were echoviruses (52.8 %), followed by coxsackie B (27.8 %), and coxsackie A viruses (8.3 %). Additional research is required to study the occurrence and persistence of enteroviruses in local surface waters and to assess the public health risk associated with an exposure to waters contaminated with these viral pathogens.

* Corresponding author: E-mail address: ivana.hrnjakovic@izjzv.org.rs, Phone: +381 21 4897 871. Fax: +381 21 6613 989.

1. Introduction

1.1. Classification of Enteroviruses

The genus *Enterovirus* belongs to the family *Picornaviridae*, order *Picornavirales*. Since 1957, when they were for the first time classified as a distinct group [1], enterovirus taxonomy has undergone numerous modifications. The original criteria for their classification were infectivity and pathogenicity in different animal species, an ability to replicate in cells of human and primate origin, and antigenic differences. According to these features, the human enteroviruses (HEV) were sub-classified into 4 subgroups: polioviruses (PV), coxsackie A viruses (CA), coxsackie B viruses (CB), and echoviruses. However, as laboratory animals are not natural hosts for human enteroviruses and because levels of pathogenicity may vary between isolates of the same serotype, beginning in 1969, newly identified types have been designated numerically, starting with enterovirus 68.

The current classification of enteroviruses takes into account their molecular properties, *i.e.*, nucleotide sequence homology, which reflects their phylogenetic relationships. The enterovirus genome contains regions, mainly in the VP1 capsid protein-encoding area, which are serotype-specific, as well as highly conserved regions common to the entire genus [2]. After amplifying and sequencing parts of the genome, the sequences of the amplicons are being compared with the sequences of prototype strains, and a virus is classified into the appropriate taxon according to genetic similarity. Strains with at least 75 % nucleotide identity and 85 % amino acid identity are considered to represent the same serotype [3].

The number of identified members of the genus *Enterovirus* has constantly been increasing [3-6]. At present, there are more than 270 types grouped into 10 species (Table 1). More than 100 serotypes of human enteroviruses are classified into 4 species, HEV A to D. Species within the genus *Enterovirus* are defined as groups of serotypes or strains, which (i) share greater than 70 % amino acid identity in the VP1, 2C, and 3CD proteins, (ii) share a limited range of natural hosts and cell receptors, (iii) have a genome base composition (G+C) that varies by no more than 1 %, and (iv) share a significant degree of compatibility in proteolytic processing, replication, encapsidation, and genetic recombination [7].

The application of molecular methods has induced extensive taxonomic changes within the family *Picornaviridae* and also within species belonging to the genus *Enterovirus*. Thus, echoviruses 22 and 23 were renamed human parechoviruses 1 and 2 and reclassified as a new genus, *Parechovirus*, in the family *Picornaviridae*. CV-A 15 and 18 were established as variants of CV-A 11 and 13, respectively; whereas CV-A 23 was proven to be essentially the same as EH 9 [7]. Furthermore, the genus *Rhinovirus* has recently been incorporated into the genus *Enterovirus* [8]. Virus classification into these two genera was based on classical criteria, such as differences in pathogenicity and stability in an acidic environment. However, genetic criteria do not justify the existence of two separate genera.

Polioviruses have previously been classified as distinct enterovirus species, which was due to their specific clinical manifestations and the cellular receptor they use. This concept was recently abandoned, and polioviruses were assigned to the species *Human enterovirus C*, which then became the type species for the entire genus [8]. Animal enteroviruses were formerly categorized into separate species based on their host specificity. At the present time, they have been reclassified into the appropriate species along with related human serotypes.

Table 1. Taxonomic structure of genus *Enterovirus* [8]

Genus *Enterovirus*		No. of serotypes:
Species:	Serotypes:	
Human enterovirus A (HEV-A)	*CV-A 2-8, 10, 12, 14, 16, EV71, 76, 89- 91 and 114,* Simian enterovirus 92, SV19, SV43, SV46, and A13	22
Human enterovirus B (HEV-B)	*CV-B 1-6, EHV 1-7, 9, 11-21, 24-27, 29-33, EV 69, 73-75, 77-88, 93, 97, 98, 100, 101, 106, 107, CV-A 9,* Simian enterovirus SA 5, and 110	60
Human enterovirus C (HEV-C)	PV 1-3, *CV-A 1, 11, 13, 17, 19-22, 24, EV 95, 96, 99, 102, 104, 105, 109, and 113*	20
Human enterovirus D (HEV-D)	*EV 68, 70, 94, and 111*	4
Human rhinovirus A (HRV-A)	HRV 1, 2, 7-13, 15, 16, 18-25, 28-34, 36, 38-41, 43-47, 49-51, 53-68, 71, 73-78, 80-82, 85, 88-90, 94-96, 98, 100, and 101	*75*
Human rhinovirus B (HRV-B)	HRV 3- 6, 14, 17, 26, 27, 35, 37, 42, 48, 52, 69, 70, 72, 79, 83, 84, 86, 91- 93, 97, and 99	*25*
Human rhinovirus C (HRV-C)	HRV 1-48	*48*
Bovine enterovirus (BEV)	*BEV 1 and 2*	2
Porcine enterovirus B (PEV-B)	*PEV 9 and 10*	2
Simian enterovirus A (SEV-A)	*A1*, A2*, SV4*, SV28*, and SA4**	5
Non-classified	Simian enterovirus SV6, EV 103, 108, 112, 115	13

*Probably represent the same serotype.

1.2. Basic Properties of Enteroviruses

Enteroviruses, *small non-enveloped single-stranded RNA viruses* with an *icosahedral symmetry, are typical members of family Picornaviridae* (Figure 1). Their virion diameter is only 24 - 30 nm, and the relative molecular weight of the virion is 8×10^6 [7].

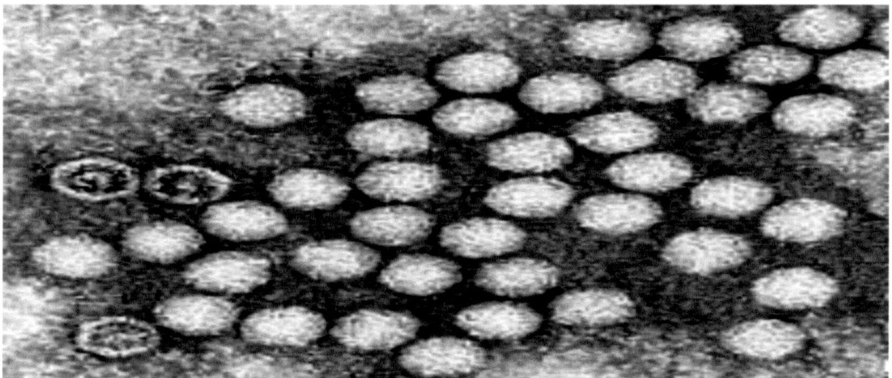

Figure 1. An enterovirus visualized by electron microscopy (by Lynda M. Stannard, Department of Medical Microbiology, University of Cape Town, http://www.web.uct.ac.za/depts/mmi/stannard/em images.html).

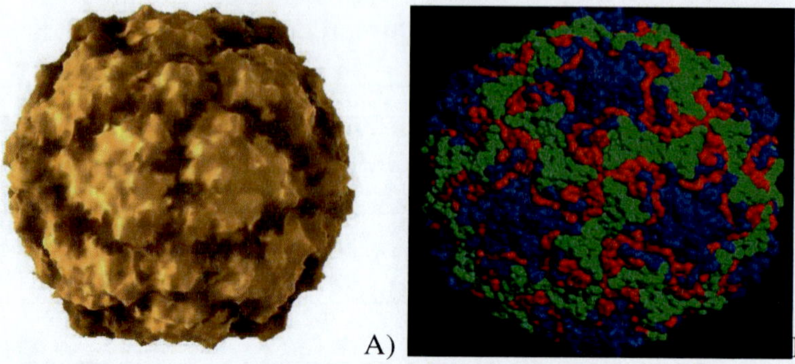

A) B)

Figure 2. A) Cryoelectronic reconstruction of CV-B3, ICTVdB - The Universal Virus Database, version 4. http://www.ncbi.nlm.nih.gov/ICTVdb/ICTVdB/; B) A computer-generated image of a picornavirus capsid (VP1 is in blue, VP2 is in green, VP3 is in red), http://www.microbiologybytes.com/virology/ Picornaviruses.html#entero.

Average capsid thickness is about 2.5 nm, and the capsid is composed of 60 identical structural units, protomeres (Figure 2). Structural units are formed from 1 copy of each of the 4 *capsid* polypeptides, which are designated VP1, VP2, VP3, and VP4 with relative molecular weights of 30 000, 27 000, 24 000, and 7 000, respectively.

The major capsid proteins VP1, VP2, and VP3 each possess a core structure comprised of an 8-stranded antiparallel β-barrel with 2 helices. Each polypeptide contains structurally unique loops connecting the β strands and unique NH_2- and COOH- termini [8]. Five structural units assemble to form each of the 12 morphological units of the capsid, the capsomers (pentamers). The protomers are oriented in such way that 5 VP1 molecules form 12 icosahedron vertices surrounded alternately by VP2 and VP3 molecules. A depression termed a "canyon", 2 nm deep and 1.2 - 1.5 nm wide, encircles each vertex of the icosahedron. VP1, VP2, and VP3 are partly exposed at the virion surface, and they are similar in structure and size, whereas significantly smaller myristylated VP4 lies in the interior of the virion. VP4 and N-terminal extensions of VP1, VP2, and VP3 form a stabilizing network on the inner surface of the poliovirus capsid [9].

The antigenicity of enteroviruses is complex and determined by epitopes localized at superficial regions of capsid proteins. Capsid protein VP1 contains the majority of serotype-specific antigenic determinants [10]. Some epitopes are composed of a continuous sequence of amino acids of a particular capsid protein (linear epitopes). Others are formed by amino acid residues from different proteins in a close proximity on the capsid surface. Genus-specific antigenic determinants are also located within the VP1, but they are not present on the surface of infectious virions. Those antigens can only be surface exposed after thermal denaturation of the viral particle, followed by the destruction of type-specific antigens [11].

The existence of a wide variety of antigenic diverse enteroviruses reflects their marked evolutionary divergence. A number of different serotypes indicate a wide range of genetic variations, which produce significant antigenic differences, *i.e.*, differences in the structure of capsid proteins while virus viability is preserved. The high degree of enterovirus diversity is due to their specific replication cycle, which is a feature of all RNA viruses. Namely, RNA viruses significantly differ from other organisms due to their extremely high mutation rate, which is some 10^3 fold higher than that of the DNA genome viruses and organisms [12]. This mutation rate is associated with the lack of proofreading ability of the RNA-dependent RNA polymerase, *i.e.*, the lack of a 3' - 5' exonuclease activity [13]. Frequent nucleotide misincorporations during viral RNA replication, being even one mutation per 10^3 - 10^4 polymerized nucleotides [14], results in a high inter- and intra-type genetic variability. Thus, antigenic variants are reported within numerous serotypes.

Enteroviruses, like RNA viruses in general, are a heterogeneous population representing a reservoir of diverse genetic variants with potentially useful phenotypes. Accordingly, even clones of enteroviruses exist as a quasispecies rather than as groups with identical viral genomes. The selective pressure induced by the immune system of the host, antiviral drugs, etc. provides favorable conditions for the rapid evolution of those viruses [15].

The enterovirus genome is single, linear, positive-stranded RNA molecule of approximately 7 400 nucleotides. The small genome size of all RNA viruses is an evolutionary consequence of a high *mutation* rate. The replication of longer genomes would result in the accumulation of too many deleterious mutations [16].

At both ends of a viral nucleic acid, there are highly conserved non-translated regions (NTR) of a genome. The *5'NTR* sequence is approximately *750-nucleotides* long and characterized by complex secondary and tertiary structures. The shorter non-translated sequence (3'NTR), which ranges from 70 - 100 nucleotides in size, is located on the 3' end of the RNA. These regions do not contain information on protein synthesis; however, they are very important in genome replication and translation processes. The 5'NTR harbors the replication initiation site-internal ribosome entry site (IRES), whereas the 3'NTR plays an important role in the initiation of replication [17]. Both genome ends are modified. A small basic protein VPg is covalently attached at the 5' end, whereas *polyadenylic* acid *sequences* are located at the *3' end* [7]. Polyadenylic acid sequences are not genetically coded, and there is a "polyadenylation signal" upstream of the 3' end of the RNA.

A single open reading frame, which is *positioned between the* 5'NTR and 3'NTR, encodes *a polyprotein of about 2200 amino acids, which is subsequently processed by virus-encoded proteases to yield 11 viral, 4 structural* (VP1, VP2, VP3, and VP4) and 7 functional proteins (2A, 2B, 2C, 3A, 3B, 3C, and 3D). Proteins 2A and 3D are proteinases while 3B is a VPg protein. Proteins 2B, 2C, 3A, and 3D (RNA-dependent RNA polymerase) are enzymes that play a role in replication [13].

The largest differences between the nucleotide sequences among types and strains of enteroviruses are within a region encoding the major capsid protein VP1, which contains the main *antigenic determinants*. An analysis of the VP1 region of the genome revealed the presence of sequences specific for the genus, individual types, and strains. However, a linkage between a particular nucleotide sequence and antigenic site responsible for serotype specificity could not be established because of the highly complex nature and discontinuity of antigenic determinants [18].

Non-translated regions and regions encoding non-structural proteins exhibit a lower rate of genetic variability. Namely, the variability of capsid proteins might be a useful strategy to escape the immune response of the host. On the other hand, changes in genomic regions encoding non-structural proteins, primarily enzymes, are mostly fatal. Thus, amino acid identity among non-structural proteins of enteroviruses is very high, particularly in regards to RNA-dependent RNA polymerase [19].

Besides mutation, genetic recombination is an important source of genetic variability for enteroviruses [14]. Genetic recombination occurs during the copying of a RNA genome, which enables the repairing of deletions and the acquisition of new genes from other viral species or from the host. Recombination is mostly limited to members of particular enterovirus species, occurs frequently within a serotype and, for still unknown reasons, almost always takes place outside the capsid-coding region of the genome [18].

1.3. Enterovirus Replication

Enterovirus, as well as all other members of the family *Picornaviridae*, replication takes place in the cytoplasm of the host cell. The kinetics of replication are rapid. The whole cycle of virus replication is completed within approximately 8 hours.

Attachment. Recognition of a susceptible cell by a virus is based on a highly specific binding between capsid protein sequence anti-receptors and the appropriate plasmalemma protein receptors. A lack or blockade of appropriate membrane receptors makes cells resistant to infection by a particular virus. The capsid plays a fundamental role in attachment to the host cell. It determines cell and tissue selectivity (tropism), host selectivity, and represents the target of the host immune response. In this respect, genetic changes in the genome region encoding capsid proteins disturb a range of viral characteristics.

Enteroviruses are adapted to use different cell membrane proteins as receptors. Variations are even observed between isolates within the same serotype [2]. The diversity of clinical syndromes, which is manifested within an affected population in epidemics caused by a particular enterovirus serotype, might be the consequence of an altered ability of the virus to bind to specific receptors. The primary sites of an interaction with cell receptors are hidden in the "canyons" of an enteroviral capsid; however, some were found at other locations on the capsid. Most likely, binding sites outside the "canyon" are only utilized for recognition without initiating the release and cell entry of the viral nucleic acid [20].

Numerous enteroviruses use immunoglobulin-like molecules as receptors. Hence, all three poliovirus serotypes use the cell adhesion glycoprotein CD155 as their receptor. He *et al.* [21] is of the opinion that quite a large contact surface of the capsid and this receptor limits the number of possible poliovirus serotypes. To preserve receptor-binding capacity, a high rate of conservation of a large number of residues on a virion's surface is essential, and mutants are "doomed to failure". The physiological function of CD155 is still unclear, and it has been suggested that CD155 plays a role in the development of the central nervous system during embryogenesis [21], as well as during cancerogenesis [20].

In contrast, CV-A 21 enters the cell via the intracellular adhesion molecule ICAM-1 [22], whereas coxsackie B viruses use the adhesion molecule named coxsackievirus and adenovirus receptor CAR [13]. Both glycoproteins are members of the large family of immunoglobulin-like molecules. In the host organism, the intracellular adhesion molecule ICAM-1 mediates *leukocyte-endothelial cell adhesion during the inflammation process* [22] while CAR represents a component of the intracellular junction in the intact epithelium [20].

Many enteroviruses recognize susceptible cells by binding to DAF (decay accelerating factor), which is a widely spread protein that regulates the activation of the complement system. Some other enteroviruses, like CV-A 9, use membrane proteins belonging to the integrin family as their receptors. Integrins are molecules involved in coupling the intracellular cytoskeleton and extracellular matrix and mediate numerous cell–cell and cell–extracellular matrix interactions [23]. Binding to DAF and integrins does not necessarily involve sites located within the "canyon" of the viral capsid and does not trigger structural changes in the capsid. These membrane proteins are believed to function as co-receptors, which enable the virus to react with the main receptor molecule [22]. Thus, CV-A 21 and CV-B 3 bind to DAF, but expression of ICAM-1 and CAR is required for triggering a productive viral infection [20].

A number of issues pertaining to the nature, distribution, variability and selection of cell-receptors for enteroviruses still need to be elucidated, as well as the influences of other cell-related factors on the mechanism for recognizing susceptible target cells.

Uncoating and genome release. Little is known about the molecular basis of the decapsidation mechanism and penetration of picornaviral nucleic acid through the cell membrane. Belnap *et al.* [9] report that during a gradual transfer through the plasmalemma poliovirus undergoes an irreversible conformational rearrangement, which is noticeable due to the decreased sedimentation coefficient of the 160S viral particle (a complete poliovirus particle) to the 135S and finally to the 80S (a particle lacking nucleic acid). Immediately after binding to membrane receptor, the 160S is transformed into the 135S particle, which has altered antigenic and proteolytic properties. These conformational changes are associated with the externalization of the VP4 protein and the N-termini of VP1 protein, which are components of the stabilizing network on the inner surface of the 160S capsid. This enables a tight attachment of the virus to the membrane. Subsequently, during the transformation from a 135S to 80S particle, the N-termini of the VP1 and parts of the VP4 emerge from the bottom of the canyon, insert into the cell membrane and undergo a rearrangement to form a transmembrane pore through which viral RNA exits the virion. RNA enters the cytoplasm after the VP3 is moved out of the way.

Translation and transcription. Enteroviruses belong to the group of positive-stranded RNA viruses, which indicates that their RNA is equivalent to the information RNA. Thus, synthesis of viral protein is initiated immediately after entering the host cell. In contrast to other RNA viruses, viruses with a positive, infectious RNA genome do not encapsidate the RNA-dependent RNA polymerase. Therefore, replication of their genome cannot begin until the genomic RNA is translated to produce polymerase and some other essential enzymes in the process of early viral protein synthesis.

Immediately after penetrating into the host cell, the virus rapidly inhibits *cellular macromolecular* synthesis. This process is termed ”shut-off” and represents the main cause of cytopathogenic changes. The “shut-off” occurs as a consequence of the inactivation of one of the proteins that composes the translation initiation complex (eIF-4G, a component of the “Cap-binding complex”), which is mediated by the A2 proteinase [17]. Consequently, the cap-binding complex is unable to recognize a 7-methylguanylate cap structure attached to the 5’ end of all eukaryotic iRNAs. In this manner, the cap-dependent translation of host iRNA discontinues and switches to the IRES-mediated translation of viral RNAs [7]. The precise *mechanism* by which the *translation apparatus* recognizes IRES at the 5’NTR of the viral genome is unknown, but the involvement of several cellular proteins is hypothesized [13]. Translation results in the creation of a unique polypeptide chain, which undergoes autocatalytic cleavage during and after translation, to generate individual viral proteins.

The process of genome replication of all RNA-positive viruses takes place within the multi-protein complex, which is associated with the cell membrane, and originates from the endoplasmic reticulum [17]. One of the basic features of RNA-positive viruses is that their nucleic acids serve as templates for both translation (synthesis of viral protein) and transcription (*synthesis* of new *complementary* RNA strands). To provide undisturbed virus replication, a balance between those two counteracting processes is necessary. Specifically, during translation the ribosome's progress along the RNA molecule is in the 5' to 3' direction, whereas ribonucleotide polymerization during viral replication commences at the 3' end of the same RNA molecule and moves in the opposite direction.

Coordination of these processes is accomplished by interactions of the appropriate viral and cellular factors with each other and with both ends of the viral RNA. The 5' NTR plays a key role in regulating translation, as well as the replication of the RNA genome [17, 24]. At the earliest stage of infection, the cellular poly(C)- and poly(A)-binding proteins (PCBP and PABP, respectively) interact with the 5' and 3' RNA ends, respectively, and stimulate translation. In addition, binding of PCBP to the 5' end protects the viral nucleic acid from cellular RNA-degrading enzymes [24]. When the amount of the viral proteins increases in the cell, the 3CD *precursors* of both the 3C *proteinase* (3C) and the polymerase (3D) bind to the 5' NTR, which represses translation and stimulates synthesis of negative RNA strands. Circularization of the genome, which is accomplished by linking the 5' and 3' ends and mediated by PABP, is the prerequisite for negative-strand synthesis [2].

RNA-dependent RNA polymerase is unable to use genomic RNA as a template for complementary strand synthesis as long as the translation of viral proteins is in progress. Thus, molecules of genomic RNA must be translated prior to RNA replication, which represents a control mechanism that prevents the replication of defective viral genomes [25].

The polymerase copies negative RNA strands to new positive RNA strands. Some of them will serve as templates for the production of new negative strands while others will be utilized as the genomes of new viruses.

Assembly and release of new virions. The molecular mechanisms of the packaging of positive RNA strands into preformed capsids are still ambiguous. It is hypothesized that RNA packaging into the picornavirus capsid is restricted to freshly synthesized RNA molecules, as the process requires the direct interaction of capsid proteins and proteins from the replication system [26]. Encapsidation of the viral genome is strongly dependent on previous successful replications, which represents an additional control mechanism ensuring elimination of defective viral RNA from packaging. Each infected cell produces approximately 10^4 - 10^5 viral particles, which are then released by cell lysis [27].

1.4. Pathogenesis of Enterovirus Infections

Enteroviruses enter the body via ingestion. Incubation periods vary between 2 and 30 days and depend on the virus type. Most frequently, it ranges from 7 to 14 days. Primary replication sites are lymphatic tissues of the throat and small intestine (tonsils, lymph glands, and *Peyer's patches). The virus then advances to the deep lymph glands and mesentherial glands via the lymphatic pathways. In some cases, after the initial stage, enteroviruses enter the bloodstream and disseminate to a variety of target organs where secondary replication is initiated. Factors that are critical for development of viremia are still unclear.* Although the actual mechanism of disease development has not yet been fully elucidated in most cases, clinical manifestations of enteroviral infection occur mainly as a result of cell destruction, *i.e.*, lysis or the immune response of the host [28].

Enteroviruses are etiological agents for a wide spectrum of diseases. No one clinical syndrome can be associated exclusively with one particular serotype. However, it has been demonstrated that some serotypes cause certain diseases more frequently than others. For example, the hand-foot-and-mouth disease and acute hemorrhagic conjunctivitis are most commonly associated with CV-A 16 and HEV 71 [27] and CV-A 24 and EV 70 [29], respectively. Alternatively, the same virus type can, under different conditions, cause various diseases. This is determined by the very nature of the RNA virus, *i.e.*, their extreme genetic heterogeneity, which excludes the existence of two identical quasispecies populations. The populations will differ even among two hosts (patients) infected from the same source, which strongly influences the nature and severity of the disease. In this respect, individual differences in symptom manifestation in particular viral diseases, different responses to vaccines and antiviral drugs, as well as different epidemiological patterns, can be explained by the presence of a variety of quasispecies populations in the patients [14].

1.5. Clinical Manifestations of Enterovirus Infections

The majority of enterovirus infections are asymptomatic or characterized by very mild symptoms while the infected person excretes the virus and produces specific antibodies. However, enterovirus infections may result in the development of a variety of acute symptoms, which affect the heart and skeletal muscles, central nervous system, pancreas, skin, and mucous membranes.

The most common manifestation of an enterovirus infection is a nonspecific febrile illness with or without a rash. Some 90 % of febrile conditions in children up to 3 months of age are of a viral etiology, and enteroviruses are the most frequent reason for the hospitalization of such children [30]. The illness is frequently associated with upper respiratory symptoms or gastrointestinal disturbances and known as “summer colds”.

Enteroviruses are responsible for a range of relatively mild diseases that spontaneously heal within several days or weeks without consequences, which include herpangina (vesicular stomatitis), exanthema, acute epidemic hemorrhagic conjunctivitis, hand-foot-and-mouth disease, and pleurodynia.

In contrast, the etiological role of numerous enteroviruses in the development of severe diseases, such as myocarditis, meningitis, neonatal sepsis, and hepatitis, is well-established. Cardiotropism is a feature predominantly attributed to coxsackie B virus, which is implicated in more than 50 % of cases of viral myocarditis [31]. The clinical course of the disease in neonates is much more severe than in older children or adults. The major symptom in all myopericarditis cases is chest pain, which is related to pericardial inflammation and may be lacking if the pericardium is not involved. Tachycardia, arrhythmia, and cardiomegaly are often observed. Complications can rarely result in lethal outcomes.

A growing body of evidence has implicated enteroviruses in the etiology and development of chronic diseases in adults, including myocarditis and dilated cardiomyopathy, which is one of the most frequent indications for heart transplantation [2]. In 10 - 20 % of patients (20 - 40 000 in the USA), enteroviral myocarditis develops into a chronic disease, which mostly progresses into dilated cardiomyopathy within a few years [2]. In cases of chronic enterovirus infections, stable double-stranded RNA virus complexes are detectable in cells from affected tissues long after termination of the acute phase. There is experimental evidence that suggests that RNA synthesis, even in the absence of infectious virus particles, has a pathological effect and may establish infections and induce cardiomyopathy [32].

It is believed that chronic fatigue syndrome (myalgic encephalomyelitis) develops as a consequence of persistent infections with particular enteroviruses [33]. Recovery from the disease may be prolonged for several months, even years, and associated with recurrent episodes of physical or psychical stress. The disease is characterized by marked fatigue, even after slightest physical activity, muscle pain, and psychical disturbance.

Emerging evidence suggests that enteroviruses are associated with the development of insulin-dependent diabetes mellitus [32]. The majority of CV-B is able to infect and damage endocrine pancreatic cells. It has been established that the presentation of antigens from damaged pancreatic β cells activates autoreactive T-cells. Such damage may trigger the autoimmune reaction underlying the development of the disease [34].

Poliomyelitis is one of the most severe enterovirosis of the CNS. Invasion of the CNS by the polioviruses is rare. Abortive poliomyelitis, which is characterized by respiratory and gastrointestinal disturbances, fatigue, fever, and headache, occurs in 4 - 8 % of affected patients and heals spontaneously within a few days. Non-paralytic poliomyelitis, which exhibits symptoms similar to aseptic meningitis, develops in approximately 1 - 2 % of the infected population. Spontaneous recovery occurs in 7 - 10 days and only progresses to its most severe form, paralytic poliomyelitis, in 0.1 - 2 % infections. The virus primarily affects the anterior horns of the spinal cord, which contains neurons that innervate the skeletal musculature. More severe cases involve the invasion of ganglia and posterior horns. A characteristic clinical manifestation is permanent flaccid paralysis and paralysis of the respiratory musculature, which can result in fatal outcomes.

Besides the polioviruses, a number of enteroviruses exhibit neurotropism. Infections of the CNS with non-polio enteroviruses are manifested as aseptic meningitis or, much more rarely, as meningoencephalitis. In some cases, paralysis, Guillain-Barre syndrome, cerebellar ataxia, and peripheral neuritis may develop [35]. About 25 - 35 % cases of meningitis are caused by enteroviruses [36], and they account for more than 90 % of all cases of viral aseptic meningitis in which a pathogen is identified [37]. In general, the disease is most common in children and young patients. Clinical signs and symptoms are non-specific and similar to those of meningitis and encephalitis caused by bacteria or other viruses. The disease presents a fever followed by meningeal symptoms and limb weakness, irritability or lethargy, anorexia, gastrointestinal and/or respiratory disturbances, and rash.

Generalized or focal enterovirus encephalitis can either be associated with meningitis, or the meninges can remain unaffected. In general, the prognosis for encephalitis is good (as in aseptic meningitis); however, neurologic sequelae or endocrine disorders, which result from damage of the *hypothalamic-hypophyseal axis,* can occur in some patients [2].

Enterovirus infections of the CNS in newborns and in immunocompromised patients present a challenge for the clinician. Meningoencephalitis is a common complication of enterovirus infections in newborns and often results in fatal outcomes [30]. Enterovirus infections in patients with hypoglobulinemia often advance to persistent states associated with the development of chronic meningitis and encephalitis. High mortality rates are reported in such groups of patients [36]. Enteroviral genomes are detectable in the liquor of immunodefficient patients with chronic meningoencephalitis during the several months after the primary infection. The mechanisms by which otherwise extremely cytolytic enteroviruses discontinue replication and establish persistence in host cells, as well as factors contributing to their reactivation, are still elusive [32].

There is no evidence on the teratogenic effects of either enteroviruses or any other picornaviruses. However, acute enterovirus infections in pregnant women, particularly at later stages of pregnancy, can cause spontaneous abortions, premature births, and disease in newborns [38]. Infections can be acquired transplacentally, but transmission during or immediately after the delivery is more likely. Onset of the disease 1 - 2 days after birth suggests an intrauterine infection and, in about 80 % of cases, the disease takes a severe course. In newborns, the clinical manifestations include sepsis, meningoencephalitis, hepatitis, myocarditis, and coagulation impairment [38]. Some recent studies indicate a correlation between infection with neurotropic coxsackie virus at the neonatal age and a variety of neuropathologies, such as schizophrenia, encephalitis lethargica, and amyotrophic lateral sclerosis [32]. Some authors emphasize the association of enteroviruses with sudden infant death syndrome [39].

Most enteroviral infections induce a stable lifelong immunity. This immunity in young children is mostly homotypic, whereas adults and older children develop a heterotypic response, *i.e.*, infection with one type induces an immunity to several serotypes. The basis for this heterotypic immune response still remains unclear [28].

1.6. Epidemiology and Prevention

The *enteroviruses* are *ubiquitous* agents with a worldwide distribution and significantly affect morbidity and mortality rates. The estimated number of enterovirus infections is approximately 50 million cases annually in the United States alone, among which some 30 - 50 000 require hospitalization [28].

Infected humans are the only natural reservoir of enterovirus infections, and feces or nasopharyngeal secretions are the most important sources of infection. Enteroviruses are primarily transmitted via the fecal-oral route. Infection is commonly acquired by direct contact with secretions from an infected person or by contact with contaminated food and/or water. Children are more susceptible to infection then the adult population, and severe forms of enterovirosis occur in individuals of poor physical condition, pregnant and parturient women, and in patients with hypo- or agammaglobulinemia.

In regions with temperate climates, enteroviral infections occur mostly during the summer and early fall when temperature and humidity reach their highest values. In the world's tropical regions, enterovirus infections occur year round or with an increased incidence during the rainy season. There are even reports on the detection of enterovirus in the feces of 40 % of children under 2 years of age in some tropical countries with poor sanitation and hygiene [2].

Some enterovirus infections reach epidemic proportions. Massive outbreaks of hemorrhagic conjunctivitis are well-established, such as the one caused by CV-A 24 in Brazil in 2003 with more than 200 000 reported cases [29]. Furthermore, EV 70 was the etiological agent of a pandemic hemorrhagic conjunctivitis that affected millions of people. The pandemic started in Ghana in 1969 and spread across Africa and then to India and southeast Asia [28]. An outbreak of hand-foot-and-mouth disease caused by enter virus 71 in Taiwan in 1998 affected more than 100 000 people with several hundred reported fatal outcomes due to complications including encephalitis, aseptic meningitis, pulmonary edema, acute flaccid paralysis, and myocarditis [10].

A universal vaccine that would provide protection against all enterovirus types is not yet available, and the isolation of infected individuals is not feasible as the majority of infections are unapparent. The only available preventative measure is following strict sanitation and hygiene precautions in children's hospitals and public nurseries, departments for immunodefficient persons and ophthalmology clinics.

Active immunization is available only in the case of polioviruses. At the present time, two polio vaccines are used throughout the world, *i.e.*, Salk's formalin-inactivated polio vaccine and Sabin's vaccine containing a live virus attenuated by *in vitro* and *in vivo* passage. Live Sabin's vaccine is more effective because it mimics natural infections that commence by viral replication in the small intestine. Its application stimulates not only humoral immunity (like Salk's vaccines do), but also a local immunity mediated by secretory IgA antibodies that neutralize viruses at each reinfection and prevent their entry into the bloodstream. The drawbacks of live vaccines include a decreased stability and potential mutation of the live attenuated virus, which may lead to a reversion to *virulence* [2]. Therefore, live attenuated vaccines have been replaced in many countries by inactivated vaccines.

The successful application of poliovirus vaccines has resulted in a *dramatic* decline in the *incidence* of poliomyelitis in regions that apply massive vaccination programs. However, according to WHO data, the 2004 vaccination program encompassed almost 94 % of children in Europe, and some 70 % in India, but only 39 % in Nigeria [40]. America (1994), the South Pacific region (2000), and Europe (2002) have been certified as polio-free regions. Poliomyelitis is expected to be declared the second completely eradicated disease, following smallpox. However, it still occurs in some countries in sub-Saharan Africa and South Asia. India, Pakistan, Nigeria, and Afghanistan remain polio-endemic regions in which wild-type poliovirus infections have been repeatedly reported. Individual cases and epidemics caused by wild strains transmitted from endemic regions are observed in various countries. Thus, during 2010, the WHO recorded 1292 wild poliovirus cases, with the highest infection rates in Tajikistan, Congo, and Pakistan (http://www.polioeradication.org/Dataandmonitoring/Polio thisweek.aspx).

Specific antiviral *therapies* for *enteroviruses, which involve the application of immunoglobulin and the antiviral drug* Pleconaril, are in development. Pleconaril prevents virus replication by incorporating into the viral capsid. *Preliminary research* reports *promising* evidence for its effect on the treatment of enteroviral respiratory infections, meningitis [36] and chronic infections in immunodefficient patients [30]. Nevertheless, some recent studies could not confirm the effectiveness of Pleconaril in the treatment of aseptic meningitis in children, whereas its application in the adult patient population only modestly shortened the course of the illness in comparison to the control group [37].

1.7. Diagnostics of Enteroviral Diseases

Due to a variety of symptoms of infection, which are not specific just for enteroviruses or for particular enteroviral serotypes, the *diagnosis of enteroviral diseases* cannot be made only on the basis of clinical manifestations. The diagnosis of enteroviral infections is mainly established by virus detection in clinical specimens by RT-PCR or isolation in cell culture.

The isolation of enteroviruses from clinical samples in cell culture, followed by an identification by indirect immunofluorescence (IFA) or neutralization test, is still regarded as the traditional diagnostic “gold standard” [6, 41, 42]. The main drawback of this approach comes from the fact that it takes an average of 3 - 14 days before the cytopathic effect (CPE) of enteroviruses is detectable [43]. Therefore, a result obtained in this manner has no influence on patient treatment but only enables a retrospective diagnosis. This disadvantage can be overcome with a method for rapid antigen detection also known as the shell-vial culture assay [43-46]. This method is based on the low speed centrifugation of inoculated cell cultures, which increases their sensitivity to the virus. In conjunction with immunofluorescence, it allows for the detection of viral antigens before the appearance of CPE, which significantly reduces the time required to obtain final results.

Molecular methods for the detection of nucleic acids have revolutionized the diagnosis of enterovirus infections [30, 47]. Although real-time reverse transcription polymerase chain reaction (real-time RT-PCR) is the most sensitive and fastest method for the diagnosis of enterovirosis [43], molecular techniques are not yet available to all laboratories. Their introduction into a routine laboratory setting requires expensive equipment, special technology, and trained personnel [48]. In many virology laboratories, cell culture is still used as the primary method for the diagnosis of enterovirus infections [48, 49]. It offers an ability to determine the type of isolates using immunofluorescence or neutralization tests. Although serotyping has no influence on the clinical management of patients, it enables the identification of the endemic circulating virus types, types responsible for an outbreak, and allow for the monitoring of changes in virulence and epidemic potential of enterovirus types.

Molecular typing is possible but not widely or commercially available [42, 49]. Diagnostic RT-PCR assays that are commonly used are based on the amplification of the highly conserved 5’ non-coding region of the enterovirus genome. These assays are very sensitive and specific and able to detect enterovirus strains that do not grow in conventional cell cultures. However, they do not allow for the identification of detected viruses beyond the genus level [42, 50]. Serotype-specific sequences are situated in the VP1 coding part of the genome, so RT-PCR and sequencing of the VP1-encoding region is applicable for virus typing [47]. However, these assays usually require cell culture prior to extraction due to a lack of sensitivity [50]. Until a reliable and sensitive nucleic acid-based typing method is commercially available, cell culture will remain necessary for the identification of enterovirus serotypes, especially for epidemiological purposes. A good alternative for molecular methods for enterovirus diagnosis is a rapid detection method of viral antigen, which represents a significant advance over a traditional isolation [41, 43, 46].

The clinical utility of serological antibody detection against the enterovirus is limited. The large number of different serotypes and the need for accurate and convalescent specimens make serology impractical for acute diagnostic purposes [30].

1.8. Enteroviruses in the Water Environment

Enteroviruses are resistant under acidic conditions (pH 3 - 5) and to inactivation by proteolytic enzymes and bile salts, which enable the viruses to pass through the stomach. Enteroviruses are excreted in the feces of infected humans (with or without clinical symptoms) into the environment, where they can persist for days under favorable conditions. According to Fong *et al.* [51], a laboratory investigation confirmed that polioviruses are detectable in water even after 60 days at 22 °C or 30 days at 30 °C, whereas the inactivation of 90 % of viral particles at 4 °C required as long as 671 days. Though susceptible to UV light (sunlight inactivates 99.9 % of poliovirus particles within 24 hours), enteroviruses are capable of retaining viability significantly longer than fecal indicator bacteria. In aggregates with solids, their persistence in natural environments and resistance to environmental factors is increased. Enteroviruses are generally more resistant to diverse physico-chemical environmental factors then vegetative bacteria [51-53]. Therefore, bacteria used as standard indicators of fecal contamination of water are not reliable indicators of a viral presence [54].

Sewage contains huge amounts of enterovirus particles [55, 56] and can contaminate groundwater used as a source of drinking water, recreational waters or shellfish-growing waters. Sewage used as fertilizer or contaminated water used for irrigation is a common source for soil viral contamination. Depending on diverse factors (rainfall, temperature, soil structure, organic carbon content, soil pore water pH, etc.), viruses can penetrate into deeper soil layers and contaminate underground sources of drinking water. Numerous authors have detected enteroviruses both in "raw" and treated drinking water that was of satisfactory bacteriological safety [52, 54, 57].

Enteroviruses are highly tolerant to temperature and salinity fluctuations [40]. Numerous studies have confirmed their presence in natural marine [58-60] and freshwater [61, 62] bathing areas. Low infective doses make recreational waters one of the major pathways of enterovirus infections [59]. The risk of infection when consuming virus-contaminated drinking water is estimated to be 10- to 10 000-fold greater compared to pathogenic bacteria at similar exposure levels [51]. Although the EU Council directive concerning the quality of bathing waters only includes bacterial indicators as microbiological quality parameters, a further study into viral indicators is recommended by a group of experts from the EU [60].

When taking into consideration food as a potential vector for spreading enterovirus infections, the consumption of shellfish grown in fecal-contaminated water presents a major hazard, as well as other food contaminated through contact with infected water [40]. Cheong *et al.* [53] proved the presence of different viruses, as well as enteroviruses, in raw vegetables and bacteriologically-safe underground water used for its irrigation.

The RT-PCR method for detecting viruses in environmental samples is sensitive and specific but is limited in that it cannot distinguish between infectious and noninfectious viral particles [57]. The integrated cell culture/RT-PCR method is recommended for the detection of enteroviruses in variable environmental samples [40, 51, 56]. This approach enables a highly sensitive and specific detection of the genome of infectious enteroviral particles. A relatively simple identification of enteroviruses and their extremely high resistance in the environment make them more reliable indicators of viral, as well as general, fecal contamination of waters, compared to bacteria and coliphages [53, 54].

2. Predominant Enterovirus Serotypes in the Vojvodina Region (Serbia)

Enteroviruses are characterized by their pronounced seasonal and geographical variability of serotypes, as well as their diversity of clinical manifestation [28]. The presence of individual antigenic types in particular geographic regions is quite constant. Some serotypes are endemic for specific regions while others occur periodically, which leads to epidemic outbreaks followed by a period of very few isolations. It is believed that such enterovirus types circulate in a particular region for some time until the majority of the population has acquired a specific immunity. Afterwards the virus recedes until a sufficient number of non-immunized individuals, *i.e.*, children born after the last period of virus activity, are present in the region. In this chapter, we will present the predominant enterovirus serotypes included in the etiology of aseptic meningitis in the Vojvodina region (Serbia).

2.1. Isolation of Enterovirus and Typing of the Obtained Isolates

To identify types of enteroviruses, a total of 70 samples (29 cerebrospinal fluid specimens, 27 stool specimens, 9 rectal swabs, and 5 throat swabs) were obtained from 52 pediatric patients suspected of having aseptic meningitis. Rectal and throat swabs were collected in viral transport medium (Earl's balanced salt solution supplemented with 1 % bovine serum albumin, penicillin 1000 U ml^{-1}, streptomycin 1000 μg ml^{-1}, and amphotericin B 5 μg ml^{-1}). Transport mediums and 10 - 20 % suspensions of fecal samples in transport mediums were centrifuged (20 min, 600 *g*), filtered (pore diameter of 0.45 μm), and inoculated into cell cultures.

Cerebrospinal fluid (CSF) samples were inoculated in cell cultures without prior treatment. If it was not possible to perform processing within 24 h of sampling, the samples were stored at -76 °C.

Three different continuous cell lines were used for isolation of enteroviruses: RD (human embryonic rhabdomyosarcoma cells), HEp-2 (human larynx carcinoma cells) and Vero (African green monkey kidney cells) (Figure 3). Cell cultures were prepared "in house" and inoculated in accordance with WHO recommended protocols [63].

Tubes were examined daily for the presence of CPE using an inverted microscope for 7 days post-inoculation. Cultures showing characteristic enterovirus CPE (Figure 4) were subpassaged for confirmation, and if positive, they were marked as CPE positive. In the absence of a CPE, two blind subpassages were performed before reporting the culture as CPE negative.

A) B)

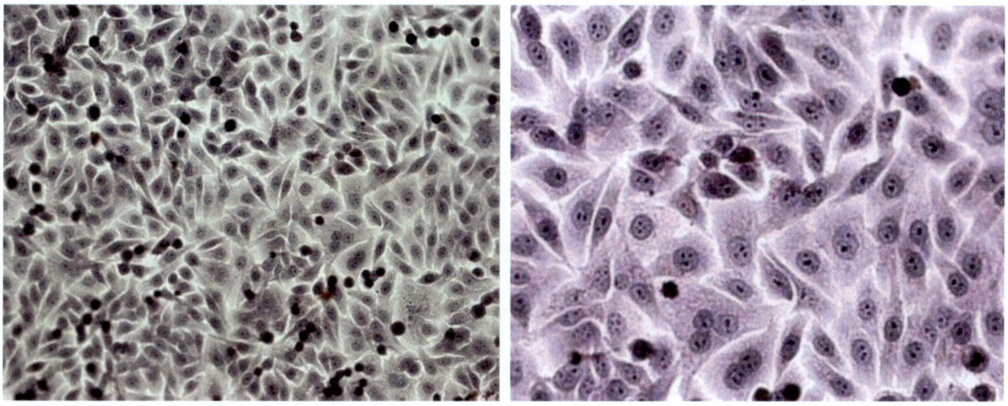

Figure 3. A non-infected HEp-2 cell culture stained with hematoxylin-eosin (HE), A) magnification 100×, B) magnification 200× (Center of Virology, Institute of Public Health of Vojvodina).

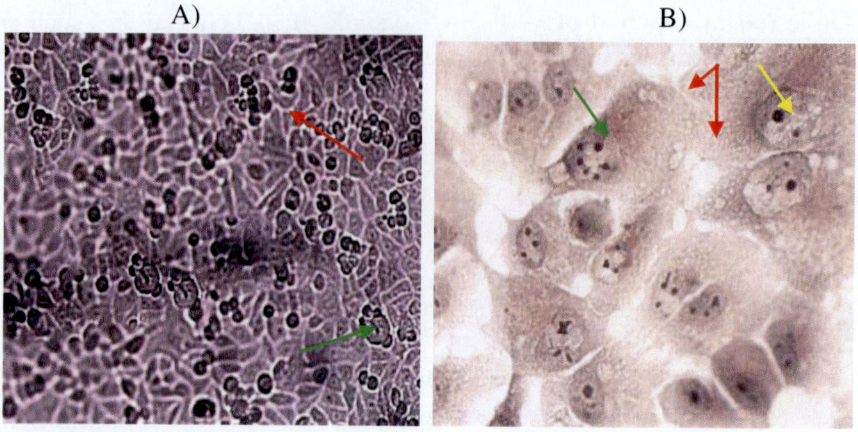

Figure 4. CPE-induced CV-B5 on Vero cell cultures stained with HE 2 days post-inoculation. A) Single or grouped, wrinkled, rounded, refractile cells (red arrow) and some enlarged due to the presence of vacuoles (green arrow), magnification 200×; B) Eccentrically positioned nuclei with clearly visible nucleoli and two intranuclear inclusions (yellow arrow), a paranuclear eosinophillic mass (green arrow), vacuoles (red arrow), magnification 400× (Center of Virology, Institute of Public Health of Vojvodina).

Simultaneously with conventional cell culturing, all samples were submitted to a rapid antigen detection (RAD) assay. Cell cultures were seeded separately in 6-well microtiter plates inoculated with 100 - 200 µl of specimens and centrifuged at 600 *g* for 60 min (36 ºC), as described by Lipson *et al.* [46]. Incubations were performed for 72 h at 36 ºC.

The presence of enteroviruses in cell cultures was confirmed using an indirect immunofluorescence assay. When approximately 75 % of cells exhibited CPE or at the end of the incubation period (for CPE negative cultures and for the RAD assay), cells were harvested and fixed in cold acetone (4 ºC) for 10 min. Uninfected cells were used as a negative control, and cells infected with a laboratory strain of coxsackievirus B3 served as a positive control. Staining was performed according to the manufacturer's instructions using Pan-Enterovirus 2E11 reagent (Millipore-Chemicon), which contains mouse monoclonal antibodies against enteroviruses. Anti-mouse immunoglobulin G FITC-labeled (Millipore-Chemicon) was used as a secondary antibody (Millipore-Chemicon). The slides were examined under a fluorescence microscope (Olympus BH-2) at 200× magnification. A bright apple-green fluorescence in the nucleus and/or cytoplasm indicated a positive result (Figure 5). A sample that gave a positive fluorescence signal for at least one cell type was considered a true positive for the presence of enteroviruses.

The usage of a Pan-Enterovirus 2E11 immunoreagent can be associated with a non-specific reaction of the antibody with reovirus 3, hepatitis A virus, some astroviruses and, to a lesser extent, with rhinoviruses.

These viruses either do not replicate in cell cultures, or their isolation requires specific conditions (a lower incubation temperature and a lower pH value for the medium for rhinoviruses). Conditions under which enterovirus isolations were attempted excluded the presence of any of aforementioned viruses. Therefore, the obtained positive immunofluorescence signals could be decidedly attributed to human enteroviruses.

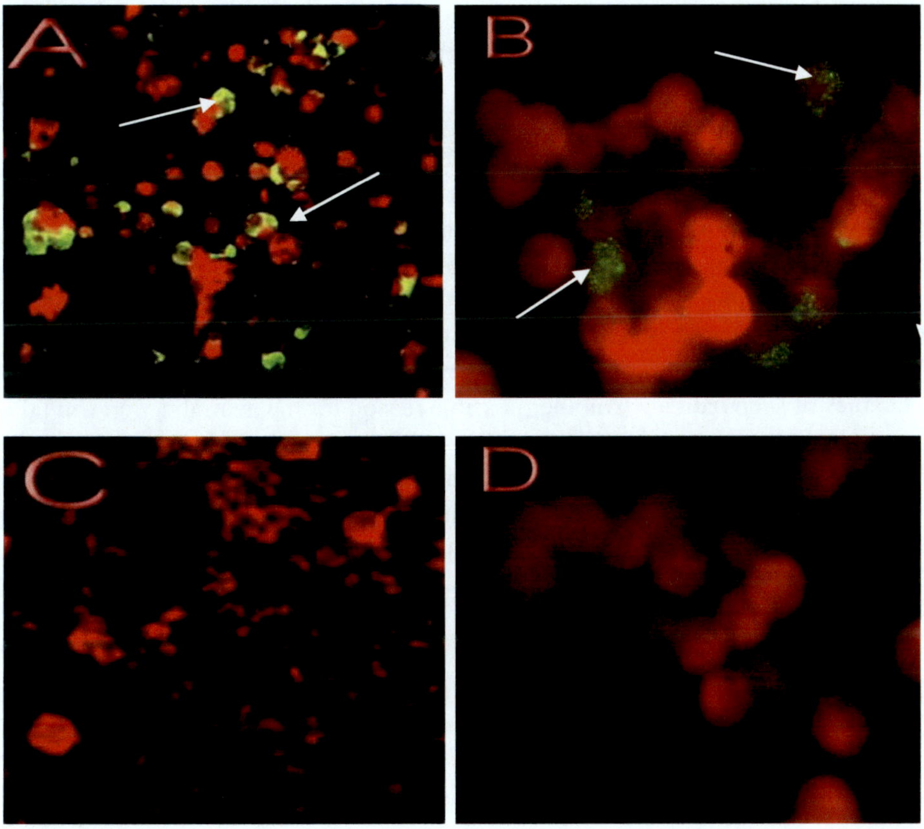

Figure 5. Immunofluorescence obtained by the use of 2E11 monoclonal antibodies. A) and B) Vero cells infected with CV-B 5 enterovirus, arrows point to red-stained cells, which cytoplasm and nucleus contain clearly visible green fluorescence; C) and D) Negative control, non-infected Vero cells without fluorescence (Center of Virology, Institute of Public Health of Vojvodina).

Serotyping of enterovirus isolates was performed by neutralization of the CPE using Lim, Benyesh-Melnick (LBM) pools of specific antisera A to H (Statens Seruminstitut, Copenhagen, Denmark). These pools allow for the identification of 42 antigenically different enteroviruses. Each isolate was tested against all the antiserum pools. Virus controls and cell controls were run for comparison. The interpretation of the results was performed by analyzing the pattern of CPE inhibition by the antiserum pools according to the scheme provided by the pools.

2.2. Analysis of the Results

Of 70 analyzed samples, enteroviruses were isolated from 36 (51.4 %). All 36 samples that proved CPE positive by classical cell culture gave a positive signal with the 2E11 monoclonal antibodies. None of CPE-negative samples displayed a positive result when IFA was used. The RAD assay gave positive results for 34 (48.6 %) samples. The sensitivity and specificity for the RAD assay compared with classical isolation were 94.4 % (34 / 36) and 100 % (34 / 34), respectively. Two CSF-positive samples by conventional cell culture were negative according to the RAD assay. This result was probably due to the low amounts of virus present in those samples, which assumes that 3 days of incubation during the RAD assay was too short for the production a sufficient number of viral particles with antigens that could be proven by IFA.

The neutralization test allowed for the typing 32 of 36 (88.9 %) enterovirus isolates. The remaining 4 (11.1 %) isolates could not be serotyped, which is a common situation in a number of studies [64, 65]. Muir *et al.* [66] offered several possible explanations for this situation. First, LBM pools enable the typing of only 42 antigenically different human enteroviruses. It is possible that our samples contained types to which no homologous antisera were available or that the samples included several diverse serotypes and, thus, prevented neutralization. According to some authors, mixed infections with diverse enterovirus types are quite frequent. Two or more types of the virus are present mainly in the fecal samples of such patients. Another possible reason for the neutralization test failure is the formation of viral aggregates, which can be neutralized only after dissociation with sodium deoxycholate or chloroform or by ultrafiltration. The occurrence of so called prime strains cannot also be excluded. Prime strains are antigenic variants of the original serotype, which were used to obtain the antiserum for the neutralization test. They are antigenically altered and, thus, can be neutralized poorly or not at all by antiserum. An isolate that is untypeable by a neutralization test could represent a new or previously unrecognized serotype. Neutralization revealed the presence of 11 different enterovirus serotypes: CV-A 16, CV-B 3, 4, 5, and EHV 2, 4, 6, 11, 13, 16, and 30 (Table 2). The results of the neutralization test indicated the dominant presence of echoviruses with an incidence rate of 52.8 % in the analyzed samples. These results correlate with literature data that suggests a wide distribution of this group of enteroviruses, which comprises the largest number of isolates obtained in numerous studies.

Table 2. Enterovirus types detected by the neutralization test

Enterovirus type	No. of positive specimens (%)				
	CSF N=29	Feces N=27	Throat swab N=5	Rectal swab N=9	Total N=70
CV-A16	2	1	-	-	3
CV-B3	2	-	-	-	2
CV-B4	-	2	-	-	2
CV-B5	3	1	2	-	6
EHV 2	1	1	-	-	2
EHV 4	1	-	-	-	1
EHV 6	1	2	1	-	4
EHV 11	2	3	-	-	5
EHV 13	-	1	-	-	1
EHV 16	1	3	-	-	4
EHV 30	1	-	1	-	2
NT	1	3	-	-	4
Total	15 (51.7)	17 (62.9)	4 (80)	0 (0)	36 (51.4)

Abbreviations: CV-A, coxsackie A virus; CV-B, coxsackie B virus; EHV, echovirus; NT, not typeable; CSF, cerebrospinal fluid; N, number of analyzed specimens.

An 8-year study conducted by Sedmak *et al.* [67] in the USA, which included the isolation and identification of human enteroviruses in clinical specimens as well as in samples of sewage waters, revealed the presence of echoviruses in 80 % of the specimens analyzed. During an epidemiological monitoring conducted by Bahri *et al.* [64] from 1992 to 2003 in Tunisia, echoviruses were the most frequently detected enteroviruses.

Echoviruses are common agents of aseptic meningitis, and outbreaks of echovirus meningitis are recorded worldwide. According to some reports, the majority of enteroviral encephalitis and aseptic meningitis is attributed to this group of enteroviruses [65]. By analyzing laboratory data on types of isolated viruses in England from 1975 to 1994, Atkinson *et al.* [68] demonstrated that 61 % of aseptic meningitis cases were associated with echoviruses.

All isolated echovirus strains have recently been reported to cause outbreaks and sporadic cases of meningitis worldwide. An outbreak of aseptic meningitis with hundreds of infected individuals occurred in Israel in 1997, and EHV 4 was the etiologic agent [69]. Echovirus 6 was identified as the viral pathogen in an epidemic of viral meningitis in Brazil in 2004, and it was also associated with numerous disease outbreaks in the USA, Australia, and Spain [70]. A small-scale epidemic of EHV 11 meningitis was reported in England in 2000 [71], whereas epidemics in Germany (2000) [72], Japan [73], and the U.S.A. [74] in 2001 were associated with echovirus type 13. During 2000, in Cuba, almost 17000 cases of meningitis caused by EHV 16 were recorded within only several months [75]. EHV 30 was responsible for over a thousand reported cases of meningitis in Taiwan in 2001 [76]. EHV 2 is considered a rare serotype and has accounted for a very small proportion of reports. According to data collected within the National Enterovirus Surveillance Program in the USA during 1970 - 2005, EHV 2 was recorded almost every year. However, it was only among the 15 most commonly identified serotypes for two years [77]. Sometimes an outbreak of meningitis can be attributed to several different enterovirus types. Thus, during 2000, epidemics of meningitis associated with EHV 6, 13 and 30 were reported in France and Belgium [78, 79].

Neutralization tests confirmed the presence of CV-B viruses in 27.8 % of samples, which corresponds with the results of a long-term study by Atkinson *et al.* [68], who proved that 29 % of all aseptic meningitis cases reported in England during a period of 20 years were caused by enteroviruses of group CV-B. Additionally, a 10-year investigation conducted in the USA by the CDC (Center for Disease Control) within the national program for enterovirus surveillance revealed that out of more than 18 000 isolates from numerous laboratories included in this research 24 % belonged to the CV-B group [80]. Out of 6 different CV-B types CV-B 3, 4, and 5, which are regarded as common causative agents of meningitis, were identified. The prominent epidemic characteristics of serotype B5 are well-known while B3 and B4 mainly produce individual meningitis cases or small-scale outbreaks. According to data collected within the enterovirus surveillance program in Japan from 1982 to1999, type B5 had the highest prevalence among all enteroviral meningitis cases associated with CV-B [81]. These same authors proved that CV-B 5 was responsible for meningitis epidemics in Japan in 2001, and the same serotype (along with several EHV) was identified as the etiological agent of large-scale epidemics in Belgium one year later [78]. Individual cases of meningitis caused by serotypes B3 and B4 were reported in Belgium in 2000 [78], Argentina from 1998 to 2003 [82], and in Spain in 2006 [50].

Enteroviruses from the coxsackie A group were identified in only 3 cases (8.3 %), and all the isolates were CV-A 16. According to a study by Atkinson *et al.* [68], CV-A was the least prevalent among all enteroviruses and accounted for only 12 % of positive liquor findings. Typing of enterovirus isolates using LBM pools (A-H) enabled the identification of only 3 CV-A serotypes, CV-A 7, 9, and 16. Although coxsackie A16 is usually associated with hand-foot-and-mouth disease, sporadic cases of meningitis, which are caused by this type, have also been reported [83].

Wild poliovirus strains are still present in some countries in Africa (Nigeria, Niger, Ivory Coast) and Asia (Pakistan, India, Afghanistan), whereas in other regions worldwide only vaccinal strains are detected. However, in 2010, the first wild poliovirus importation into the European region because it was declared polio free in 2002, resulted in over 450 confirmed cases in Tajikistan, Russia, Turkmenistan, and Kazakhstan. This investigation confirmed that the virus was transmitted from another endemic region outside of Europe (the Global Polio Eradication Initiative: http://www.polioeradication.org). The adoption of a new polio vaccination policy, which implicates the immunization of children exclusively with a dead inactivated vaccine, significantly reduced the rate of isolation of vaccinal poliovirus strains from clinical and environmental samples. Thus, Sedmak *et al.* [67], during 2000, 2001, and 2002, did not isolate any poliovirus from examined clinical specimens and sewage water samples. Additionally, continual monitoring of enteroviruses, as a part of a comprehensive poliomyelitis eradication program, has been conducted in France since 2000. During the first five years of enterovirus monitoring, more than 190 000 different clinical specimens were examined, and vaccinal poliovirus strains were detected in only two of them [79]. In Republic of Serbia, a live oral poliomyelitis vaccine is still in use; however, in examined samples, no poliovirus was identified. This may be due to the fact that none of the patients, from whom the samples were obtained, had been immunized against poliomyelitis for several weeks before sampling.

Although isolation by cell culture is considered the "gold standard" in diagnostics of human enterovirus infections, there is no universal consensus on the optimal combination and number of cell cultures required for isolation. The susceptibility of a particular cell line to specific enterovirus serotypes was determined according to data obtained during virus isolation in cell cultures, which indicated the type of cell culture in which a particular sample produced CPE, as well as based on the results of a neutralization test, which detected the type of enterovirus that was present in particular samples (Table 3).

Data from the literature pertaining to the sensitivities of particular cell cultures to various enterovirus serotypes are inconsistent. It is difficult to compare the results of different studies as they refer to diverse cell systems and serotypes. In addition, data on the susceptibility of particular cell cultures to specific serotypes are often contradictory. Van Loon *et al.* [84] performed the quality assessments of 12 laboratories with respect to the isolation and typing of different enteroviruses. Although all laboratories were provided with the same set containing 8 isolates of a known serotype, the results did not reveal any obvious pattern pertaining to the susceptibility of a particular cell line to a specific serotype. Specifically, the same isolates were reported to produce CPE in a cell culture in one laboratory but were CPE negative in the same culture in another laboratory.

The results presented in this chapter confirm that a single cell line is not sufficient for the isolation of all enteroviruses as even 17 (47.2 %) isolates could be detected in only one cell line. The use of multiple cell lines increases enterovirus recovery. Only in some particular cases can isolations be performed utilizing a reduced set of cell cultures.

Table 3. Sensitivity of Vero, HEp-2, and RD cell cultures to different enterovirus types

Enterovirus Type	Specimen	Cell culture		
		Vero	HEp-2	RD
1. CV-A16	CSF	-	-	+
2. CV-A16	CSF	-	-	+
3. CV-A16	Feces	-	-	+
1. CV-B3	CSF	+		-
2. CV-B3	CSF	+	+	-
3. CV-B4	Feces	+	+	-
4. CV-B4	Feces	+	+	-
5. CV-B5	Throat swab	+	+	-
6. CV-B5	Throat swab	+	+	-
7. CV-B5	CSF	+	-	-
8. CV-B5	CSF	+	-	-
9. CV-B5	CSF	+	-	-
10. CV-B5	Feces	+	-	-
1. EHV 2	CSF	+	-	+
2. EHV 2	Feces	+	-	-
3. EHV 4	CSF	+	+	+
4. EHV 6	CSF	-	+	-
5. EHV 6	Feces	-	+	-
6. EHV 6	Feces	-	+	-
7. EHV 6	Throat swab	-	+	-
8. EHV 11	Feces	+	+	+
9. EHV 11	Feces	+	+	+
10. EHV 11	CSF	+	+	+
11. EHV 11	CSF	+	-	+
12. EHV 11	Feces	+	-	+
13.EHV 13	Feces	-	+	-
14.EHV 16	CSF	+	-	+
15.EHV 16	Feces	+	-	+
16.EHV 16	Feces	+	-	+
17.EHV 16	Feces	+	-	+
18.EHV 30	CSF	-	+	+
19.EHV30	Throat swab	-	+	+
1.	CSF	-	-	+
2.	Feces	-	-	+
3.	Feces	-	-	+
4.	Feces	-	-	+
Total		22	16	20

Abbreviations: CV-A, coxsackie A virus; CV-B, coxsackie B virus; EHV, echovirus; NT, not typeable; CSF, cerebrospinal fluid.

Thus, if the presence of coxsackie B virus is proved in the samples analyzed during epidemics, the application of RD cell culture can be discontinued. However, sometimes such a decision can be made according to a patient's clinical symptoms. For example, if the symptoms suggest the hand-foot-and-mouth disease most commonly caused by CV-A 16, the isolation procedure should include only the RD cell line. However, none of the enterovirus serotypes is related to only one particular disease, which lessens the number of cell cultures that could result in a misdetection of the pathogen. Moreover, outbreaks can be caused by several enteroviruses belonging to various groups and thus decreasing the number of cell lines would significantly reduce the spectrum of identified types.

RD cell culture is generally considered the culture of choice for the isolation of coxsackie A viruses [67] as well as for the majority of echoviruses [70, 72], whereas it is less convenient for the detection of coxsackie B viruses [67, 84]. The results presented in this chapter strongly correspond with this statement. All 3 CV-A 16 isolates were obtained only in RD cell culture, and the majority of echoviruses (13 / 19) produced CPE in this culture. In contrast, enteroviruses of the coxsackie B group could not be isolated by RD cell culture.

Based on the obtained results, as well as the findings of other authors [85], Vero cell culture proved to be the most efficient in isolating coxsackie B serotypes. All 10 isolates from this group successfully replicated in this cell line. Moreover, the majority of echoviruses (12 / 19) produced CPE in this culture.

In general, the HEp-2 cell line is considered less suitable for the recovery of echoviruses [67]; however, its sensitivity towards echoviruses (11 / 19) was not significantly different if compared with the other two cell lines. HEp-2 proved to be the most useful cell line for the isolation of coxsackie B serotypes (5 / 10).

It is well-demonstrated that enteroviral particles are shed in large numbers in the feces of infected persons, which makes stool samples most convenient for virus isolation in cell cultures. Therefore, the isolation rate for stool samples (17 / 27, 62.9 %) was expected to be higher than for CSF (15 / 29, 51.7 %, Table 2). However, the identification of enterovirus in a stool samples is less significant because it may be the result of an asymptomatic infection, preceding infection or polio vaccination. On the other hand, detection of an enterovirus in a CSF sample is reliable evidence of its etiological role in this particular CNS disease.

Enteroviruses are present in large amounts in throat swabs only during the early stage of infection. Thus, a successful isolation from these samples is highly determined by the time of sampling. Out of 5 examined throat swabs, enteroviruses were detected in 4 (80 %). Considering the small number of examined throat swabs, such a high isolation rate might be accidental and must be considered with some reservations.

Rectal swabs proved least suitable for enterovirus isolation as enteroviruses were not detected in any of the 9 analyzed samples. Because rectal swabs contain only small amounts of viruses, they are only utilized for isolations when stool samples are not available. Moreover, the presence of enteroviruses in the throat and on rectal swabs is of a limited diagnostic value as it may be due to an asymptomatic infection.

Though some epidemics of enterovirus meningitis have been characterized by a significant number of affected adults [76], the disease predominantly occurs in children. The majority of patients are under 15 years old [70, 72, 75], and children of pre-school age are the most susceptible population for enterovirus infections, which can be attributed to the fecal-oral route of transmission as hygiene habits have *not* fully *developed* by this age. The presence of enterovirus in at least one sample was proved in 30 of 52 (57.7 %) examined children.

Table 4. Isolated HEV serotypes based on patient age categories

<table>
<tr><th rowspan="2">Age group</th><th colspan="4">No. of positive patients (%)</th><th colspan="2" rowspan="2">No. positive/
No. analyzed (%)</th></tr>
<tr><th>CV-A</th><th>CV-B</th><th>EHV</th><th>NT</th></tr>
<tr><td>0-2</td><td>-</td><td>6 (27.3)</td><td>8 (36.4)</td><td>1 (4.5)</td><td>15/22 (68.2)</td><td rowspan="2">24/37
(64.1)</td></tr>
<tr><td>3-5</td><td>2 (13.3)</td><td>3 (20.0)</td><td>4 (26.6)</td><td>-</td><td>9/15 (60.0)</td></tr>
<tr><td>6-8</td><td>-</td><td>-</td><td>3 (33.3)</td><td>2 (22.2)</td><td>5/9 (44.4)</td><td rowspan="2">6/15
(30.5)</td></tr>
<tr><td>9-11</td><td>1 (16.7)</td><td>-</td><td>-</td><td>-</td><td>1/6 (16.7)</td></tr>
</table>

Abbreviations: CV-A, coxsackie A virus; CV-B, coxsackie B virus; EHV, echovirus; NT, not typeable.

The rate of infection decreased with age and dropped from 68.2 % in patients age 0 - 2 years old to 17.7 % in children 9 - 11 years old (Table 4). The rate of positive patients below 5 years old (64.1 %) was significantly higher in relation to the group of older children (30.5 %; $t = 2.27$; $p > 0.05$).

Conclusion

Genus *Enterovirus* is highly dynamic and possesses a variety of serotypes that are susceptible to genetic changes, which result in the formation of antigenic variants characterized by altered tropisms and virulence.

Monitoring the presence of enterovirus serotypes enables the identification of the types that are endemic for a particular region, and the types that are responsible for outbreaks, as well as control of the pathogenic potential and the virulence and epidemic potential of various enteroviral serotypes. The results presented in this chapter demonstrate the importance of human enteroviruses (coxsackie virus A16, coxsackie virus B3, 4 and 5, and echovirus 2, 4, 6, 11, 13, 16, and 30) as the etiological agents of aseptic meningitis in the autonomous province of Vojvodina.

Additional research is needed to study the occurrence and persistence of enteroviruses in local surface waters and to assess the public health risk associated with exposure to waters contaminated with these viral pathogens. In addition, it is essential to investigate the influence of different environmental factors, especially climate factors, on the abundance and survival of enteroviruses in natural waters. These data would provide better knowledge about the effects of climate variability and global climate change on the epidemiology of enterovirus infection.

Acknowledgements

The research presented in this chapter was realized as a part of the project "Studying climate change and its influence on the environment: impacts, adaptation and mitigation" (43007) financed by the Ministry of Education and Science of the Republic of Serbia within the framework of integrated and interdisciplinary research for the period 2011-2014.

REFERENCES

[1] J.L. Melnick, H.A. Wenner, C.A. Phillips, In *Diagnostic procedures for viral, rickettsial and chlamydial infections,* E.H. Lennette, J.C. Schmidt, Ed.; 5th edition; American Public Health Association: Washington, D.C., 1979; pp 471-534.

[2] P.D. Minor, P. Muir, In *Prinicples and practice of clinical virology,* A.J. Zuckerman, J.E. Banatvala, P.D. Griffiths, B.D. Schoub, P. Mortimer, Ed.; 6th edition; John Wiley and Sons Ltd.: Chichester, UK, 2009; pp 601-624.

[3] B.A. Brown, K. Maher, M.R. Flemister, P. Naraghi-Arani, M. Uddin, M.S. Oberste, M.A. Pallansch, *J. Gen. Virol.*, 90, 1713-1723 (2009).

[4] M.S. Oberste, K. Maher, W.A. Nix, S.M. Michele, M. Uddin, D. Schnurr, S. al-Busaidy, C. Akoua-Koffi, M.A. Pallansch, *Virus. Res.*, 128, 34-42 (2007).

[5] N. Junttila, N. Leveque, J.P. Kabue, G. Cartet, F. Mushiya, J.J. Muyembe-Tamfum, A. Trompette, B. Lina, L.O. Magnius, and other autors, *J. Med. Virol.*, 79, 393-400 (2007).

[6] T. Yamashita, M. Ito, H. Tsuzuki, K. Sakae, H. Minagawa, *J. Gen. Virol.*, 91, 1063-1066 (2010).

[7] G. Stanway, F. Brown, P. Christian, T. Hovi, T. Hyypia, A.M.Q. King, N.J. Knowles, S. M. Lemon, P.D. Minor, M.A. Pallansch, A.C. Palmenberg, T. Skern, In *Virus Taxonomy, Eighth Report of the International Committee on Taxonomy of Viruses*; C.M. Fauquet, M.A. Mayo, J. Maniloff, U. Desselberger, L.A. Ball, Ed.; Elsevier Academic Press: London, UK, 2005; pp 757-778.

[8] N.J. Knowles (2011) The picornavirus home page. On line: http://www.picornastudygroup.com

[9] D.M. Belnap, D.J. Filman, B.L. Trus, N. Cheng, F.P. Booy, J.F. Conway, S. Curry, C.N. Hiremath, S.K. Tsang, A.C. Steven, J.M. Hogle, *J. Virol.*, 74, 1342-1354 (2000).

[10] V. Caro, S. Guillot, F. Delpeyroux, R. Crainic, *J. Gen.Virol.*, 82, 79-91 (2001).

[11] X. Zhang, Y. Li, T. Peng, M. Aasa, L. Zhang, Y. Yang, L. Archard, *J. Histochem. Cytochem.*, 48, 579-584 (2000).

[12] E. Domingo, *J. Nutr.*, 127, 958-961 (1997).

[13] V. R. Racaniello, In *Fields Virology.* D.M. Knipe, P.M. Howley, D.E. Griffin, R.A. Lamb, M.A. Martin, B. Roizman, S.E. Strauss, Ed.; 5th edition; Lippincott Williams and Wilkins: USA, 2007; pp 795–838.

[14] N.L. Yozwiak, P. Skewes-Cox, A. Gordon, S. Saborio, G. Kuan, A. Balmaseda, D. Ganem, E. Harris, J.L. De Risi, *J. Virol.*, 84, 9047-9058 (2010).

[15] P. Ahlquist, *Science,* 5571, 1270-1273 (2002).

[16] E.C. Holmes, *J. Virol.*, 77, 3893-3897 (2003).

[17] N.H. De Jesus (2007). Epidemics to eradication: the modern history of poliomyelitis. *Virol. J. On line:* http://www.virologyj.com/content/4/1/70

[18] M.S. Oberste, K. Maher, M.A. Pallansch. *J. Virol.*, 78, 855-867 (2004).

[19] A.M. Lindberg, P. Andersson, C. Savolainen, M.N. Mulders, T. Hovi, *J. Gen. Virol.*, 84, 1223-1235 (2003).

[20] T.J. Tuthill, E. Groppelli, J.M. Hogle, D.J. Rowlands, *Curr. Top. Microbiol. Immunol.*, 343, 43–89 (2010).

[21] Y. He, S. Mueller, P.R. Chipman, C.M. Bator, X. Peng, V.D. Bowman, S. Mukhopadhyay, E. Wimmer, R.J. Kuhn, M.G. Rossmann, *J. Virol.*, 77, 4827-4835 (2003).

[22] C. Xiao, C.M. Bator-Kelly, E. Rieder, P.R. Chipman, A. Craig, R.J. Kuhn, E. Wimmer, M.G. Rossmann, *Structure*, 13, 1019-1033 (2005).

[23] C.H. Williams, T. Kajander, T. Hyypia, T. Jackson, D. Sheppard, G. Stanway, *J. Virol.,* 78, 6967-6973 (2004).

[24] P. Ahlquist, A.O. Noueiry, W.M. Lee, D.B. Kushner, B.T. Dye, *J. Virol.*, 77, 8181-8186 (2003).

[25] A.V. Gamarnik, R. Andino, *Gene. Dev.*, 12, 2293-2304 (1998).

[26] Y. Liu, C. Wang, S. Mueller, A.V. Paul, E. Wimmer, P. Jiang, *PLoS. Pathog.*, 6:e1001066, (2010).

[27] T. Hyypia, T. Hovi, N.J. Knowles, G. Stanway, *J. Gen. Virol.*, 78, 1-11 (1997).

[28] G. Palacios, M. S. Oberste, *J. Neuro. Virol.*, 11, 424-433 (2005).

[29] F.E.A. Moura, D.C.S. Ribeiro, N. Gurgel, A.C. da Silva Mendes, F.N. Tavares, C.N.G. Timoteo, E.E. da Silva, *Br. J. Ophtalmol.*, 90, 1091-1093 (2006).

[30] M.T. Hawkes, W. Vaundry, *Paediatr. Child. Health.*, 10, 383-388 (2005).

[31] K. Pond, *Water recreation and disease: plausibility of associated infections: acute effects, sequele and mortality*; IWA Publishing: London, p. 239 (2005).

[32] R. Feuer, C.M. Ruller, N. An, J.M. Tabor-Godwin, R.E. Rhoades, S. Maciejewski, R.R. Pagarigan, C.T. Cornell, S.J. Crocker, W.B. Kiosses, N. Pham-Mitchell, I.I. Campbell, J.L. Whitton, *J. Virol.*, 83, 9356-9369 (2009).

[33] J.K.S. Chia, *J.Clin.Pathol.*, 58, 1126-1132 (2005).

[34] V. Jerant Patic, *Imunology*; Faculty of Medicine, University of Novi Sad: Novi Sad, p. 695 (2000). (In Serbian).

[35] T. Chonmaitree, C.D. Baldwin, H.L. Lucia, *Clin. Microbiol. Rev.*, 2, 1-14 (1989).

[36] H A. Rotbart and F. G. Hayden, *Arch. Fam. Med.*, 9, 913-920 (2000).

[37] N. David and M.D. Irani, *Neurol. Clin.*, 26, 635-viii (2008).

[38] G. Enders, In *Microbiology and microbial infections,* L. Collier, A. Balows, M. Sussman, Ed.; 9th edition; Arnold: London, UK, 1998; Vol. 1 Virology, pp 874-915.

[39] R. Dettmeyer, A. Baasner, M. Schlamann, S.A. Padosch, C. Haag, R. Kandolf, B. Madea, *Pediatr. Res.*, 55, 947-952 (2004).

[40] B. Rajtar, M. Majek, L. Polanski, M. Polz-Dacewicz, *Ann. Agric. Environ. Med.*, 15, 199-203 (2008).

[41] M. Perez-Ruiz, J.M. Navarro-Mari, P. del Valle, M. Rosa-Fraile, *J. Med. Microbiol.*, 52, 789-91 (2003).

[42] S.S. Afifi, S.A. Zaki, A.F. Mohamed, H.E. Hosseiny, *Aust. J. Basic. Appl. Sci.*, 3, 3230-3238 (2009).

[43] E. Terletskaia-Ladwig, S. Meier, R. Hahn, M. Leinmuller, F. Schneider, M.A. Enders, *J. Med. Microbiol.*, 57, 1000-1006 (2008).

[44] S L. Klespies, D.E. Cebula, C.L. Kelley, D. Galehouse, C.C. Maurer, *J. Clin. Microbiol.*, 34, 1465-1467 (1996).

[45] G.J.J. Van Doornum, J.C. De Jong, *J. Clin. Microbiol.*, 36, 2865-2868 (1998).

[46] S.M. Lipson, K. David, F. Shaikh, L. Qian, *J. Clin. Microbiol.*, 39, 2755-2759 (2001).

[47] V. Milošević, V. Jerant-Patić, I. Hrnjaković Cvjetković, J. Radovanov Tadić, G. Kovačević, *Med. Today.*, 5, 442-451 (2006). (In Serbian).

[48] T.L. Lin, Y.S. Li, C.W. Huang, C.C. Hsu, H.S. Wu, T.C. Tseng, C.F. Yang, *J. Clin. Microbiol.*, 46, 785-788 (2008).

[49] R C. She, G. Crist, E. Billetdeaux, J. Langer, C.A. Petti, *J. Clin. Virol.*, 37, 151-155 (2006).

[50] M. Cabrerizo, J.E. Echevarria, I. González, T. de Miguel, G. Trallero, *J. Med. Virol.*, 80, 1018–1024 (2008).

[51] T.T. Fong, E. K. Lipp, *Microbiol. Mol. Biol. Rev.*, 69, 357-371 (2005).

[52] Bosch, *Int. Microbiol.*, 1, 191-196 (1998).

[53] S. Cheong, C. Lee, S. W. Song, W. C. Choi, C. H. Lee, S.J. Kim, *Appl. Environ. Microbiol.*, 75, 7745-7751 (2009).

[54] M.M. Ehlers, W.O.K. Grabow, D N. Pavlov, *Water. Res.*, 39, 2253-2258 (2005).

[55] W.J. Lodder, H.H.J.L. Van den Berg, S.A. Rutjes, A.M. de Roda Husman, *Appl. Environ. Microbiol.*, 76, 5965-5971 (2010).

[56] G. La Rosa, M. Pourshaban, M. Iaconelli, M. Muscillo, *Ann. Ist. Super. Sanita.*, 46, 266-273 (2010).

[57] M.A. Borchardt, P.D. Bertz, S.K. Spencer, D.A. Battigelli, *Appl. Environ. Microbiol.,* 69, 1172-1180 (2003).

[58] D.W. Griffin, K.A. Donaldson, J.H. Paul, J.B. Rose, *Clin. Microbiol. Rev.,* 16, 129-143 (2003).

[59] H. Katayama, A. Shimasaki, S. Ohgaki, *Appl. Environ. Microbiol.*, 68, 1033-1039 (2002).

[60] L. Moce-Llivina, F. Lucena, J. Jofre, *Appl. Environ. Microbiol.*, 71, 6838-684, (2005).

[61] W.O.K. Grabow, K.L. Botma, J.C. de Villiers, C.G. Clay, B. Erasmus, *B. World Health Organ.*, 77, 973-980 (1999).

[62] C.D. Chapron, N.A. Ballester, J.H. Fontaine, C.N. Frades, A.B. Margolin, *App. Environ. Microbiol.*, 66, 2520-2525 (2000).

[63] World Health Organization, *Polio laboratory manual, 4th edition*; WHO/IVB/04.10; World Health Organization: Geneva, p. 157 (2004).

[64] O. Bahri, D. Rezig, B. Nejma-Oueslati, B. Ben Yahia, J. Ben Sassi, N. Hogga, A. Sadraoui, H. Triki, *J. Med. Microbiol.*, 54, 63-69 (2005).

[65] C.H. Joo, J. Ahn, I. Seo, Y.K. Kim, D.H. Kim, H.N. Hong, H. Lee, *Intervirology*, 48, 97-103 (2005).

[66] P. Muir, U. Kammerer, K. Korn, M.N. Mulders, T. Poyry, B. Weissbrich, R. Kandolf, G. M. Cleator, A.M. van Loon, *Clin. Microbiol. Rev.,* 11, 202-227 (1998).

[67] G. Sedmak, D. Bina, J. MacDonald, *Appl. Environ. Microbiol*, 69, 7181-7187 (2003).

[68] P.J. Atkinson, M. Sharland, H. Maguire, *Arch. Dis. Child.*, 78, 373-374 (1998).

[69] R. Handsher, L.M. Shulman, B. Abramovitz, I. Silberstein, M. Meuman, M. Topperberg-Oikawa, T. Fisher, E. Mendelson, *J. Clin Virol.*, 13, 29-36 (1999).

[70] Luchs, D.H. Russo, A. Cilli, F.F. Costa, S.G. Morillo, B.C. Machado, A.C.G. Pellini, R.C.C. Carmona, M.C.S.T. Timenetsky, *Bras. J. Microbiol.,* 39, 28-31 (2008).

[71] E. Somekh, T. Shohat, R. Handsher, F. Serour, *Epidemiol. Infect.*, 126, 441-444 (2001).

[72] S. Diedrich, and E. Schreier (2001). Aseptic meningitis in Germany associated with echovirus type 13. *BMC Infect.Dis.* On line: http://www.biomedcentral.com/1471-2334/14.

[73] M. Keino, M. Kanno, K. Hirasawa, T. Watari, M. Mikawa, K. Saito, K. Kato, M. Katayose, H. Yoshida, *JPN. J. Infect. Dis.*, 22, 317-318 (2001).

[74] CDC. *M.M.W.R. Weekly*, 50, 777-780 (2001).

[75] L. Sarmiento, P. Mas, A. Goyenechea, R. Palomera, L. Morier, V. Capo, I. Quintana, M. Santin, *Emerg. Infect. Dis.*, 7, 887-889. (2001).

[76] C.H. Kao, S.S.J. Lee, Y.C. Liu, M.Y. Yen, Y.S. Chen, S.R. Wan, H.H. Lin, W.R. Lin, C.K. Huang, C. Chin, *J. Microbiol. Immunol. Infect.*, 367, 192-196 (2003).

[77] N. Khetsuriani, A. LaMonte-Fowlkes, S. Oberste, M.A. Pallansch, *M.M.W.R. Weekly.*, 55, 1-20 (2006).

[78] Thoelen, P. Lemey, I. Van der Donck, K. Beuselinck, A. M. Lindberg, M. Van Ranst, *J. Med. Virol.*, 70, 420-429 (2003).

[79] D. Antona, N. Leveque, J.J. Chomel, S. Dubrou, D. Levy-Bruhl, B. Lina, *Eur. J. Clin. Microbiol.*, 26, 403-412 (2007).

[80] M. Moore, M.H. Kaplan, J. McPhee, D.J. Bregman, S.W. Klein, *Public Health Rep.*, 99, 515-522 (1984).

[81] Y. Kitahori, Y. Inoue, Y. Maruhashi, O. Adachi, S. Imai, *JPN. J. Infect. Dis.*, 56, 75-76 (2003).

[82] A.S. Mistchenko, M. Viegas, M.P. Latta, P.R. Barrero, *J. Clin Virol.*, 37, 293-299 (2006).

[83] D. Brunel, N. Leveque, J. Jacques, F. Renois, J. Motte, L. Andreoletti, *J. Clin. Virol.*, 42, 225-228 (2008).

[84] A.M. Van Loon, G.C. Cleator, A. Ras, *B. World Health Organ.*, 77, 217-223 (1999).

[85] H. Kubo, N. Iritani, Y. Seto, *Emerg. Infect. Dis.*, 8, 298-305 (2002).

In: Essays on Fundamental and Applied Environmental Topics ISBN 978-1-61942-522-4
Editor: Dragutin Mihailovic

Chapter 12

MONITORING OF FOREST ECOSYSTEMS IN SERBIA

S. Orlović**, *Z. Galić, S. Stojnić and B. Klašnja
Institute of Lowland Forestry and Environment,
University of Novi Sad, Novi Sad, Serbia

ABSTRACT

The importance of forests in soil protection, carbon sequestration, and the development of a number of activities that are significant to the population of and rural development in the Republic of Serbia is immeasurable. Forest monitoring programs have been established due to many factors that disturb the conditions of forests throughout Serbia, such as the (i) National Forest Inventory (NFI) and (ii) International Co-operative Program on the Assessment and Monitoring of Air Pollution Effects on Forests (ICP – Forests). In addition, the effects of climate changes on forests have been monitored. The areas, ownerships and capacities of the major tree species from the forests were evaluated by the NFI. It was noted that forest areas increased by 465 611 ha from 1979 to 2007 and that forest conditions also improved. Two levels of monitoring of air pollution effects on forests (ICP – Forests) were established. Level I has been implemented since 2003 at 130 observation plots and level II since 2009 at 2 observation plots. Results that were obtained during several years of monitoring at level I have revealed deteriorated crown conditions as the first signs of weakness and potential drying. Level II monitoring that was conducted over the course of two years showed differences between measurements in tree crown conditions, foliage, soil and soil solution chemical composition, tree growth, ground vegetation, atmospheric deposition, meteorological data, leaf phenology and forest litter composition, which were comparable to previous results. However, the durations of the studies are still not long enough to draw reliable conclusions.

The sites, where microclimatic conditions and oxidative drought stresses were observed, were set up to monitor the impacts of climate changes in the most important forests and plantations in Serbia. These analyses revealed significant differences in

* Corresponding author: E-mail address: sasao@uns.ac.rs; Phone: +381 21 540 382. Fax: +381 21 540 385.

microclimate parameters (air temperatures and relative humidity). To gather information on the adaptability of the most important tree species in the area, the International beech provenance tests were established, in which the most significant morphological, anatomical and physiological parameters were observed. Differing heights of juvenile stage plants corresponding to different growth locations are presented. The monitoring of forest ecosystems in Serbia are in the beginning stages. The collected data are insufficient for establishing a database for forest ecosystem monitoring, and the results cannot be systematized at this time. In the future, studies should focus on long-term monitoring and simulate ecosystem dynamics and functions.

1. INTRODUCTION

At present, forests cover approximately 31 % of the total land area or 4 billion hectares. In addition, 12 % of forests or 95 million hectares have been set aside for the conservation and preservation of biological diversity [1]. They provide raw materials for renewable and environmentally sound products and play important roles in the economy, biological diversity, the global carbon cycle and water balance [2]. Forests are important sources of environmental safety and social and recreational services, especially considering the increasingly urbanized societies [3]. They are important resources for rural development and jobs, local communities, forest owners, and forest-related companies [4]. They are of immense significance to the protection of agricultural and forest lands against erosion and are the most significant forms of vegetation in terms of net resources and the sequestration and retention of carbon, ensuring that activities, such as beekeeping, hunting, sports, recreation, and tourism, are available to rural populations and for rural development [5]. Forests also provide timber as the raw material for the wood industry and biomass for energy production.

There are numerous factors that threaten forest survival. Humans directly impact the reduction of forest areas by converting them into agricultural land, especially in the tropics where 13 million hectares of forest are lost annually; therefore, environmental and productive forest functions are also lost [1]. Forest conditions are also threatened by numerous abiotic and biotic factors, the most significant of which are air and water pollution, climate changes, and insect and disease infestations [3]. All of these factors lead to the weakening of the plants, and in some cases to forest drying, causing environmental changes. Long term studies and observations of forest ecosystems can provide valuable indicators of ecosystem functioning and are key factors for improving ecosystems and natural resource management [6]. For this reason, several monitoring programs have been introduced at the international level, of which the best known are the International Co-operative Program on the Assessment and Monitoring of Air Pollution Effects on Forests (ICP - Forests) and the International Long-Term Ecological Research Network (ILTER). The most comprehensive European scale for the monitoring of forest ecosystems is provided by the UN Economic Commission and the European Union [7]. By the implementation of forest environmental monitoring, the basis for a Europe-wide standard guideline (ICP - Forests) set by the United Nations Economic Commission for Europe (UNECE) Convention on Long-range Transboundary Air Pollution for a Level II Program has been initiated for data recording and research of cause-and-effect chains with the goal of publishing scientifically proven recommendations for forestry management and environmental policies. Unlike ICP - Forests, the ILTER program is not only dedicated to forests but also to terrestrial ecosystems in which long-term environmental and socio-economic research is conducted. In addition to pollution, global climate changes significantly undermine forest conditions, which was the primary reason for the initiation of climate monitoring. This monitoring has attained much public attention due to the possible global warming that has taken place in recent years. The available observational data indicate the existence of persistent global trends during the last century [8]. The last report from the Intergovernmental Panel of Climate Change (IPCC) revealed the existence of increased air temperatures that are caused by anthropogenic factors [9]. Increased mean air temperatures can lead to extreme climatic conditions, such as fluctuations in spring temperatures; i.e., more frequent and longer intervals of drought [9]. These climatic extremes of increased mean air temperatures may adversely affect the vitalities of trees, increasing the likelihood of secondary damage that is caused by diseases and pests. The effects of climatic extremes on forest ecosystems have been described for forests in the boreal and alpic belts [10] and those in the lowlands [11]. Given the variety of scenarios for climate change, it can be expected that they will significantly affect all natural systems, including forests [12]. The main task of the current research is to ensure the survival of forests under the existing and modified conditions. It is also expected that such forest management systems will be developed to ensure for the mitigation of climatic changes by increasing the levels of the existing carbon absorbents through forest management, adaptive management and improved sustainability and multifunctionality [5]. To ensure the sustainable management of forests worldwide, including those in Serbia, and to mitigate any adverse effects, the continuous monitoring of forest conditions is necessary. According to a recent study, harmful insects caused the devastation of 35 million ha of forests in a temperate climatic zone in 2010 and thus significantly decreased the vitality of forest trees, making it possible for other noxious factors to cause the drying of forests [1]. The monitoring of the health status and vitality of forests should ensure for the determination of new harmful insects and diseases that were previously not present to allow for adequate preparatory measures and prevent the devastation of forests.

2. Monitoring of Forests in Serbia

Forest monitoring networks in Serbia have been organized under the NFI, the ICP - Forests system of monitoring level I and level II forest conditions, and from the experimental monitoring of the influences of climate changes on forests.

2.1. National Forest Inventory

National forest inventories have been conducted in many countries with the aim of obtaining information for the preparation of appropriate forest management measures. Only inventories of forest stands were conducted in Serbia until 2004. More recent information that was gathered on forest stocks was directly used in the formation of general and private forest management programs, while their reductions and sublimations were the bases for the creation of the fundamentals of forest management, micro-economic planning, and correspondence with international communities.

Problems arising from the lack of quantitative information regarding stand inventory, which have originated from great variabilities in current volume increments and often unreliable records on forest management among other things, make such information unreliable in terms of real multi-functional planning at higher levels, especially at the national level.

Information regarding the forest stock in Serbia was previously obtained using other procedures that were even less acceptable than that which was mentioned above and from which some highly incompatible basic data on forest conditions can be encountered in the expert and scientific literature. Reliable information that is used for the higher level planning of forest areas has been successfully provided by the NFI (for large areas) [13]. The first data on the national inventory of forests in Serbia were released in 2009. According to these data, the total forested area was 2 252 400 ha or 29.1 %. Thus, Serbia is considered to be a moderately forested country.

This level of forest coverage is similar to that worldwide, which is 30 %, and is significantly lower than the European levels of 46 %. Other forested lands, according to the international definition, include bushes and shrubs, which cover 4.9 % of the territory and encompass a total of 36.3 % in relation to the area of productive land. Forest coverage increased by 5.2 % since 1979, and this certainly had positive impacts on the condition and quality of the environment in general [13].

According to ownership, 1 194 000 ha or 53 % of Serbian forests belong to the state, and 1 058 400 ha or 47 % are privately owned. A study that was conducted in 2000 reported that the relationship between state and private forests in the following European countries is as follows: Austria 17.5 / 82.5 %, Croatia 75.5 / 24.5 %, Bosnia and Herzegovina 78.4 / 21.6 %, Czech Republic 84.1 / 15.9 %, France 26.2 / 73.8 %, Romania 94.6 / 5.4 %, Slovenia 30.0 / 70.0 %, and Finland 28.9 / 71.1 %. Compared to previous reference reports that described 45.2 % (state forests) and 54.8 % (private forests) ownership, there was a change of 1.8 % in favor of private forests. Compared to the study that was performed in 1979, the increase in forest area was 356 863 ha for the state owned and 108 748 ha for the privately owned forests (Figure 1) [13].

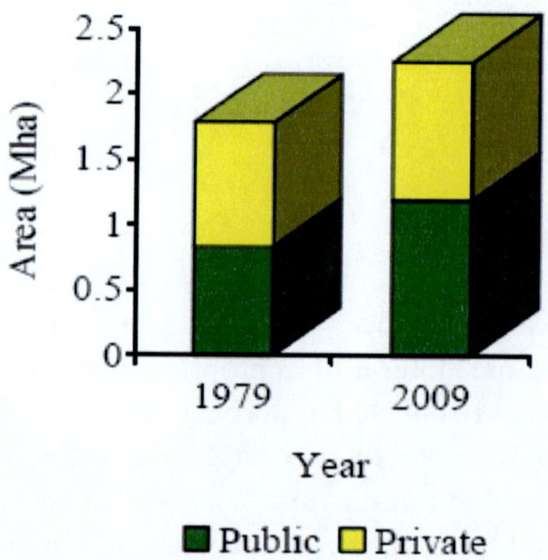

Figure 1. Forest area in Serbia [13].

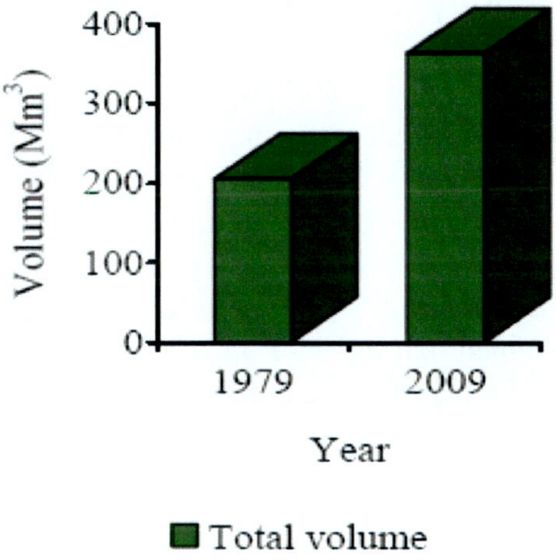

Figure 2. Forest volume in Serbia [13].

In 1979, the total estimated timber volume was 205 480 735 m^3, and in 2009, it was 362 487 417 m^3 (Figure 2). The average volume in 1979 was 115 $m^3\ ha^{-1}$, which was less than that observed in Austria, Bulgaria, and Croatia, and more than that observed in Bosnia and Herzegovina. In general, the state forests are considered to be satisfactory because their average volume in 2009 was 185 $m^3\ ha^{-1.}$ In quantitative terms, privately owned forests are less satisfactory with an average volume of 133 $m^3\ ha^{-1}$ because of their origins of dominant emergence and age structures [13].

According to the Serbian National Forest Inventory, 49 tree species were found to be dominant deciduous species (40) as opposed to coniferous (9). Uneven wood volumes were recorded for the tree species from the total forest stock. Beech is a dominant species in Serbian forests, which had a total volume of 40.5 %, followed by Turkey oak (13.0 %), sessile oak (5.9 %), Hungarian oak (5.8 %), hornbeam (4.2 %), black locust (3.1 %), common oak (2.5 %), and ash (1.6 %). Norway spruce is the most common coniferous tree species and had a volume of 5.2 %; pine and Austrian pine and European silver fir showed volumes of 4.5 % and 2.3 %, respectively. Clones of Euramerican poplars showed a total volume of 1.7 % (Figure 3). The largest decrease in volume over the past 20 years was observed for beech [13].

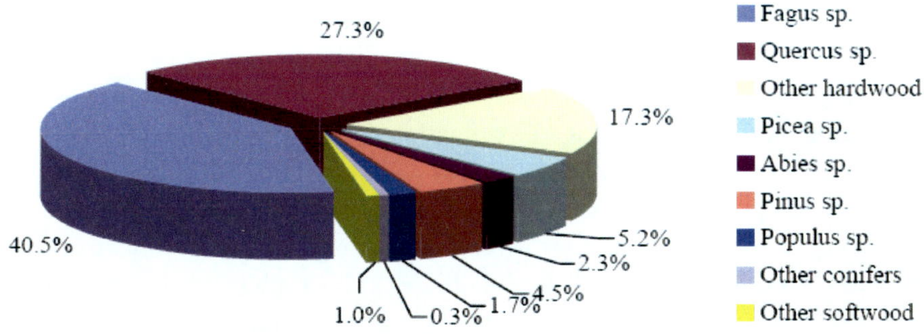

Figure 3. Percentage of tree species in Serbia by volume [13].

2.2. Monitoring of Air Pollution Effects on Forests

Monitoring of forests is an essential tool for the detection of the negative impacts and creation of mitigation measures. Because the monitoring system is only fully efficient if it acts upon large areas throughout the network, it is necessary to monitor the effectiveness of the implemented measures [14]. Damage to forests in Europe, especially to coniferous tree species became noticeable during the 1980s over large areas. The first visible symptoms of reduced vitality were defoliation and pronounced crown discoloration regardless of the vegetation period. It was assumed that the damage occurred as the result of air pollution, and thus the ICP - Forests was established because it is global and recognizes no country borders [15].

The program is implemented at two levels, I and II. It has been conducted in Serbia since 2003.

2.2.1. Monitoring at Level I

This monitoring system covers the most important European forests, and conditions of both vitality and ample health are observed at the 6000 observation plots [15].

The level I network in Serbia contains 130 observation plots (Figure 4) uniformly distributed over the network with distances of 16 km between points. Within them, the following parameters were observed: crown conditions, soil chemistry and forest tree nutrition. According to the defoliation data that was collected from 2003 - 2010, uniform values were present in both the deciduous and coniferous tree species (Figure 5). In conifers, the years 2004 and 2005 showed the highest percentages of low to moderate intensity defoliation. After the first two extremely substandard years, defoliation in the coniferous species became uniform, while in the broadleaved species it varied greatly. In 2007, the broadleaved species showed higher levels of defoliation compared to the coniferous species [15].

The discoloration of the Serbian forests was the most pronounced in the coniferous species in 2005 (Figure 6), while in 2009 and 2010, the conditions improved. The observational periods can be separated as follows: (i) the period from 2004 to 2007 in which discoloration was more severe and (ii) the period from 2009 - 2010 in which the percentage of discoloration of the affected tress did not fall below 90 %.

Data in Figure 7 depict the total damage of all of the trees originating from both of the categories of defoliation and discoloration. In this figure (Figure 7), similar trends in both damage and discoloration were observed.

2.2.2. Monitoring at Level II

This level of monitoring, which is known as intensive monitoring, is carried out on the plots that are located in the dominant forest ecosystems to study the cases and consequential relationships of anthropogenic and natural factors that cause damage to forest ecosystems. It is conducted using fewer plots than does level I monitoring. The main objective of intensive monitoring is to obtain better insight into the effects of air pollution and other stress factors on forest ecosystems. The level II monitoring program includes the following group of parameters: crown conditions, analysis of assimilation organs, ground vegetation, phenological observation, forest litter composition, analysis of deposition, growth and yield, ozone injury, soil and soil solution analysis and meteorological measurements. Because all of the observations were not every year, the number of plots for data collection varies depending on the year [15].

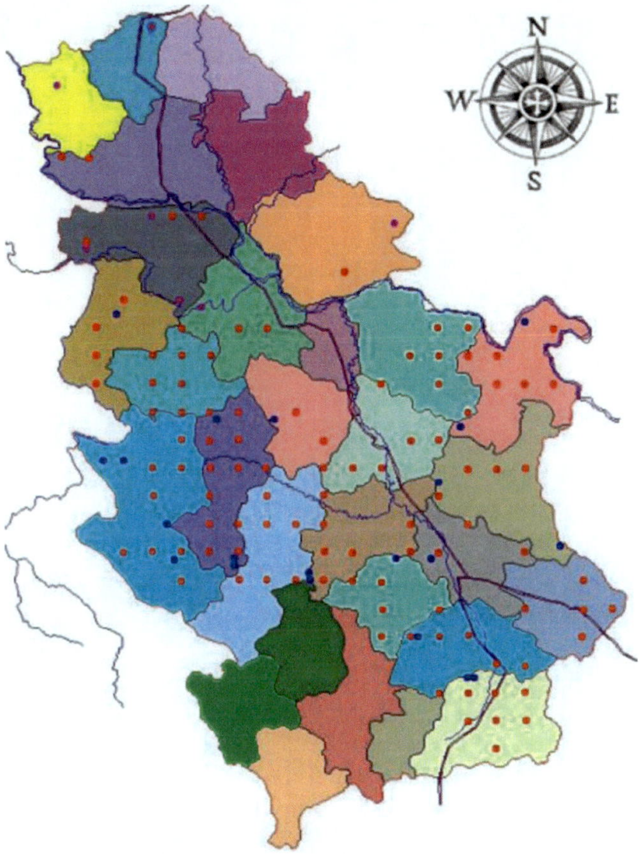

Figure 4. Observation plots in Serbia (level I and level II) [15].

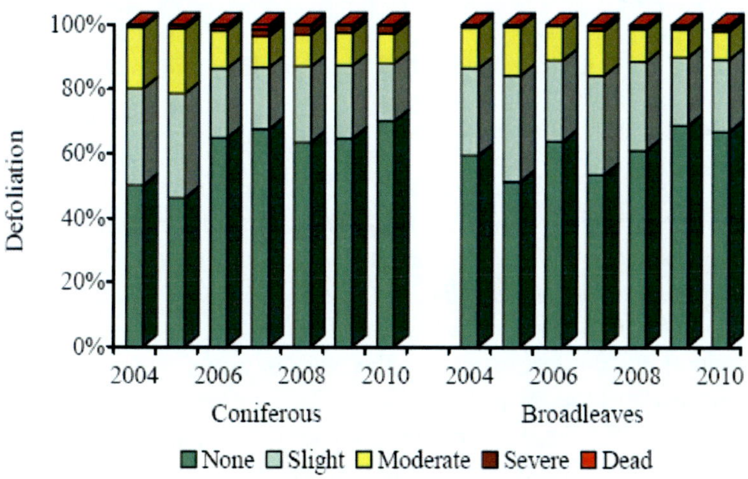

Figure 5. Defoliation of forests in Serbia in the period 2004 - 2010 [15].

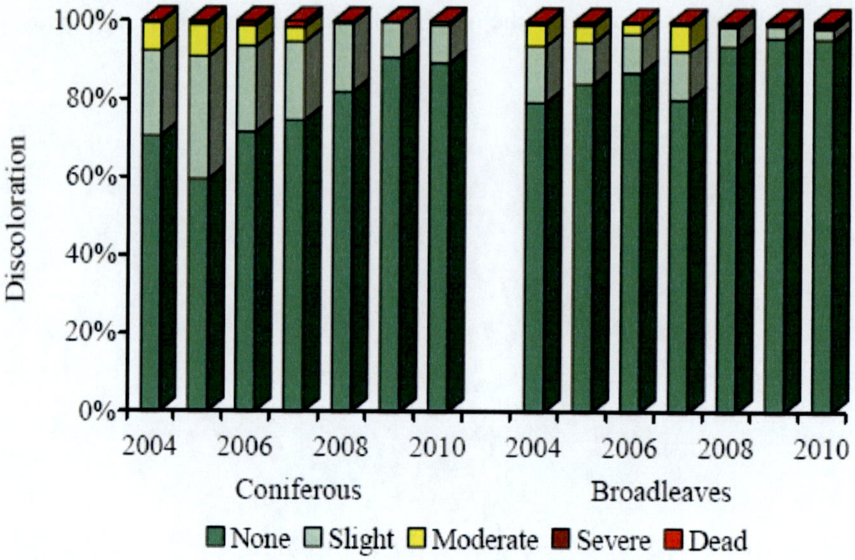

Figure 6. Discoloration of forests in Serbia in the period 2004 - 2010 [15].

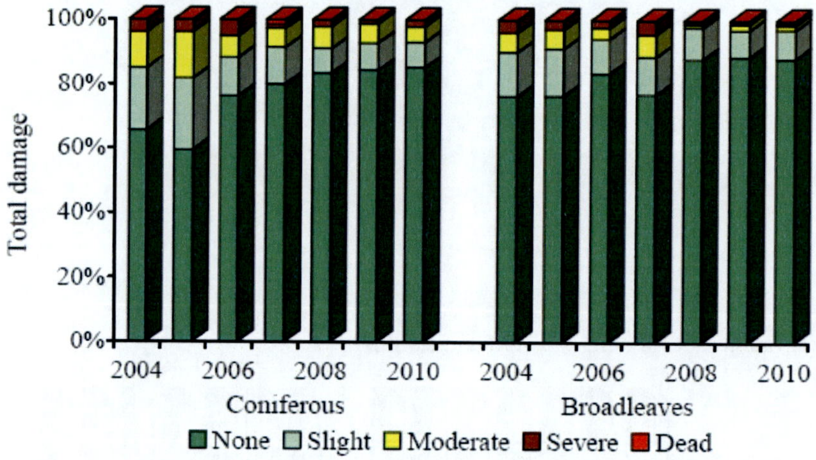

Figure 7. Damages of forests in Serbia in the period 2004 - 2010 [15].

Level II monitoring in Serbia was initiated in 2009 for the Fruška Gora National Park and in 2010 for the Kopaonik National Park. The observation plot at the Fruška Gora National Park is situated in association with the dominant sessile oak (*Quercus petrea*), and there are also large-leaved linden (*Tilia grandifolia*) and beech (*Fagus sylvatica*) trees present (Figure 8). The observation plot at the Kopaonik National Park is situated in association with *Picetum excelsae,* where the dominant tree species is the Norway spruce. Monitoring was carried out at both plots throughout the respective years. The following factors were monitored: conditions of tree crowns, chemical compositions of tree leaves, tree growth rates, leaf phenology, chemical compositions of precipitation, chemical compositions of leaf litter, soil chemical composition, leaf ozone damage, ground vegetation and meteorogical parameters.

Figure 8. Area of observation plot of monitoring at the level II (Fruška Gora National park).

Table 1. Chemical analysis of leaf litterfall (fallen leaves and branches) from Fruška Gora National Park [15]

Type of tree	Part of plant	C (g/100g)	N (mg/g)	S (mg/g)	P (mg/g)	Ca (mg/g)	Mg (mg/g)	K (mg/g)
Sessile oak	Leaves	32.97	7.29	1.002	0.648	3.672	2.827	1.044
Sessile oak	Branches	34.96	7.50	0.648	0.438	6.490	0.088	0.774

Because the results from the two-year monitoring were insufficient to draw reliable conclusions, only selected results will be reported from this point forward.

Data on fallen leaves and branches that were obtained during the 2010 vegetation period from the Fruška Gora National Park are presented in Table 1, including the analyses of carbon, nitrogen, sulfur, phosphorous, calcium, magnesium and potassium.

3. Monitoring of Forest Ecosystems under Climate Changes

In the 2007 report from the ICCP Working Group 1, it was concluded that the global atmospheric concentrations of CO_2, CH_4, and NO_2 have significantly increased as a result of human activities since 1750 and now far exceed preindustrial values [9]. The global increase in CO_2 was primarily due to the use of fossil fuels and changes in land usage, while the increased concentrations of CH_4 and NO_2 were primarily caused by agricultural activities [9].

In the 2007 report from the Intergovernmental Panel on Climate Change Working Group 2, the responsibilities of natural and human systems for the observed and projected changes in natural and environmental factors were analyzed. This report estimated that in the late 21st century, increased numbers of hot days and nights can be expected, the result of which will be increased numbers of insects, increased energy production and decreased air quality in urban areas. Increasing frequencies and durations of heat waves are likely to lead to reduced harvest yields in warmer regions, increased risks of forest fires, increased water demands, problems with water quality and increased human mortality. It is also likely that with heavier precipitation, major damage to crops, increased soil erosion and reduced qualities of surface and ground water may occur [9].

Forest ecosystems are sensitive to climate changes due to the long life-span of tree species [16]. Previous investigations have suggested that in the future, forest ecosystems will be greatly influenced by increased air temperatures, decreased precipitation levels, increased CO_2 levels, fires, increases in the intensity and duration of droughts, storm damages, extremely high precipitation levels, and insect and pathogen infestations [17].

The impacts of the increased air temperatures and decreased precipitation levels on the durations and intensities of droughts could be the most prominent for forests. Changes in mean annual air temperatures at the global level in forest ecosystems will impact the intensities of photosynthesis and respiration, the decomposition of organic matter in soil and its mineralization, phenology, the onset of frost, the distribution, adaptation and evolution of species [17].

The most severe drought occurred in Europe during a heat wave in 2003 [16, 18], and the most extreme happened in Germany and France and lasted for six months [17]. During the dry period of 2003, a direct influence on tree growth was confirmed with reductions in leaf area being determined after several months [19].

In Serbia and throughout Europe, a trend towards increasing air temperatures and decreased precipitation levels has been reported [20].

Figure 9 shows the existence of an evident trend of decreased precipitation levels in specific locations in Serbia. The most pronounced decrease in precipitation occurred in eastern and southeastern Serbia and in part of Srem in the vicinity of Sremska Mitrovica [21], while the most markedly increased precipitation occurred in the eastern part of Serbia and the Prizren (south), Novi Sad and Vršac (north) municipalities.

An important trend of increasing air temperatures is present throughout Serbia. The highest increase in air temperature was present in the wider area of Loznica and Belgrade, and in Negotinska Krajina (Figure 9). A slightly negative trend towards decreased air temperatures was recorded only in the southeastern part of the country. Estimates based on climate modeling according to moderate scenarios suggest that annual temperatures in Serbia will increase by 2.6 ^{o}C until the end of the century. According to climate model projections, heat waves will occur more frequently and will be of higher intensity and longer durations [20].

The intensity of droughts were increased compared to the previous period due to precipitation deficits. The most pronounced droughts were observed in the north, east, and southeast parts of Serbia and the lowest drought intensity occurred in the west (Figure 10). Climate change projections indicate that the durations and frequencies of dry periods will increase in eastern and southeastern Serbia [20]. There are numerous drought indices that are currently in use worldwide, and one of the most frequently used is SPI (standardized precipitation index) [22]. As occurs in other parts of the world, SPI can also be used in Serbia as the primary drought index. Figure 10 shows that in 1990, there was a significant drought in southwestern Serbia, while in 2000, the entire country suffered from extreme drought with the exceptions of the Zlatibor and Raska regions [20].

The increasing air temperatures, reductions in precipitation and the resulting increased drought intensity can all negatively impact the forest sector. Areas with the most intense trends of increasing air temperatures, decreasing precipitation levels, and severe drought in Serbia include the beech forests. In addition, the climate change have significantly negative effects on forests in the Vojvodina region (north part of Serbia).

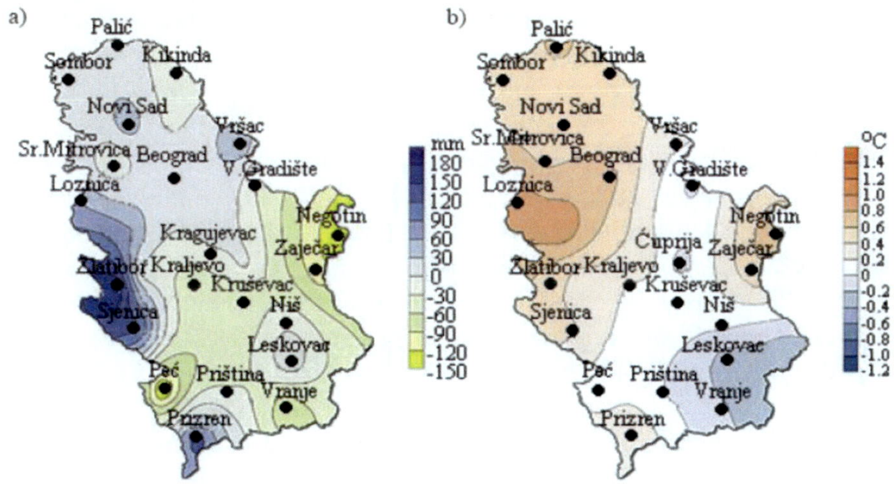

Figure 9. Trends of precipitation (left a panel) and air temperature (right b panel) in Serbia for the period (1950 - 2004) [20].

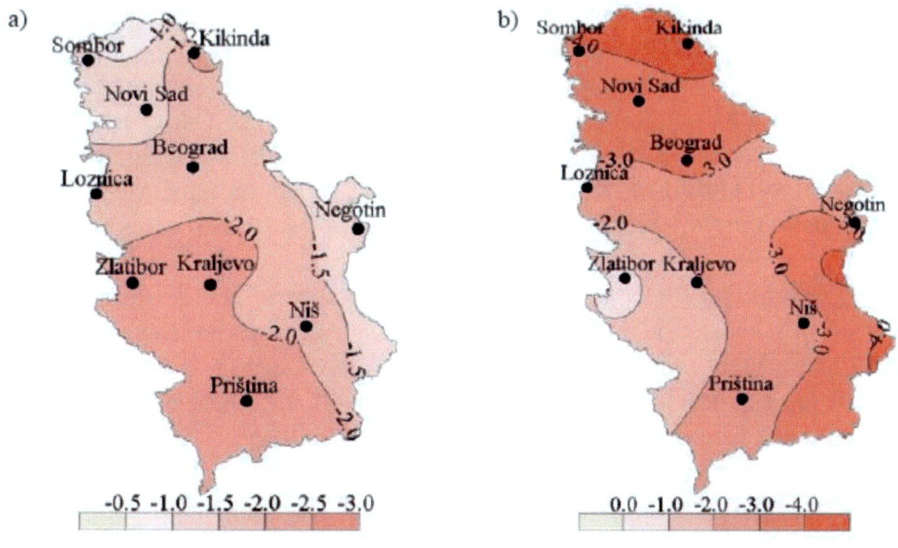

Figure 10. Spatial distribution of drought in 1990(a) and 2000(b) and its severity expressed in SPI values [20].

3.1. Monitoring of Micro-Climatic Conditions

Global climate models allow for future climate projections. They include forests as the most important factors contributing the most extensively to exchanges in energy fluxes with the lower atmosphere.

Thus, it is crucial to investigate how climate change has specifically impacted forests. However, climate models cannot provide adequately comprehensive information to elucidate this problem; in particular, to reveal processes that occur inside the forests that are highly sensitive to climate change. For that reason, the monitoring of the effects of climate change was initiated in poplar stands in 2006, while it was established in 2009 for the remaining forest types. The sites, in which observations of micro-climate conditions and the impacts of drought stress were observed, were established in the following natural stands: (1) *Fraxineto - Querceto roboris* Jov. - common oak and ash forests; (2) *Querceto frainetto - cerris* Rud.- Hungarian and Turkey oak forests; (3) *Quercetum montanum* Cer. et Jov. - Sessile oak forest with beech; (4) *Fagenion moesiace montanum* Jov.- mountain beech forest; (5) *Blechneto - Piceto - Abieto - Fagetum sphagnetosum* Col. – beech; (6) *Picetum excelsae Serbicum* Rud. - Norway spruce forest, and artificial stands: (1) *Populus x euramericana* (Dode) Guinier - Euramerican poplar; (2) *Pinus nigra* Arn. and *Pseudotsuga menziessi* (Mirb.) Franco.- Austrian pine and Douglas fir. In addition, special studies were conducted in the international provenance trials of beech, which is the most common forest species in Serbia, with the aim of estimating their potential for adaptation to altered conditions.

Data on microclimate characteristics are used for explaining the distribution, development and movement of the organisms in natural ecosystems. Major ecological processes, such as productivity, mineralization, the decomposition of leaf litter, the distribution of diseases and insects and the possibility of the occurrence of natural disasters are all related to microclimatic conditions [23, 24].

By monitoring air temperatures in three time intervals (hourly, daily and monthly), a spatial variation in microclimate conditions [25], indicating changes in vegetation zones, i.e., natural distribution of species, can be obtained [5].

Air temperatures and relative air humidity were recorded hourly at the above mentioned stands as part of the monitoring of microclimate conditions. Daily amplitudes of air temperatures that were recorded for the stands in August 2009 for the eight different forest species are given in Figure 11. The observed values are shown only for August because that was the driest month in the vegetation period according to results that were obtained by other authors [26].

Figure 11 shows that the largest differences in the daily amplitudes of air temperatures were observed in poplar stand, while the smallest ones were measured inside of the sessile oak stand. Amplitudes of the monthly air temperatures were also the greatest in poplar stand (14.8 °C) and the smallest in sessile oak stands (5.4 °C). In other studied stands, greater monthly amplitudes were observed for the common oak and ash (10.4 °C), Hungarian and Turkey oak (9.7 °C), and artificial Austrian pine and Douglas fir stands (9.2 °C).

The largest amplitudes of daily temperatures were observed for the poplar stand. Air temperature measurements that were taken in August 2010, which was when the drought was the most severe, revealed that the mean daily air temperatures for the poplar stand ranged from 13.3 do 25.6 °C.

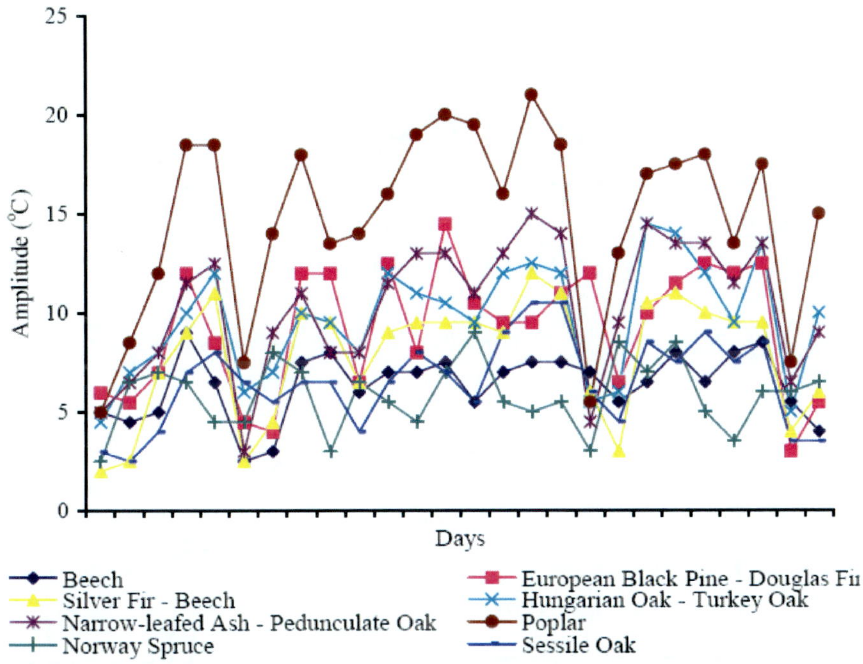

Figure 11. Daily amplitudes of air temperatures for the eight forest species during August at different experimental plots.

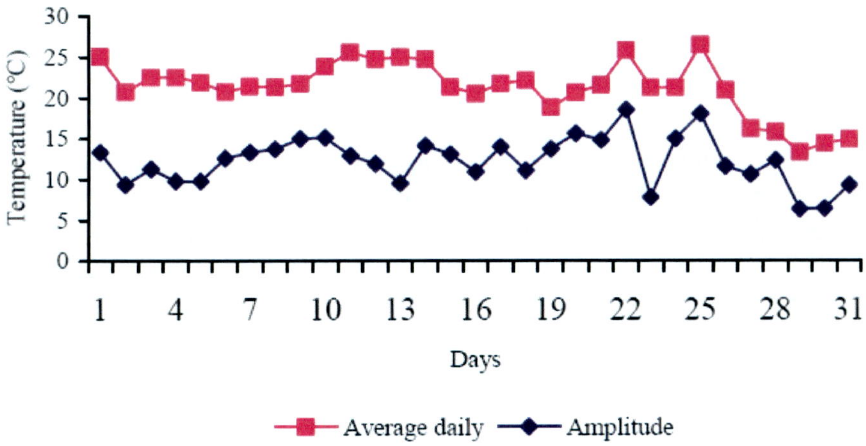

Figure 12. Amplitudes and mean daily air temperatures for poplar stand in August 2010.

The smallest daily amplitude of air temperature was 6.3 °C (August 29th) and the largest was 18.5 °C (August 22nd) as shown in Figure 12. It was previously mentioned that increased air temperatures may co-occur with larger amplitudes of daily air temperature variation, indicating that temperature conditions will not tend to favor poplar stand growth. The parameter of relative humidity may mitigate the unfavorable influence of high air temperatures in a stand to some extent. With elevated levels of air humidity, greater tolerances to increased air temperatures can be expected.

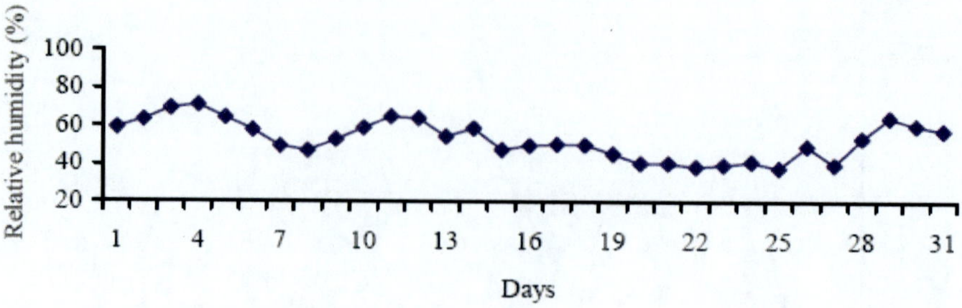

Figure 13. Relative humidity levels % of poplar stand in August 2010 in Kaćka suma (near Novi Sad).

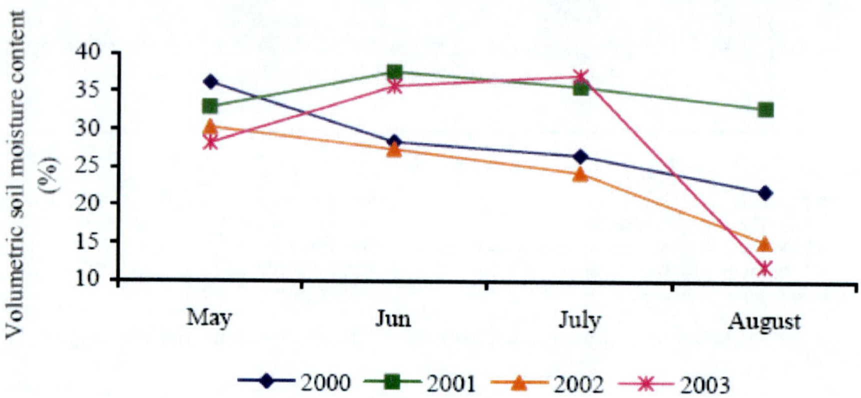

Figure 14. Volumetric soil moisture content (%) in *Populus x euramericana,* the fluvisol f. sandy loam, from 2000-2003 in Kaćka suma (near Novi Sad).

Air humidity is normally measured in the afternoon hours when it is at its lowest levels. Relative humidity in poplar stands that were measured at 2 pm during the month of August ranged from 39.8 to 71 % (Figure 13). The smallest relative humidity value was recorded during the hottest day of the vegetation period, which was 39.8 %. Considering the fact that increased air temperatures led to decreased relative humidity levels and that in the future, more severe drought periods can be expected, reduced vitalities can also be expected. Consequently, the decreased productivity of poplar stands can be expected.

The observed differences in the microclimatic conditions clearly indicate the deterioration of habitat conditions for the poplar stand. However, additional wetting with underground water may allow for the survival of these stands.

The simulation of the effects of possible factors on the development of poplar stand in altered climatic conditions can be derived based on research that was conducted before and during the heat wave in 2003. The quantity of available water depended on hydrological conditions throughout the year (Figure 14). If hydrological conditions were closer to normal, the water soil supply was higher. Long periods of drought led to relatively uniform reductions in the quantities of available water in the soil, and thereby caused unfavorable conditions for plant growth.

The occurrence of heat waves is a special case that can cause reductions in total soil water by up to 1/3 in only one month and can affect the growth and development of plants [27].

Because more frequent and higher intensity drought periods are expected, the survival of forests in the lowlands of Serbia requires remarkable scientific and practical attention.

There is no sufficient information available regarding the sensitivity of ecosystems to climate changes over the long-term, and thus it is necessary to conduct monitoring over longer periods [19]. Strategies of forest ecosystem management in terms of climate change should take into consideration the biological diversity of forests and the impact of drought on stands of tolerant tree species [17] because studies have indicated the different adaptability of species depending on the specific climatic change [28 - 30].

3.2. Studies Dealing with Drought-Induced Oxidative Stress in Poplar Stands

Drought-induced oxidative stress has been studied in poplar stands grown in different forms of fluvisol for several years. Drought-induced oxidative stress and antioxidant activities were examined and compared in leaves of *Populus x euramericana* Pannonia that were grown in two different forms of fluvisol (sandy and loamy). According to results that were obtained using the DPPH (2,2-diphenyl-1-picrylhydrazyl) method, poplars that were cultivated in sandy fluvisol showed higher scavenger capacities than those cultivated in loamy fluvisol, with maximum levels being reached in June (RSC=90.171 ± 0.906 %). FRAP (Ferric Reducing Antioxidant Power) test results followed the same trends as those obtained using the DPPH method, and the antioxidant activities of both dramatically decreased after the month of June. The highest rate of lipid peroxidation (31.027 ± 0.449 nmol MDA per mg proteins) was observed in June, while the largest amount of free proline was detected in July (121.782 ± 0.651 nmol per mg proteins), which were both observed in poplars that were grown in sandy fluvisol. All of the results regarding the higher oxidative and drought stresses and the resulting higher antioxidant activities in the poplars that were cultivated in sandy fluvisol may be addressed to the lower humidity levels of the sandy compared to the loamy fluvisol, in addition to other stressful environmental conditions, particularly during the summer [31].

3.3. Studies of the International Beech Provenance Tests

The first provenance trial in the world involving forest trees was established in 1820, and to date, they have been carried out on more than 50 tree species from the temperate climate zone [32]. Several provenance trials encompassing different tree species, including walnut, wild cherry, common oak, Norway spruce, Austrian pine, Norway maple, and beech, have been established in Serbia [33].

According to Gömöry (2010) [34], the main aim of provenance trials is to identify populations that are characterized by excellent growth and adaptability so that they can be used as seed sources for future afforestations. Numerous investigations into beech provenance trials, mostly involving observations of phenology and survival and plant growth, have been carried out to date.

The leaf flushing phenology that were observed in several provenance trials proved the existence of clinal patterns moving from the east to the west; i.e., populations from the Alps and the southeastern part of Europe have tendency towards early flushing in comparison to those from regions that are under the influence of the Atlantic Ocean [34 - 36]. In addition, by observing height growth in provenance trials, it was determined that provenances often showed genetic variations of ecotypic character [37 - 39].

More recent investigations into beech provenance trials were mainly focused on the impact of impending climate changes on the species. In the last decade, numerous studies have been conducted in which the variability between provenances were observed with the goal of selecting for drought resistance and assessing the drought-induced effects on the physiological and morphological characteristics of different provenances [40 - 42]. The international beech provenance trials are the first trials of this kind in Serbia. They were established in the spring of 2007 within the European network of beech provenance trials. Within this network, beech provenance trials have been established on six separate occasions in Europe (1986, ‘87, ‘88, ‘95, ‘98 and ‘07).

The last series of beech provenance trials was established in 2007 within *COST Action E52: "Evaluation of the Genetic Resources of Beech for Sustainable Forestry"*.

On that occasion, a total of 7 provenance trials were established (2 in Serbia and Germany and one in each of the following countries: Croatia, Bosnia and Herzegovina, and Italy), in which 20 provenances were represented by the Balkan region (Serbia, Croatia, Bosnia and Herzegovina) and 12 by Austria, Germany, Hungary, Italy, Switzerland and Romania). The trials were designed in such a way that 15 provenances were common to all sites [43].

Provenance trials in Serbia require special attention for the following reasons: (a) beech is the most important tree species in Serbia because pure beech stands cover approximately 29.3 % of the total forest area and account for 40.5 % of the total wood volume [7]; (b) climate scenarios for the southern European region predicted a reduction in precipitation levels during the vegetation period and an increase in average air temperatures [16], which will result in the endangerment of the beech forest within the natural range [44]; (c) the results from the obtained provenance trials would further elucidate knowledge regarding which provenances should be grown in the future in Serbian habitats to ensure for the sustainable management of forest ecosystems.

After the establishment of provenance trials, large scale studies of anatomical, morphological, physiological and biochemical parameters were initiated to obtain data on the variability of the observed parameters and differences in the adaptability potentials of the studied beech provenances.

The main objective was the development of new bases for the determination of strategies for the gene pool preservation of this important species. This would create the basis for the assessment of the adaptability of beech to changes in environmental conditions and would make optimal choices of adequate provenances for reforestation and afforestation possible.

Initial studies regarding beech provenance trials in Serbia were related to observations of variabilities in the morphological and physiological characteristics of the observed provenances under nursery (field) conditions. Results that were obtained from these studies revealed significant variabilities, both inside the same and among different provenances. Also in some cases the effects of habitat conditions masked differences in provenance-associated genetic characteristics [45 - 47].

Table 2. Two-way ANOVA[1] for net photosynthesis, transpiration and stomatal conductance [46]

	Net photosynthesis ($\mu mol\ m^{-2}\ s^{-1}$)	Transpiration ($mmol\ m^{-2}\ s^{-1}$)	Stomatal conductance ($mol\ m^{-2}\ s^{-1}$)
Provenance (P)	0.77 ^{ns}	115.54***	180.31***
Locality (L)	5.38***	50.31***	15.86***
P x L	10.98***	73.75***	26.20***

ns- not significant, higly significant (***) P<0.001

The measurements of the physiological parameters that were obtained in June 2009 (net photosynthesis, transpiration and stomatal conductance) showed that differences between provenances were caused by differences in the their genetic constitutions in addition to environmental factors (Table 2). The existence of genetic variability within and among analyzed provenances is a necessary prerequisite for the preservation and directed usage of the gene pool of chosen provenances. Statistically significant effects of the interactions between provenances and localities revealed the existence of potentials for adaptation.

Average height values for 12 provenances (5 and 6 years of age) are given in Figs. 15 and 16.

Measurements conducted on 5-year-old nursery seedlings revealed that plants from provenances at the locality of Fruška Gora were taller than plants from Debeli Lug. Mean heights taken for the provenances on Fruška Gora and Debeli Lug were 114.8 and 96.5 cm, respectively (horizontal lines in Figure 15).

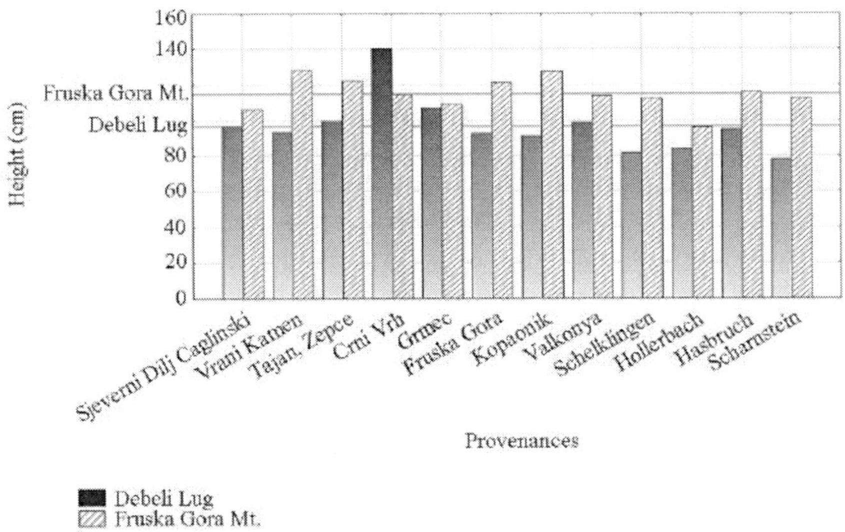

Figure 15. Average heights of five-year-old plants in provenance tests performed in 2009.

[1] ANOVA – Analysis of variance.

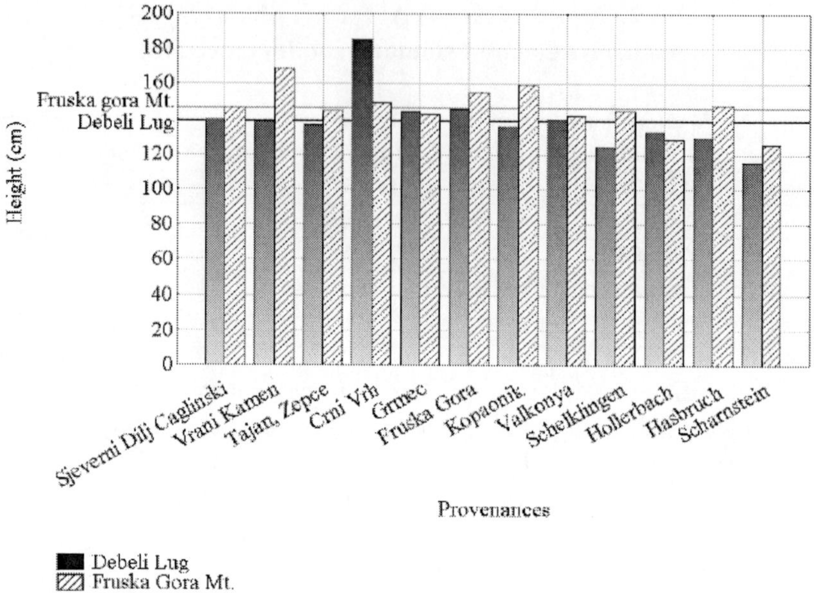

Figure 16. Average heights of five-year-old plants in provenance tests performed in 2010.

The following provenances at Fruška Gora were considered to be the most optimal: Vrani Kamen (Croatia), Kopaonik (Serbia), Fruška Gora (Serbia) and Tajan (Bosnia and Herzegovina), while in Debeli Lug, the most optimal were Crni Vrh, Grmec, Tajan from Bosnia and Herzegovina and Valkonya (Hungary). As stated by Wuehlisch [44], during the first years after the establishment of the field trials, the height growth of the plants did not reflect the full potentials of the provenances; this was also the result of the influence of habitat on the genotypes to a lesser extent. The successes of the rooting abilities and stress caused by planting are the primary influences on height growth at this stage.

The measurements that were obtained a year later when plants were 6 years old are shown in Figure 16. The average heights measured for trees from the provenances Fruška Gora and Debeli Lug were 146.1 and 138.9 cm, respectively (horizontal lines in Figure 16). The height order of the provenances in the tests also changed. The greatest change occurred in the provenances Tajan and Zepce (Bosnia and Herzegovina), which had above-average heights in 2009, while in 2010, the heights were below average for both localities. Studies that were conducted by Kleinschmit and Svolba [48] revealed that the beech provenances changed rank even above 42 years of age. Thus, Muhs [49] considered that the assessment of the adaptabilities of individual provenance could only be reliably made when the test was conducted on tress that were aged 60 years and older.

When observing provenances that had achieved lower sizes during 2010, it was noted that they were geographically distant from the site of the trial establishment. This was particularly evident at Debeli Lug, where the 4 lowest provenances were also the most distant. In Fruška Gora, the lowest height values were determined for the provenances Höllerbach (Germany) and Scharnstein (Austria), while the provenance Schelklingen (Germany) had an average height that was below the average for this locality. Figure 17 shows average provenance heights depending on the latitude and longitude of the locality of origination.

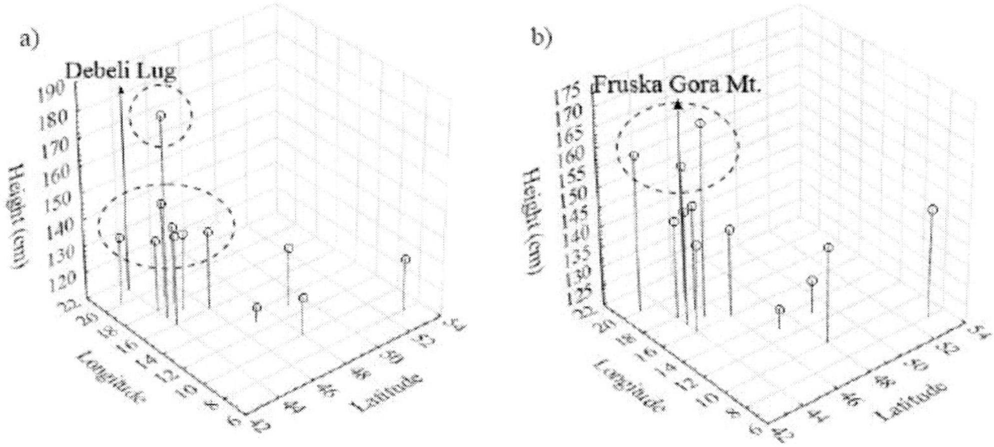

Figure 17. Mean height (cm) for provenances depending on latitude and longitude of origin at the following sites: a) Debeli Lug, b) Fruška Gora.

Dependence was shown separately for the two localities from which the provenances trials were established. This chart shows that the trees from the provenances originating from the localities that were geographically closest to the sites of trail establishment (on Figure 17, they are marked by dashed circular lines) were generally higher compared to those of the more distant provenances. The same data from the walnut provenance test (*Juglans regia* L.) was presented by Hemery et al. [50], who found clear clinal patterns for the declining heights that occurred with increasing distances between localities from which the provenances originated and the localities where the trials were established.

Because studies involving beech provenance trials in Serbia are still at the initial stages, it is too soon to draw any conclusions on the adaptabilities of individual provenances. Previous research results may be used to aid in the understanding of the reactions of plants to changed site conditions at juvenile stages. The precise data that is capable of elucidating the pre-set goals can only be obtained after many years or even decades of research. These data will aid in the determination of the following: (1) the variability of anatomical, physiological, and morphological beech characteristics that would allow for the further improvement of this species in Serbia; (2) the structural-functional relations that are based on detailed anatomical, physiological and morphological studies (relations between anatomical properties and physiological processes); i.e., the adaptability of individual provenances based on the measurements of morphological parameters, and 3) the eco-physiological characteristics of provenances based on the observations of parameters involving the plant water regime, which are significant for the cultivation of certain soil types.

CONCLUSIONS

There are many factors that threaten the survival of forests and lead to reductions in forest areas and the loss of environmental and productive functions. For these reasons, the monitoring of forests is conducted to detect disturbances and to ensure for the appropriate measures for adaptation and the mitigation of negative effects to be implemented over time. A national forest inventory was conducted in Serbia for the first time using new methodologies that gave reliable data regarding the most important characteristics of forest ecosystems. It clearly showed that the area of Serbian forests has increased since the last inventory by 356 863 ha for state and 108 748 ha for private property. It also showed that beech is still the most widespread tree species, which clearly indicates the necessity of focusing on monitoring and protective measures for these types of forests. Due to noticeable changes that were detected in the 1980s, the International Cooperation Program for the Evaluation and Monitoring of the Effects of Air Pollution on Forests (ICP - Forests) was established. It has been implemented in Serbia since 2003 at level I and from 2009 at level II. The results of the monitoring clearly showed the deteriorated conditions of the forests as indicated by their discoloration and defoliation. The levels of these processes were uniform in the coniferous species and varied in the deciduous species depending on the year. Intensive monitoring at level II was carried out at observation plots in the Fruška Gora and Kopaonik, where the evaluations of the conditions of the tree crowns, foliage, soil and soil solution chemical composition, growth rates, ground vegetation, atmospheric deposition, meteorological data, phenology and forest litter components were conducted. Over the course of 2 years, the results revealed significant differences with regards to the observed parameters and the importance of continuing this work. Due to intensive global climate changes and future scenarios, experimental fields were set up in Serbian forests with the most important tree species, on which microclimatic conditions and oxidative stresses to drought were monitored. Measurements of microclimatic parameters (air temperature and relative humidity) that were conducted over the course of 2 years revealed significant differences that were species dependent. The greatest differences in terms of minimum and maximum daily temperatures were found in the poplar stand and the least in the sessile oak stand. Differences between average monthly minimum and maximum air temperatures in August were also the greatest in the poplar (14 °C) and the least in the sessile oak stand (5.4 °C). In other studied stands, the greatest difference between average minimum and maximum air temperatures was also found (10.4 °C for sessile oak and ash, 9.7 °C for Hungarian oak and Turkey oak, and 9.2 °C for the artificial stands of Austrian pine and Douglas fir). Because the monitoring of the poplar stand was conducted over the last 5 years, the relative humidity has also been given in the results. The results revealed that increased temperatures led to decreased relative humidity. Because of the fact that more frequent drought periods can be expected in the future, decreased vitality and lower productivity of poplar stands can also be expected.

A higher antioxidative activity was found for the poplar stand growing in the fluvisol sandy soil. For the international provenance beech trials, the adaptations to site conditions were studied. By measuring the heights of the plants in the juvenile stage, it was determined that the provenances originating from the localities that were geographically closer to the sites of trial establishment were generally higher compared to the more distant provenances, indicating a clinal pattern.

The monitoring of forest ecosystems in Serbia are in the initial stages. The collected data are insufficient for the establishment of a database for forest ecosystem monitoring, and the results cannot be systematized at this time. In the future, the scope and key results of research studies on long-term monitoring have to involve the simulation of ecosystem dynamics and functioning. To identify new requirements for forest monitoring under climate changes in actual forest conditions and their changes over time, measurements have to be obtained from different forest types because of their long life cycles. In the future, the monitoring of forest ecosystems could elucidate the following issues: forest productivity, carbon sequestration, the potential distribution of species and the possible adaptation to extreme events (forest fires, droughts, heat waves, storms, floods, diseases and pest outbreaks). The main goal of the monitoring of forest ecosystems could be the adaptive forest management in response to climate changes.

ACKNOWLEDGEMENTS

This chapter was realized as a part of the project "Studying climate change and its influence on the environment: impacts, adaptation and mitigation" (No. 43007), which is financed by the Ministry of Education and Science of the Republic of Serbia within the framework of integrated and interdisciplinary research over the period 2011 - 2014.

REFERENCES

[1] Global Forest Ressources Assessment, *Key Findings,* 1-11 (2010) [www.fao.org].

[2] S. Orlović et al., *Status and Potential of Forests in the Autonomous Province of Vojvodina*, Novi Sad, RS, 2005; p. 139 (In Serbian).

[3] E. Paoletti, A. Bytnerowicz, C. Andersen, A. Augustaitis, M. Ferretti, N. Grulke, N. et all. Gunthardt-Goerg, S. Madeleine, J. Innes, D. Johnson, D. Karnosky, J. Luangjame, R. Matyssek, S. McNulty, G. Muller-Starck, R. Musselman, K. Percy, *The Scientific World Journal,* 1-8 (2007).

[4] Spatial Plan of Republic of Serbia, 1-75 (In Serbian).

[5] R. Kadović, M. Medarević, *Proceedings of the Forests and Climate Change*, Belgrade, RS, 17-26 (2007).

[6] D. Lindenmayer, G. Likens, *Trend. Ecol. Evol.*, 24(9), 482-486 (2009).

[7] M. Lorenz, G. Becher, V. Mues, E. Ulrich, *Int. J. Environ. Stud.*, 65(3), 299-309 (2008).

[8] T. Amoriello, A. Consantini, In *Ecological Condition of Selected Forest Ecosystem in Italy, Annali – Special Issue. Status and changes 1995 - 2005;* M. Ferretti, B. Petriccione, F. Bussotti, G. Fabbio, Eds.; report of the Task Force on Integrated and Combined (IandC) evaluation of the CONECOFOR programme. Annali C.R.A. - Centro di Ricerca per la Selvicoltura, IT, 2005; pp. 73-83.

[9] IPCC Fourth Assessment Report: Climate Change (2007).

[10] J. Fuhrer, M. Beniston, A. Fischlin, C. Frei, S. Goyette, K. Jasper, C. Pfister, *Clim. Change,* 79, 79-102 (2006).

[11] C. Dorland, R. Tol, J. Palutikof, *Clim. Change*, 43, 513-535 (1999).

[12] S. Kostadinov, In *Global Environmental Change: Challanges to Science and Society in Southeastern Europe*; V. Alexandrov, M.F. Gajdusek, C.G. Knight, A. Yotova, Eds.; Springer Science, 2010; pp. 177-190.

[13] S. Banković, M. Medarević, D. Pantić, N. Petrović, *The National Forest Inventory of the Republic of Serbia*; Belgrade, RS, 2009; p. 237 (In Serbian).

[14] P. Lasch, F. Suckow, F.W. Badeck, *Symposium of Forest in a Changing Environment - results of 20 years ICP Forests Monitoring*, Göttingen, DE, 136-141 (2007).

[15] R. Nevenić, Lj. Rakonjac, S. Orlović, *Monitoring and Impact Assessment of Air Pollution and its Effects in Forest Ecosystems on Territory of the Republic of Serbia – Monitoring of Forest Condition*, Belgrade, RS, 2011; p. 295.

[16] M. Lindner, M. Maroschek, S. Netherer, A. Kremer, A. Barbati, J. Garcia-Gonzalo et all. *Forest Ecol. Manag.*, 259, 698-709 (2010).

[17] G. Hemery, *Short-Term Scientific Mission Report for Working Group 1, COST Action E42*, 2007; p. 73.

[18] P. Ciais, M. Reichstein, N. Viovy, A. Granier, J. Ogee, V. Allard, et all. *Science,* 437, 529-533 (2005).

[19] M. Dobbertin, *Eur. J. Forest Res.*, 124, 319-333 (2005).

[20] P. Spasov, D. Spasov, P. Petrović, *Book of Abstracts of International Conference on Drought Mitigation and Prevention of Land Desertification,* ICID, EWG of ICID, Full paper on CD, Bled, SI (2002).

[21] R. Jevtić, M. Telečki, B. Lalić, D.T. Mihailović, M., Malešević, In *Advances in Environmental Modeling and Measurements;* D.T. Mihailovic, B. Lalić, Eds.; Nova Science Publishers Inc: New York, US, 2010; pp. 209-222.

[22] T.B. McKee, N.J. Doesken, J. Kleist, *Proceedings of the Eighth Conference on Applied Climatology*, American Meteorological Society, Boston , MA, 179-184 (1993).

[23] J. Chen, S. Saunders, T. Crow, R. Naiman, K. Brosofske, G. Mroz, B. Brookshire, J. Franklin, *BioScience,* 49(4), 288-297 (1999).

[24] S. Wang, H. Ruan, Y. Han, *Ecol. Res.,* 25, 1113-1120 (2010).

[25] T. Vanwalleghem, R. Meentemeyer, *Ecosystems* 12, 1158-1172 (2009).

[26] L. Kolendowicz, E. Bendroz, *Quaestiones Geographicae*, 28, 31-39, (2008).

[27] Galic Z., Orlovic S., Klasnja B. *Proceedings International Scientific Conference Forest Ecosystems and Climate Changes,* Belgrade, RS, 213-219 (2010).

[28] H. Davi, E. Dufrene, C. Francois, G. Le Maire, D. Loustau, A. Bosc, et all. *Agric. Forest Meteorol.,* 141, 35-56 (2006).

[29] M. Macias, L. Andreu, O. Bosch, J.J. Camarero, E. Gutierrez, *Clim. Change,* 79, 289-313 (2006).

[30] S. Gartner, A. Reif, F. Xystrakis, U. Sayer, N. Bendagha, A. Matzarakis, *J. Veg. Sci.,* 19, 757-768 (2008).

[31] M. Kebert, S. Orlović, D. Štajner, B. Popović, Z. Galić, *Book of Abstracts of 5th International Poplar Symposium (IPS V),* Orvieto, IT, p. 128 (2010).

[32] J.W. Wright, *Introduction to Forest Genetics*. Academic Press: London, GB, 1976; p. 463.

[33] V. Isajev, V. Ivetić, A. Lučić, Lj. Rakonjac, *Genetics,* 41(3), 309-327 (2009).

[34] D. Gömöry, *Proceedings of the Workshop and MC Meeting of the COST Action E52 Evaluation of Beech Genetic Resources for Sustainable Forestry,* Thessaloniki, GR, 90-97 (2010).

[35] D. Gömöry, L. Paule, J. Vysny, *Bot. J. Linn. Soc.,* 154, 165-174 (2007).

[36] G. Von Wuehlisch, H. Krusche, J. Muhs, *Silvae Genet.,* 44(5-6), 343-346 (1995).

[37] M. Ivanković, S. Bogdan, G. Božić, *Šumarski list,* 11-12, 529-541 (2008) (In Croatian).

[38] W. Barzdajn, *Leoene Prace Badawcze,* 70(2),101-111 (2009).

[39] A. Jazbec, K. Šegotić, M. Ivanković, H. Marjanović, S. Perić, *Forestry,* 80(2), 151-162 (2007).
[40] L. Rose, C. Leuchner, B. Köckemann, H. Buschmann, *Eur. J. Forest Res.*, 128(4), 335-343 (2009).
[41] A.D. Peuke, C. Schraml, W. Hartung, H. Rennenberg, *New Phytol.,* 154, 373-387 (2002).
[42] A.D. Peuke, H. Rennenberg, *Tree Physiol.,* 31, 196-207 (2011).
[43] G. von Wuehlisch, D. Ballian, S. Bogdan, M. Forstreuter, R. Giannini, B. Götz, M. Ivankovic, S. Orlovic, A. Pilipovic, M. Sijacic Nikolic, *Book of abstracts*, Burgos, ES, 2010, p. 21.
[44] G. Von Wuehlisch, *Proceedings from the 7th International Beech Symposium IUFRO Research Group 1.10.00, Improvement and Silviculture of Beech.* Tehran, IR, 135-144, (2004).
[45] M. Šijačić-Nikolić, J. Milovanović, R. Knežević, *Glasnik Šumarskog fakulteta u Banjoj Luci*, 6, 61-71 (2006) (In Serbian).
[46] M. Šijačić-Nikolić, V. Ivetić, R. Knežević, J. Milovanović, *Acta herbologica,* 16(1), 15-27 (2007) (In Serbian).
[47] S. Stojnić, S. Orlović, A. Pilipović, M. Kebert, M. Šijačić-Nikolić, D. Vilotić, *Acta Silvatica and Lignaria Hungarica*, 6, 135-142 (2010).
[48] J. Kleinschmit, J. Svolba, In: *Genetics and silviculture of beech,* Madsen, S. (Ed.): *Procidings of the 5th IUFRO Beech Symposium 1994,* Hørsholm, DK, 15-34 (1995).
[49] H.J. Muhs, L. Paule, L. Ionita, G. von Wuehlisch, *Book of abstracts*, Burgos, ES, 2010, p. 19.
[50] G.E. Hemery, P.S. Savill, A. Thakur, *Forestry,* 78(2),121-133 (2005).

In: Essays on Fundamental and Applied Environmental Topics ISBN 978-1-61942-522-4
Editor: Dragutin Mihailovic

Chapter 13

THE IMPACT OF CLIMATE CHANGE ON DISEASES AND PESTS OF SMALL GRAINS AND SUNFLOWERS IN THE VOJVODINA REGION (SERBIA)

R. Jevtić[1,*], S. Maširević[2] and D. Vajgand[3]
[1]Small Grains Department, Institute of Field and Vegetable Crops, Novi Sad, Serbia
[2]Department for Phytomedicine and Environmental Protection, Faculty of Agriculture, University of Novi Sad, Novi Sad, Serbia
[3]Agroprotekt, Sombor, Serbia

ABSTRACT

The risk of disease and pest damage to agricultural crops has increased significantly as a result of climatic changes. The climatic change has resulted in the dominance of pathogens that require higher temperatures for their development or are better able to adapt to drought conditions. This is the reason why the fungal diseases of the genus *Fusarium* and *Septoria* spp. that affect small grains have played a dominant role in causing significant damage and why the causal agents of sunflower spots develop very well and rapidly at high temperatures. The diseased plants may suffer from complete defoliation and premature ripening. Fungi of the genus *Alternaria* are favored by high temperatures accompanied by a short period of wet weather. The parasite develops best and most rapidly at temperatures between 20 and 30 °C.

During the 2009 - 2010 growing season, there was a significant *Fusarium* infestation on wheat, barley and triticale in the Vojvodina region (Serbia). In the locality of Rimski Šančevi near Novi Sad, the percentage of *Fusarium*-damaged kernels in some commercial fields (the spring-planted variety Natasha) reached up to 33.3 % per m^2. In the same growing season and locality, the levels of deoxynivalenol (DON) in cereal samples were very high.

The effectiveness of the *Pm* resistance genes is correlated with the air temperature conditions. During the 1990 - 2009 period, numerous genotypes had the V7, V6 and V8

[*] Corresponding author: E-mail address: radivoje.jevtic@ifvcns.ns.ac.rs; Phone: +381 21 489 8207. Fax: +381 21 489 8222.

virulence alleles. The most efficient among the sexual populations of the parasite was the gene combination Pm5+6 from the variety Coker 983. Compared with the previous period, the rust (*Puccinia* spp.) pathogens have recently formed fewer generations and have transitioned quickly from the uredo to the teleuto stage, which is a direct result of the climate change impact.

Orobanche cumana is a serious problem in sunflowers. Since the 1990s, broomrape has caused significant damage to the susceptible hybrids in Serbia. In the north of the Bačka region alone, 15 000 ha have been endangered. The yield losses depend on the intensity of the attack and can range from 5 to 100 %.

Besides having a great impact on pathogens, the climate change also influences pests. Moths are insects and thus do not have a constant body temperature. The speed of their development is directly correlated to the air temperature. Increasing temperature leads to changes in the cycle of development and the diversity of the individual species. A species that we can safely say has become dominant due to the increase in air temperature is the cotton bollworm (*Helicoverpa armigera*). In years with mass outbreaks of the cotton bollworm in the north of Vojvodina, 60 to 100 % of maize ears and sunflower heads and 20 to 94 % of string bean pods were reportedly damaged.

1. INTRODUCTION

Climate change not only influences cultivated plants, but it can also cause changes in the development of parasitic organisms. Climate change also influences the resistance of the plant hosts and leads to the modification of host-pathogen interactions. The most frequent effects are changes to the geographic distribution of the pest organisms, namely the appearance of thermophilic species in moderate climate zones, which could endanger efficient control of the pests. In the absence of barriers, it may be possible for a species or for communities to migrate in response to the changing conditions. Vegetation zones may move towards higher latitudes or higher altitudes following shifts in mean air temperatures. The movements will be more pronounced at higher latitudes, where the temperatures further from the equator are expected to increase. In the mid-latitude regions (45 to 60°), the present temperature zones could shift by 150 - 550 km [1]. There will be a northward expansion of warmer-season seed crops (e.g., soybean and sunflower) [2]. On the other hand, the yield of spring crops, most importantly sunflowers, will prospectively decrease in Central and South Europe [3]. The pathogens that are normally less aggressive in natural plant communities could devastate the crop monocultures that grow in close proximity [4]. The occurrence of new diseases, pests and weeds is a direct consequence of the climate changes in Austria, Italy, Greece, Poland, Russia and Serbia [5]. In the Vojvodina region (Serbia), the climatic changes have increased the incidence of small-grain pathogens and influenced their prevalence, adaptability and development cycle. The pathogen control principles used so far have shown weaknesses in certain areas, indicating a need for new research methods under the new set of circumstances.

The keystone factor in determining the rate of physiological development in sunflowers is the accumulated air temperature (mean daily air temperature over 6 ^{o}C). The projections from the UKCIP02-United Kingdom Climate Impact Programme [6] data indicate that the area suitable for sunflower production will increase to approximately 79 % of the land in England by 2050. There will be a substantial increase in the number of varieties that will be suitable for use in the United Kingdom. Rainfall levels will decrease, but these conditions will favor the drought-tolerant sunflower [6]. The climate change implicates the shifts in sunflower production in northern countries as well as in southern ones, where this plant is commonly found in the fields. These changes will cause some diseases to become more important in sunflower production than they were previously.

The studies conducted over a number of years in the Vojvodina region have shown that a large number of biotic and abiotic factors may lead to significant yield losses. It is therefore necessary to understand their prevalence and their direct correlation to the climatic changes [7]. This Chapter includes an analysis of the severity of the diseases caused by the parasitic fungi and pests in the Vojvodina region in Serbia.

2. Climate Change: Diseases and Pests in the Vojvodina Region (Serbia)

The global warming is now evident from observations of the increases in global mean air and ocean temperatures, widespread melting of snow and ice and rising global average sea level [6]. The global mean temperature has increased from 1906 to 2005 (0.74 °C ± 0.19 °C), and over the past 25 years, it has been rising at a rate of approximately 0.2 °C every ten years (± 0.05 °C) [6]. The changes can be expected to occur on different continents, expanding deserts and raising sea levels.

The climatic change is most often explained by a rise in air temperature. Regardless of whether they are high or low, excessive air temperatures negatively affect plants and plant processes in the form of stress. In some years in the Vojvodina region, air temperatures had a significant effect on the overall small-grain production. Those years are often characterized as being unfavorable for small-grain production, which may mean that the climatic change had the most prominent effect during those years [7].

Between 1961 and 1990, the mean annual air temperature in Serbia in areas that are up to 300 m in altitude was 10.9 °C. According to trend data values for a period of 35 years (1966 - 2000), the annual air temperature in Serbia increased in intensity by 1 °C over a 100-year period. However, when looking at periods shorter than 100 years, the shorter periods show a higher increase in temperature. This fact indicates that the warming has intensified in recent decades on an annual basis [8]. Global warming has been noticed in the Vojvodina region as well. Presently, the annual mean temperature in Vojvodina is 11 °C [9]. Annual temperatures in Vojvodina are expected to rise by approximately 1.3 °C by 2040 or thereabouts and by 2.4 °C by 2080. [10]. There are no significant differences among the climate models in the projected relative change in the annual temperature [10].

The highest mean annual air temperature is found in the southeastern part of Vojvodina. The gradient of mean annual temperature increases from the northwest (Palić, 10.7 °C) to the southeast (Vršac, 11.7 °C) with the exception of Mount Fruška Gora (10 °C). The mean annual air temperature is 10.8 °C in Srem, 10.9 °C in Bačka and 11.2 °C in Banat [11]. The trends in the annual air temperature (°C) measured at the 12 meteorological stations in the Vojvodina region during the period from 1949 - 1999 are shown in Figure 1. [12].

On the average, the annual amount of precipitation in Serbia increases with altitude. In lower-lying regions such as Vojvodina, the annual precipitation levels range from 540 to 820 mm [13].

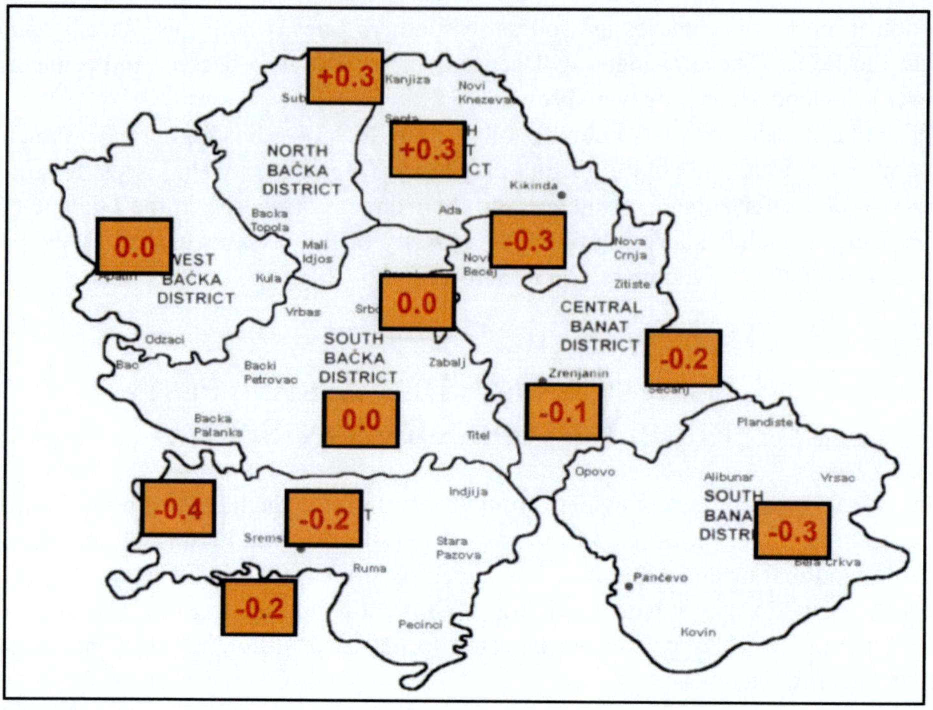

Figure 1. Trends in the annual air temperature (°C) in the Vojvodina region during the period from 1949 - 1999 (Source: [12]).

In the Vojvodina region, the mean annual precipitation is 602 mm [9]. The trends in the annual precipitation (mm) in the Vojvodina region for the period from 1949 - 1999 are shown in Figure 2 [12]. The annual sum of the precipitation in Serbia over the past 52 years has tended to decline [8]. According to Djurdjevic and Rajkovic [14], in the first 30 years of the 21st century, the expected change in the amount of precipitation in Serbia is positive (20 to 30 mm $year^{-1}$), while for the last 30 years, the whole country has been far drier (with a decrease in precipitation of up to 30 mm $year^{-1}$) with the exception of the Vojvodina region [10].

In the most recent 10-year period, increasing air temperatures during the winter were recorded both in Serbia and worldwide. Warmer winters favor the overwintering of many types of disease-causing agents. Based on results obtained from questionnaires, Mihailović et al. [15] concluded that over a 15-year period, the following diseases occurred frequently in the Vojvodina region as a result of the impact of climate change: powdery mildew of cereals (*Erysiphe graminis*), Fusarium head blight, Cercospora leaf spot (*Cercospora beticola*), sunflower blight (*Plasmopara halsltedii*) and potato and tomato Alternaria leaf spot (*Alternaria solani*). These diseases occurred because of the dry and warm springs and dry summers with occasional showers.

Insects do not have a constant body temperature; instead, their body temperature is directly correlated to the air temperature. The data in Table 1 show that the egg stage within the same insect species, for example, may last anywhere from 2 to 14 days depending on the ambient temperature. Therefore, temperature fluctuations within a given area greatly impact insect development. Moth flight data in our study were obtained by the daily collection of moths using an RO Agrobečej-type light trap that uses a 250 W mercury bulb as the source of light.

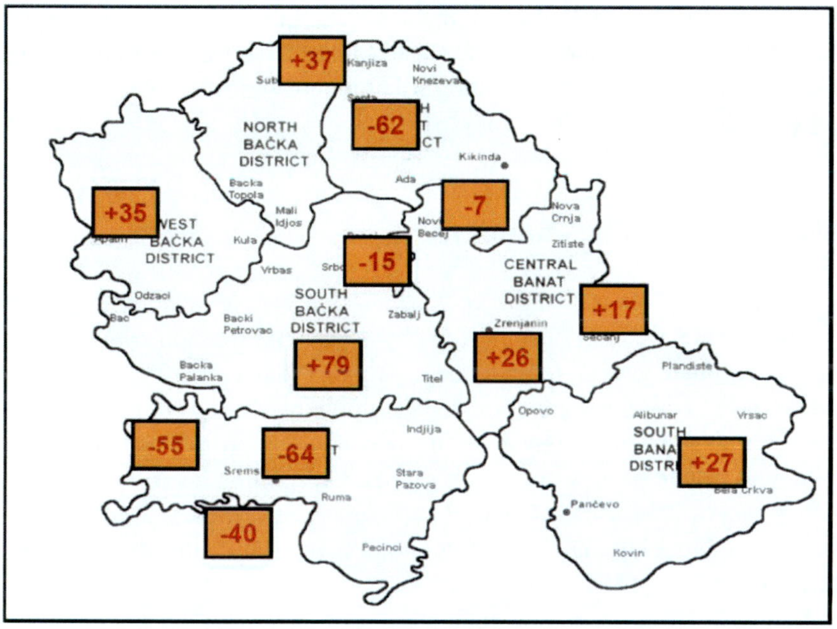

Figure 2. Trends in the annual precipitation (mm) in the Vojvodina region during the period from 1949 - 1999 (Source: [12]).

Table 1. Duration of *Agrotis segetum* development at various temperatures

Stage of development	Duration of development in days at various temperatures							
	15 °C	18 °C	20 °C	24 °C	27 °C	29 °C	34 °C	36 °C
Egg	14	9	8	5	4	2	3	3
Caterpillar	70	42	40	28	23	21	23	26
Pupa	41	24	22	15	10	10	12	11

The changes in the occurrence of several moth species that have been observed in the area of the town of Sombor in the Vojvodina region are indicative of how the climate change has affected the behavior of these species in terms of the duration of their development stages, their abundance, and the increasing number of generations that they now produce.

3. Small Grain Diseases

Small-grain pathogens can be divided into three groups based on the symptoms and changes they cause in the plant. These are: obligate parasites (*Puccinia* sp. and *Blumeria* sp.), organisms causing spot-type diseases (*Septoria* sp., *Pyrenophora* sp. and *Rhynchosporium*), and agents causing grain destruction (*Fusarium* sp.). According to the mechanisms established between the plant and the pathogen in the course of the infection, the parasites can be classified as either obligate or facultative [7].

In the period from 2009 - 2010, there was a significant decrease in the grain yields of wheat, barley and triticale, regardless of the time of sowing. Large amounts of precipitation and the alternation of warm and cold periods during April, May, and June significantly favored the pathogens of small grains, most notably the causal organisms of powdery mildews, leaf rust, leaf spots, and Fusarium head blight. In low-lying fields, where the water is retained for a long period of time, the lodging of plants was found due to the appearance of the pathogens that cause root and foot rot [16].

3.1. Obligate or Biotrophic Pathogens

Powdery Mildew (*Blumeria /Erysiphe/ graminis*). Powdery mildew is a disease that occurs in warmer weather conditions preceded by occasional wet and cool spells. The conidia of this parasite germinate at high temperatures ranging from 1 to 30 °C without the presence of water, while the infection occurs at air temperatures between 5 and 30 °C. Warm springs and summers promote the occurrence of the parasite. According to Duveiller et al. [17], powdery mildew is important in China, where at least 12 million hectares can be affected, and also in South America [18]. Because natural epidemics of powdery mildew do not occur in Mexico, most of the breeding materials of grain are being sent from Mexico to the regions of China and South America [17].

In small grains, changes in the population and virulence of *Blumeria* (= *Erysiphe*) *graminis tritici* have been monitored by tracking the sexual and asexual parts of the population. In Vojvodina, the causal agent of powdery mildew in wheat develops a total of 15 - 18 generations, including the transitional period between the harvest and the new planting when the so-called green bridge is formed [19]. After 30 asexual cycles of *Erysiphe graminis*, some cultivars may adapt to the pathogen [20, 21].

The durability of resistance may be threatened, however, if the number of infection cycles within a growing season increases because of one or more of the following factors: increased fecundity, more pathogen generations per season, or a more suitable microclimate for disease development. This may lead to a more rapid evolution of aggressive pathogen races [22]. During the period from 1990 - 2009, according to Jevtic et al. [23], numerous genotypes had the V7, V6 and V8 virulence alleles. The most efficient among the sexual populations of the parasite was the gene combination Pm5+6 from the variety Coker 983 (Figure 3). A large number of the pathotypes produced by sexual reproduction either cannot exist in nature or their frequencies in the population are very low.

The effectiveness of the *Pm* resistance genes is correlated to the air temperature conditions. The wheat plant response to the pathogen is reflected in the number and the size of the pustules forming on the leaves, resulting in the existence of different infection types.

Race-specific resistance often provides complete control, but most race-specific genes begin to lose effectiveness within 2 - 4 years [24]. Unlike race-specific resistance, non-race resistance is incomplete but durable. It is characterized by a reduction in the rate of the disease development and is also called partial resistance. Partial resistance to powdery mildew may involve one or several of the following components [25]: (i) increased incubation or latent period (from inoculation to the appearance of symptoms), (ii) reduced infection frequency (the number of colonies per unit of leaf area), (iii) reduced infectious period (the length of time during which the colony produces viable spores), (iv) reduced infection lesion size (colony size), and (v) reduced spore production (the number of spores produced per unit of leaf area over a particular length of time).

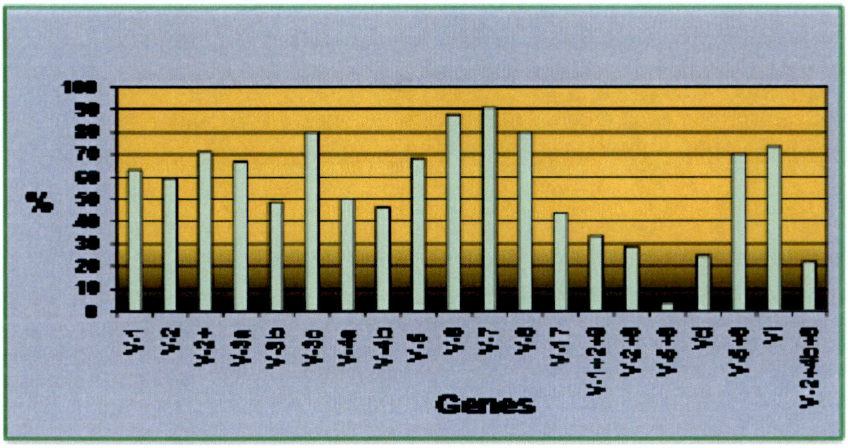

Figure 3. The frequency of the virulence genes (%) of *Blumeria graminis tritici* from 1990 to 2009.

The identification of the wheat varieties with partial resistance is of great importance for wheat breeding programs because this type of resistance has provided durable control of powdery mildew in cereals in several countries [26].

Multi-year data have shown that *B. graminis tritici* reaches a second peak of infection during the last few days of May in the growing conditions of Vojvodina. After that, the fungus begins to develop cleistothecia as a result of its sexual cycle, and the risk of any major damage to the crops practically vanishes. There is a direct link between the occurrence and the severity of leaf rust and the causal agent of powdery mildew. The more severe the outbreak of powdery mildew, the lower the incidence of leaf rust will be and vice versa.

Rusts of Small Grains (*Puccinia* spp.). Rusts are caused by *Puccinia triticina* in wheat, *Puccinia hordei* in barley, and *Puccinia coronata* in oats. Wheat rust fungi differ in their requirements from 2 - 15 ^{o}C for stripe rust, 10 - 30 ^{o}C for leaf rust, and 15 - 35 ^{o}C for stem rust [27, 28].

In the Vojvodina region, the occurrence of the rust-causing pathogens in small grains has stagnated in recent years compared with the previous period. This is attributed to the earlier occurrence of spot-type diseases, resulting in their earlier expansion across the leaf area, which accelerates aging and reduces leaf area duration. Compared with the previous period, these pathogens have recently formed fewer generations and have transitioned quickly from the uredo to the teleuto stage, which is a direct result of the impact of the climate change.

The varying effectiveness of the resistance genes based on the changes in temperature (temperature sensitive genes) is pronounced in the case of the causal agents of cereal rusts. The resistance genes in wheat against the rust fungi *P. recondita* and *P. striiformis* have benefited from high temperature, according to Legreve and Duveiller [29]. According to Milus et al. [30], *P. striiformis* has demonstrated that wheat rust fungi can adapt to and benefit from higher air temperatures to cause severe disease in previously unfavorable environmental conditions.

Examples include the *Pg3* and *Pg4* genes for resistance to the causal agent of oat stem rust [22], the *Lr26* gene linked to resistance against the causal organism of wheat leaf rust (*P. triticina*), and the generally increased occurrence of the causal agent of wheat yellow rust, *P. striiformis* [31].

Oat crops suffer significant damage from crown rust (*P. coronata*) and stem rust (*P. graminis avenae*). Infections by these pathogens appear later in oat crops than in wheat and barley and are therefore more severe in the spring genotypes of the crop. However, the effectiveness of the resistance genes (Pm and Lr, Pc ones) in a population is dependant on the temperature fluctuations. The severity of the attacks by the pathogens is reflected in the different number and the size of the pustules forming on the leaves, based on which different infection types are identified. By making a rapid transition from the stage of urediniospores to that of teliospores, leaf rust causes less damage than expected.

Jevtić et al. [7] have introduced the concept of the Uredo-Teleuto Spore Coefficient (UTC). The uredo-teleuto spore coefficient (UTC) is calculated based on the ratio of the total leaf coverage by *P. coronata* to the coverage by the teleutospores. The UTC shows the rapidity of transition from the uredo to the teleuto stage under the influence of temperature on the reproductive organs of the pathogen [7].

Disease resistance can be differentially affected by the air temperature. For example, cultivars of wheat on the United Kingdom Recommended List [32] were routinely tested against isolates of brown rust (leaf rust) (*P. recondita*) at 10 °C and 25 °C for 20 years [33, 34]. Generally, about half showed a differential resistance expression, either effective at 10 °C and not at 25 °C or vice versa [35]. Browder and Eversmeyer [36] also reported similar differential temperature responses in the same host-pathogen system, and both groups also demonstrated that the effect was not necessarily attributable simply to the resistance gene expression, as isolates showed differential responses, too. Similar temperature sensitivities have been reported previously for the same rust [37], *P. striiformis* [38], and *P. tritici* in oats [39].

In 1993 - 1994 in Serbia and Vojvodina, an epidemic of leaf rust occurred in wheat. According to Jevtic et al. [40], the main reasons were 59 days with rainfall and favorable temperatures (15 - 20 °C), susceptible genotypes, and volunteer wheat plants that enabled the maintenance and reproduction of the pathogen. In 2007 in the Vojvodina region, there was a severe outbreak of yellow rust that was attributed to the inoculum arriving by air current from Romania to Vojvodina via the Carpathians [40].

Based on the empirical climate-disease models established with a limited set of observational data, Jahn et al. [41] predicted that an increase in temperature combined with less precipitation could increase leaf rust in wheat and barley (*P. triticina*), decrease infestation levels caused by foot rot and leaf blotch diseases, and cause no change in powdery mildew on small grains.

3.2. Necrotrophic Pathogens

Unlike rust, necrotrophic pathogens have always been a major constraint to wheat production worldwide. Some of the most economically important pathogens in Australia are *Pyrenophora tritici repentis* and *Phaeospheria nodorum* [42]. In the Vojvodina region, the most economically important necrotrophic pathogens are fungi of the genus *Septoria* and *Pyrenophora tritici-repentis*.

Leaf and glume Blotches (*Septoria* spp.). The climatic changes have resulted in the dominance of the pathogens that require higher temperatures for their development or are better adapted to drought conditions. This is the reason why fungi of the genus *Septoria* have assumed the dominant role, causing significant damage.

The causal agent of leaf and glume blotches is currently considered to be one of the pathogens with the greatest impact on global wheat production and is the main wheat pathogen in most of Europe, North Africa, South America, and several other parts of the world. The occurrence of these diseases is enabled by warm and wet conditions. The reproductive organs of the fungi, pycnidiospores, germinate at 2 - 37 °C. Major outbreaks of leaf and glume blotches in wheat occur during warm and wet springs and summers, causing significant yield losses. The minimum temperature for germination is 2 - 3 °C, while the optimum and maximum temperatures are 20 - 25 °C and 33 - 37 °C, respectively. Infection in the field will stop if the temperature drops below 7 °C for two consecutive days. The pycnospores are released from the pycnides when the leaf is moistened for 30 minutes or more. The spores form in a thick, sticky substance with high sugar and protein concentrations. These reserve substances enable the pycnospores to maintain moistness during dry spells.

During the wheat growing season, a number of secondary infections occur (seven to nine generations of the pathogen). The number of generations can be easily calculated based on the incubation period length by month. The longest incubations occur in the autumn: 36, 28 and 25 days. In the spring, the length of the incubation period gradually decreases with increasing temperature until it reaches approximately 11 days in June.

In 2006 at Rimski Šančevi near Novi Sad, we studied 3800 wheat genotypes and found that 2064 (or 55.6 %) were severely infected at a level of 30 % or more [43]. This represents at least one third of the leaf area being affected by the spots caused by the parasite, which has a direct influence on yields. The severity of the *S. tritici* attack was greater in winter than in spring genotypes, leading to the conclusion that the primary infection of the winter genotypes occurred in the autumn.

In 2009 in several locations in the Vojvodina region, there were reports of deterioration of the lower wheat leaves due to early infections by the fungus *S. tritici,* resulting in a multitude of pycnides with pycnospores. This means that the primary infections occurred in the autumn of 2008, when the air temperature conditions were optimal. The largest number of pycnides formed on the leaves closest to the ground (Figure 4). At phenophase 29 - 30, a correlation was found between the number of pycnides formed and the infection severity in the first seven leaves (Figure 5). The correlation was complete in the variety Simonida (r = 0.93) and very high in the variety Pobeda (r = 0.88). *S. tritici* infection severity is most commonly estimated based on the ratio of the necrotic leaf area to the green leaf area [44].

Tan Spot of Wheat (*Pyrenophora tritici-repentis*). This pathogen was described in Serbia in 1997 [45], when it occurred sporadically in a small number of locations in the province of Vojvodina. In recent years, the occurrence of this spot has become more severe, and has been attributed directly to the influence of climatic changes. The fungus needs high temperatures for growth. Depending on the type of toxin it exudes, *P. tritici-repentis* can produce three kinds of symptoms, ranging from small spots to chlorosis to complete necrosis of the leaves. More severe infections may result in a complete devastation of the leaf mass, with the necrosis spreading from the tip of the lamina down to the base. After analyzing samples from several sites, it has been determined that the pathogen caused significant yield losses in the susceptible varieties of wheat in 2008, with the hard (durum) cultivars being the most susceptible. Nikolaev [46] presented data on the occurrence of the new disease *pyrenosphoriosis* in northwest Russia.

Figure 4. *Septoria tritici* showing a number of pycnides formed on the lower leaves.

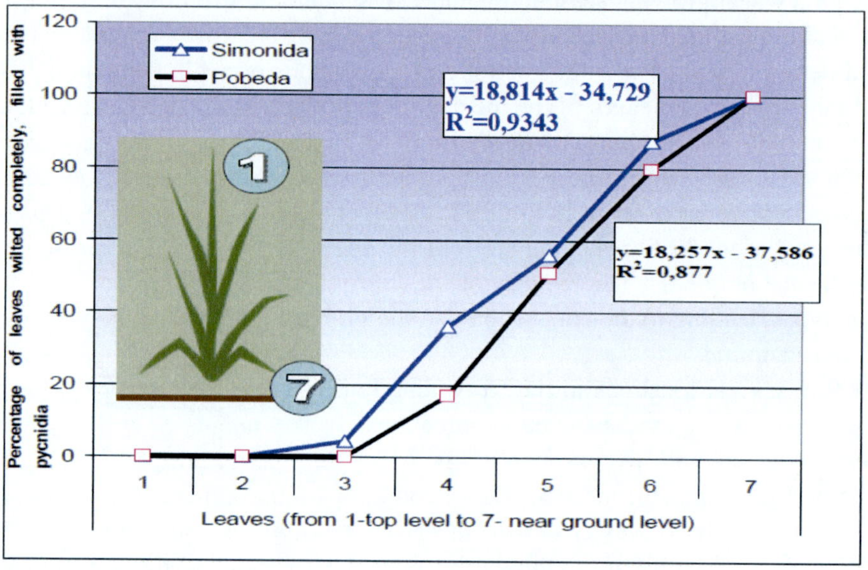

Figure 5. The correlation between the number of pycnides formed and the infection severity in the first seven leaves for the varieties Simonida (blue line) and Pobeda (black line).

3.3. Grain Pathogens

Fusarium Head Blight (*Fusarium* spp.), *F. graminearum,* perfect stage *Gibberella zeae; F. moniliforme,* Perfect Stage *G. fujikuroi*, etc. The polyphagous species of the genus *Fusarium* are ubiquitous and attack small grains and maize as well as many other cultivated species and plants of the spontaneous flora. These parasitic fungi are favored by warm and wet weather and grow successfully at temperatures of 3 ~ 30 °C. They are typical parasites of weakness, as they successfully infect plants weakened by prolonged periods of drought and warm weather. Severe outbreaks of Fusarium disease and major damages occur during warm and dry summers that were preceded by a wet spring.

All the above wheat diseases have been occurring in the Vojvodina region with significantly increased severity in recent years, which was not previously the case. This can be partially attributed to the changes in climatic factors in the province, most notably the increased temperatures in spring and summer.

Head blight, caused by the fungi of the genus *Fusarium*, was found in large numbers in the varieties having awns, regardless of whether they were winter, facultative, or spring varieties. The percentage of diseased spikes was directly correlated to the planting date and preceding crop as well as to the coincidence of the ascospore-release phase in the parasite and the flowering stage in the plant. Because of this, the control of this parasite did not produce the expected results in the commercial crops, as head blight appeared in some localities despite the fact that the crops were treated against the disease.

In the 2009 - 2010 growing season in the locality of Rimski Šančevi near Novi Sad, the percentage of *Fusarium*-damaged kernels in some commercial fields (the spring-sown variety Natasha) reached up to 33.3 % per m^2. However, in a collection of genotypes originating from different countries of the world, out of 461 genotypes, only seven, or 1.5 %, had more than 5 % of the *Fusarium*-damaged kernels. The largest number of genotypes, 381 or 82.6 %, had 0 - 1 % of the ears infected per m^2 [47].

With the fungi of the genus *Fusarium*, Telečki and Jevtić [48] reported finding a very strong negative correlation between 1000-grain weight and the percentage of grains infected by *F. graminearum* in the hard wheat cultivar Dušan (r = - 88.9) as well as a strong correlation of the same kind in the cultivar Durumko (r = - 74.0). In 70 % of cases, there is a greater risk of the occurrence of Fusarium diseases in fields that had maize or a monoculture as the preceding crop. This is explained by the fungus entering the perfect stage on harvest residues and releasing ascospores at the peak of the wheat flowering. However, if the crop phenophase fails to coincide with favorable conditions for the development of the pathogen (warm and wet weather), individual varieties often manage to avoid the attack (escape-type resistance). This is why the disease prognostication system is of great importance from the economic and environmental points of view.

Changes in crop rotations in response to the climate change may influence the future importance of specific pathogens. For example, if the warming in the northern latitudes enables forage maize to be grown in the rotation, the maize will leave residues in which pathogens such as Fusarium Head Blight (FHB) could build up high levels of inoculum for subsequent wheat and barley crops [49].

Deoxynivalenol (DON). Cereals crops are infected by fungi from the genus *Fusarium* that can contaminate the edible parts with mycotoxins, the toxic secondary metabolites of fungi [50]. These can significantly affect human and animal health [50].

The Commission of the European Communities [51] has established the following tolerance values for DON in cereals and cereal-based products: (i) unprocessed cereals other than durum wheat, oats and maize - 1250 $\mu g\ kg^{-1}$, (ii) unprocessed durum wheat and oats - 1750 $\mu g\ kg^{-1}$, (iii) unprocessed maize - 1750 $\mu g\ kg^{-1}$, (iv) cereal flour, including maize flour, maize grits and maize meal - 750 $\mu g\ kg^{-1}$, (v) bread, pastries, biscuits, cereal snacks and breakfast cereals - 500 $\mu g\ kg^{-1}$, (vi) pasta, dry - 750 $\mu g\ kg^{-1}$ and (vii) processed cereal-based food for infants and young children and baby food - 200 $\mu g\ kg^{-1}$. The maximum permitted level of DON has been set in Serbia only for feed and complete feeding stuffs for hogs, with the permitted level being 0.5 $mg\ kg^{-1}$ [52].

According to Jajić et al. [52], during 2009 - 2010 there was a significant Fusarium infestation on wheat, barley and triticale in the Vojvodina region. The levels of deoxynivalenol (DON) in cereal samples were measured after the 2009 - 2010 harvest season.

Table 2. The occurrence of deoxynivalenol in small-grain cereal samples from the 2010 harvest season

Cereal	No. of samples	No. of positive samples (%)	Concentration in samples	
			Average (mg kg^{-1} ± SD)	Range (mg kg^{-1})
Wheat	17	14 (82.4)	0.435 ± 0.43	0.068 - 1.572
Barley	4	-	-	-
Triticale	1	1 (100)	1.952	1.952
Total	22	15 (68.2)	0.537 ± 0.57	0.068 - 1.952

The presence of DON in 22 samples of small-grain cereals is shown in Table 2. Out of the 17 wheat samples analyzed, as many as 14 (82.4 %) tested positive for the presence of DON. The concentration range of DON was 0.068 - 1.572 mg kg^{-1}, with an average content of 0.435 mg kg^{-1}. None of the four barley samples analyzed was positive for the presence of DON. The highest concentration of DON, 1.952 mg kg^{-1}, was found in a sample of triticale. The overall presence of DON in all the analyzed samples was as high as 68.2 %, with the average DON content of 0.537 mg kg^{-1}.

A number of productive models have been developed worldwide for *Fusarium* head blight and the risk of DON contamination in small-grain cereals [50]. The DONCAST model is based on the weather conditions just prior to and during the wheat head emergence [50]. Rossi et al. [50, 53] have developed a risk index (TOX-risk). The risk is calculated based on the following equation:

$$\text{TOX-risk} = SPO \times DIS \times INF \times GS \times INV$$

where SPO is the sporulation rate, DIS is the dispersal rate, INF is the infection rate, GS is the host growing stage and INV is the invasion rate [50].

Van der Fels-Klerx et al. [54], demonstrated the possibility of using the GEMIS (Geographical Emerging Mycotoxin Identification System) model as an example, on a European scale, of the effects of changing temperature (+ 2 °C) and rainfall (+ 3 mm) on heading date and cereal productivity areas. However, this does not consider the interaction between the factors on specific mycotoxins or the impact that CO_2 × temperature × drought impacts might have [50].

Bunts (*Tilletia* spp.). Wheat bunt is an old, almost forgotten disease that caused huge damage in 1993 - 1997. In 1997, it infected 100 000 - 150 000 tons of wheat seed, which, given the average yield of four tons recorded that year on state-owned land in the country, corresponds to 25 000 - 37 500 hectares worth of production [55].

The domestic scientific community had long believed that the teliospores of the causal agents of wheat bunt (*Tilletia* spp.) that are found on the seeds were the main source of the inoculum and that the critical period for the infection ended with the moment of emergence [56, 57]. Recently, however, significant new results have been obtained by Koprivica [58] in a study involving the daily measurements of the soil temperature during the growing of wheat under the conditions of artificial grain and soil infection and the daily monitoring of the conditions needed for the occurrence of infection. When wheat was infected by teliospores present on the seeds and in the soil, the critical period lasted 1 - 120 days after planting ($r > +0.42$). The temperatures during that period ranged from - 8 to 13.1 °C. The maximum infection levels were achieved when the temperatures were within the 2.0 - 12.3 °C range, or 4.0 - 5.0 °C on average. The most plants were infected during the fourth ten-day period after planting, while a somewhat smaller number were infected in the sixth ten-day period. These findings have shown that, in Serbian conditions, the soil is the more important source of the inoculum and that infection by *Tilletia* sp. takes place during emergence.

During the 2007 - 2008 harvest season, Župunski et al. [59] analyzed 167 non-processed seed samples of wheat. Wheat grain was collected from 39 municipalities in seven districts of Vojvodina and six municipalities in five districts of Central Serbia. It was found that 145 of the samples (86.8 %) were contaminated with *Tilletia* species, while the remaining 22 (13.2 %) were not. By using the rep-PCR fingerprinting technique, it was found that the DNA isolates of *T. tritici* originating from the Serbian wheat samples had 80 % similarity with the positive control for *T. tritici*. One isolate shared a similarity of 60 % with *T. tritici, T. controversa* and *T. laevis*. It was presumed that this isolate belonged to *T. bromi*. The isolate of *T. laevis* shared a similarity of 70 % with the isolates of *T. tritici* and *T. controversa,* while *T. walkeri* was more than 10 % similar to *T. tritici, T. controversa* and *T. laevis*. Although *T. controversa* and *T. tritici* had a high percentage of genetic similarity, they were not clustered together. Our results suggest that rep-PCR fingerprinting could be a useful tool to monitor the presence of morphologically similar *Tilletia* species in wheat production areas

4. Sunflower Diseases

More than 50 diseases (fungal, bacterial, and viral) have been reported in sunflowers worldwide. Among them, only 10 are sufficiently significant on the global scale to warrant attention and breeding for resistance. The appearance of sunflower disease and broomrape has been observed in Serbia over a number of years. In this chapter, the main sunflower diseases will be discussed with special emphasis on broomrape.

Broomrapes (*Orobanche* spp.) are obligate parasitic flowering plants that belong to the family Orobanchaceae. Broomrapes are native primarily to the Mediterranean region (i.e., North Africa, the Middle East, and southern Europe) and western Asia, where they cause significant crop damage [60]. With the anticipated climatic changes taking the form of higher temperatures and drought in many areas of the world, the *Orobanche* species could pose an increased threat to agriculture by expanding their ranges farther to the north in Europe as well as elsewhere [61]. Scenario analyses of the expected future climate revealed a likelihood of the potential range of *Orobanche crenata* expanding into temperate Europe [62]. Global warming could also influence broomrape distribution in Serbia [63, 64].

Orobanche cumana has been a serious problem in sunflowers (Figure 6) in the warm and arid areas between the Mediterranean and the Black Sea, but it has become more widespread in temperate climatic zones as well. The yield losses depend on the severity of the attack and can range from 5 to 100 % [65]. Besides sunflowers, broomrape can be important in tobacco and tomato crops as well.

The broomrape genus (*Orobanche* L.) is characterized by pronounced biodiversity. Of its 99 species identified so far throughout the world, 26 have been found in Serbia.

Figure 6. *Orobanche cumana* in sunflowers.

The species of the *Orobanche* genus are highly variable and include 56 varieties, 226 forms (a total of 308 taxons), and a number of physiological races adjusted to specific crops. *Orobanche cumana* Wallr. (*O. cernua* Loefl.) is especially important as a parasitic species in sunflowers, but it is also frequently found on the roots of tomato, tobacco and some weed plants. Other species of the parasite on sunflower are: *O. coerulescens*, *O. minor, O. ramosa, O. brassicae O. aegyptica*, and *O. crenata*. The last two have not been described in Serbia [66, 67]. Broomrape seeds remain viable for 8 - 12 years. In addition to the huge infectious potential of the pest, another great problem is the heterogeneity of its population, i.e., the existence of multiple physiological races, each specific to a particular sunflower region. The appearance of new broomrape races has been reported in recent years in a host of countries (Romania, Bulgaria, Turkey, Spain, and Russia). The analysis of the populations showed that broomrape in Serbia belongs to race E. Most sunflower hybrids are susceptible to race E [68].

Since the 1990s, broomrape has caused significant damage to the susceptible hybrids in Serbia. In the north of the Bačka region alone, 15 000 ha have been endangered. In this area, 75 broomrapes per plant have been observed in a susceptible sunflower hybrid [64]. It is significant to underline that broomrape has appeared in several new localities in the north and central parts of Vojvodina where it had not been registered previously. Until several years ago, the region of central Banat (the eastern part of the Vojvodina) was considered an area with low broomrape infection levels. Broomrape appeared for the first time in the area in 2001. Until 2007, the appearance of this parasitic plant there was sporadic. From that year onwards, however, the presence of broomrape plants became more frequent and intensive, and this pest has come to cover more fields. Field infestation has ranged from five to as many as 10 or more broomrapes per sunflower plant. In the southeast part of Banat near the border with Romania, the appearance of broomrape was observed for the first time in 2004. In this region, the problem of broomrape exists on the slopes of the Deliblato Sands almost every year. The main focus of the spread occurs on sandy soils, and broomrape is present in almost every field. Infested sunflower fields have been found at the Vatin border crossing very close to the border with Romania. This parasitic plant was spreading in 2009, and for the first time, it was observed in the southern part of the region. The attack severity has ranged from 2 - 10 broomrape plants per sunflower plant. Areas of broomrape infection can also be found in the Bor County near the state border with Romania and Bulgaria (according to the data submitted to the Directorate for Plant Protection). This parasitic plant had been unknown to the farmers in this region until it was noticed there three years ago. The area of broomrape-infected fields is spreading in this region due to the presence of susceptible hosts, the parasite's huge seed potential, and a favorable climate change (warm and dry weather).

Sclerotinia sclerotiorum, the causal agent of white rot, is a worldwide pathogen [69]. The fungus attacks approximately 400 plant species including the most important crops and many weed species [70]. The disease is highly important in the temperate climates that form the main region of sunflower production [69]. Sclerotinia white rot is one of the most important sunflower diseases in Serbia, where it appears almost every year with varying intensity [71]. In favorable climate conditions, this fungus can cause severe economic damage with significant yield losses [72]. The main problem is that there are no Sclerotinia-resistant sunflower hybrids; there are only differences in the susceptibility of different sunflower genotypes to the pathogen.

Sclerotinia sclerotiorum causes three main types of diseases on sunflowers: the root, stem and head forms of white rot. The root form usually occurs at the budding stage, causing the plants to wilt. Primary infections are usually caused by the mycelium from the sclerotia, which can preserve their vitality in the soil for 8 - 10 years. The stem and head infections are caused by ascospores later in the growing season. Temperatures between 15 - 20 °C are optimal for ascospore germinations, and the ascospores also require water for 16 - 24 hours. Higher temperatures are expected to increase the diseases' severity; Sclerotinia apothecial activity could extend throughout the year with a 2 °C temperature rise. The incidence of the disease is correlated to the number of rainy days during the flowering period, so the more critical factor is likely to be the rainfall distribution. The soils must be moist for several days for the sclerotia of *S. sclerotiorum* to germinate and produce apothecia. Dry soils and high temperatures in summer can inhibit spore production, with temperatures > 26 °C inhibiting ascospore infections [73].

During 2005 in Serbia, there was 50 % more growing-season rainfall than the average. In July and August, moreover, there were 80 and 153 % more precipitation than the average, respectively. The number of days with rainfall was also higher than the long-term average during the three summer months. Such conditions had a strong influence on the appearance of white head rot, and the severity of the disease was the highest since sunflower growing started in Serbia. The number of diseased plants varied from 2 to more than 60 %. A similar situation was recorded that year in the neighboring countries. In Croatia, there were fields with 100 % infected heads [74]. In Hungary, many fields had 90 % infected plants and very low yields. In Romania, the situation was similar [72]. Such a dramatic outbreak of Sclerotinia could be repeated in the upcoming years due to the climatic conditions with a highly unequal rainfall distribution during the vegetation period.

Climate change can produce changes in the strains of *S. sclerotiorum* that affect crops if some types adapt to the higher temperature conditions. The ability of the sclerotia formed in cool (10 °C) or warm (25 °C) temperatures to germinate carpogenically varies with the isolate and, more specifically, with its geographic origin [75]. Rainfall is expected to be less frequent but more intense. Infections that do occur may develop more rapidly [6, 76].

Plasmopara helianthii, the causal agent of sunflower downy mildew, is a widespread obligate sunflower pathogen [77, 78]. It exists in the form of many physiological races controlled by 15 dominant resistance genes. The main host is the sunflower, but more than 100 host species from the family Asteraceae can be present in the life-cycle of the fungus.

P. halstedii is seed-borne and soil-borne, with its oospores capable of surviving for as long as 8 - 10 years in the soil. In the book *New and Priority Seed-borne Diseases in Western Europe*, among the other seed pathogens listed under the first category are *Plasmopara halstedii* of sunflowers (with 2 new races in France) and *Diaporthe helianthi* of sunflowers [79]. *P. halstedii* may induce two types of disease symptoms.

The most important and typical symptoms appear after primary infection. In such cases, systemically infected dwarf plants with chlorotic mottling of the leaves along the main veins can be observed. Under moist conditions, a white downy growth composed of sporangiophores and sporangia of the fungus appears on the lower leaf surface. The primary infection can considerably affect the yield because the majority of the systemically infected plants either die prematurely or hardly produce any viable seeds. Moisture and temperature are the most important environmental factors affecting infection and spread. Consequently, rainfall or intensive irrigation will be prerequisites for the initiation of the primary infection, particularly during the critical first 2 - 3 weeks after sowing [80, 81]. The earlier the infection occurs in the season, the more severe the disease will be in the plant. The second type of symptom occurs in the secondary infection with manifestations such as leaf spots with irregular shape and different sizes. If cool and wet weather prevails during the vegetation period, there are more infected plants [82].

In the U.S.A., 11 races of the pathogen have been identified (2000 - 2008). A new race of the fungus was reported in 2002 in Spain [83]. The authors pointed out the importance of climatic conditions in the spring becoming favorable for plant infection and disease development.

The climatic change during the recent years and the wide use of resistant varieties and hybrids in agricultural practice are the main reasons for the limited distribution of downy mildew on sunflowers in Bulgaria. The disease has occurred only in separate locations and with a low percentage of attack (up to 10 %). The current situation stimulates the pathogen to develop new races. Over a period of 12 years (1988 - 2000), only two downy mildew races were identified in Bulgaria. Now, however, there are five. Race 700 has the largest area of occurrence [84].

Sunflower downy mildew was one of the most important sunflower diseases in Serbia during 1960 - 1970, when mostly Russian varieties of sunflower were grown in the country. In that period, there were 5 - 10 % diseased plants on average, and yield losses were proportional to that. In the years that were favorable for disease development, 50 % of the plants in some fields were infected. Now, the dominant race in Serbia is race 730 [85]. In recent years, however, the disease could not spread significantly because of the frequent occurrence of very dry and warm springs that are not favorable for the development of sunflower downy mildew in Serbia.

The disease is extremely difficult or impossible to eradicate once it is established in an area. Sunflower hybrids resistant to downy mildew are available, but new pathogenic races of the fungus are being formed in nature, making questionable the use of formerly resistant cultivars in a particular area [86]. High soil temperature (above 25 °C) and/or drought are limiting factors.

Phomopsis helianthi, the causal agent of stem canker in sunflowers, was described for the first time in Yugoslavia in 1980 [87]. After that, the disease was noticed in Romania, Hungary, France and the U.S.A. It is thought that the Diaporthe/Phomopsis complex in sunflowers consists of more than one species. Sunflower stem canker can cause yield losses and the decrease of seed oil content [88]. In the period from 1983 to 1985, new, more resistant sunflower hybrids were introduced into production. The disease has always posed a big problem in sunflower production in Europe, but today it is also a major problem for sunflower growers across the globe.

The fungus maintains itself in infected plant debris as a mycelium capable of forming perithecia in favorable temperature and humidity conditions. The perithecia can form only in conditions of high humidity, and the ascospores are released after the perithecia absorb enough water. The release of the ascospores occurs at temperatures between 15 and 20 °C and stops at 30 °C. The disease development is favored by wet and warm weather. Infections successfully occur at temperatures between 15 and 35 °C. The disease intensity is dependent on the frequency and amount of rainfall. Frequent rainfall during the budding stage leads to severe infections. It has been pointed out that as the changing climate affects parts of northern and eastern France, diseases such as Phomopsis are likely to become more important [6]. *P. helianthi* affects sunflowers most intensively in those places where there is short-term humidification, and the mean daily air temperature exceeds 20 °C during a period of two months [88]. Debaekea and Moinardb [89] have concluded that the microclimate resulting from crop management and cultivar architecture is responsible for the early Phomopsis attacks that are the most detrimental to sunflower yields [89].

Phoma macdonaldi, the causal agent of Phoma black stem, was noticed for the first time in Serbia during the 1960s. The pathogen is widely spread in Europe, but it is also widespread in Canada, U.S.A. and Argentina, where it is also of significant importance. Infected plants have a smaller head diameter and reduced seed and oil yields. At the beginning of the 1980s, 80 % of the plants in some field were infected [90]. After that period, the disease became less important, but in recent years it has regained significance.

The most typical symptoms appear on the sunflower stem in the form of black spots. In conditions favorable for disease development, the stem base becomes covered with such spots. The infected sunflower debris in the field is the basic source of inoculums in which the fungus persists in the form of mycelia or pycnidia. The picnospores serve to spread the infection. Warm and wet weather favors the development of the disease, and the optimal temperature is about 25 °C, in which case the incubation lasts for 4 days. The intensity of the disease can be affected by smaller amounts and poorer distribution of rainfall, which has been the case in recent years. Such weather conditions lead to a reduction of plant vitality and to an increase in susceptibility to the disease. One of the indicators that more attention should be paid to this fungus is the reports of its higher aggressiveness in France [91]. Because of the existence of the teleomorph stadium, whose presence has been confirmed in Serbia, a variability of aggressiveness in the fungus population [92] can be assumed. *Phoma* is developing very fast at high temperatures. Due to global warming, it appears more intensively and causes significant damages in Serbia.

Alternaria helianthi is the causal agent of sunflower Alternaria blight. This fungus was first reported in 1934 in Africa, and now it has become widespread. A severe disease attack can cause yield losses of up to 80 %, especially in wet and warm weather conditions. In the infected seeds, fungi of the genus *Alternaria* can produce mycotoxins that affect seed quality.

The pathogen is maintained in the infected plant debris and, at a low percentage, in the infected seeds as well. The infections in sunflowers occur across a wide temperature range, but the optimal temperature for disease development is above 20 °C. At higher air humidity, the disease severity is higher. The incubation period at temperatures between 20 and 30 °C is two days. Frequent rainfalls favor disease development [93]. Because of the climate change (especially the warming) and the features of the fungus (its thermophilic origin), there has been a higher incidence of Alternaria in Serbia in recent years. In the eastern part of Croatia, Loncaric [94] observed that *Alternaria helianthi* on sunflower occurred at higher intensity than before due to the climate change [94].

Sunflower powdery mildew can be caused by S*phaerotheca fuliginea*, *Erysiphe cichoracearum* and *Leveillula taurica*. The last of these pathogens has not yet been described in Serbia. Powdery mildew of lower intensity was observed in the second part of the growing season without influencing the yield in the Vojvodina region. Favorable conditions for disease development are warm and dry weather with short wet periods. Conidia germinations can be observed at temperatures between 5 and 35 °C, with 25 °C being optimal.

Macrophomina phaseolina is a plant pathogen that causes charcoal rot in sunflowers. The pathogen prefers very high temperatures (above 30 °C) and dry weather without rainfall. The fungus maintains itself with sclerotia in the soil and in the seeds. Macrophomina is a thermophilic fungus that appears more frequently in tropical regions, but due to the temperature increase, it has spread to the northern part of Europe as well. Charcoal rot has been a big problem in the Vojvodina region in the years with very hot and arid periods after the flowering of the sunflowers. It has been reported recently in sunflowers in the Czech Republic and Slovakia [95, 96]. An incidence of unusual disease adaptation to the changed ecological circumstances (e.g., hot weather and dry soils, drought and heat tolerance) has occurred with the charcoal rot disease caused by the polyphagous fungus *Macrophomina phaseolina* in sunflowers, maize, potatos, peppers, beans, faba beans, etc. in Hungary since 1970 [97, 98].

In the conditions that are favorable for disease development, the number of infected plants can reach 80 %, and yield loses are approximately 25 %. In the warm and droughty years, the total infection of all plants in the field has been reported. Under drought stress conditions for sunflowers, the fungus can cause significant damage in Serbia and the Balkan Peninsula.

Verticillium dahlie and *Verticilium albo-atrum* are the causal agents of sunflower wilt. This disease is important in the warmer climates (most notably Argentina), where it mainly causes economic losses. The diseased plants show the typical symptoms of wilt. On the cross sections of roots and stems, the phloem and vascular bundle necrosis can be observed. In contrast to Sclerotinia and its dependence on moist soils and days with rainfall, Verticillium wilt is likely to be more damaging when summer temperatures are higher, and crops are under moisture stress. The host infection requires moist soils and soil temperatures above 15 °C. Beyond that threshold, the disease has an optimum soil temperature of 24 °C and is damaging when the rainfall is low [99].

Verticilium albo-atrum and *Verticillium dahlie* are maintained in the infected plant debris and soil with resting mycelium and microsclerotia. Warmer soil (22 - 27 °C) favors the development of *V. dahlie,* while *V. albo-atrum* expresses its pathogenicity at somewhat lower temperatures (16 - 27 °C). Despite the fact that Verticillium wilt was not important in Serbia in the past, now due to the climate change, it might increase in significance.

Botrytis cinerea, the causal agent of sunflower grey rot, has been the most damaging when causing severe head rot on late-harvested crops. The causal agent of grey mold can appear any time from germination to harvesting. The most prevalent symptom in the Vojvodina region is the grey head rot. The characteristic powdery grey cover can be seen under high humidity conditions. The pathogen is a problem in the UK and parts of France. Due to the climate change, the adoption of early maturing varieties and the completion of the harvest during September should reduce the severity of attacks in the Vojvodina region. The rainfall distribution will be influential, as heavy rain over 2 to 3-day periods is likely to be sufficient for sporulation, infection and lesion development to occur.

Abundant sporulation appears from mid-April to the end of May. The optimal temperature for conidia germination is between 17 and 27 °C.

Sunflower dry rot is caused by the fungi of the genus *Rhizopus*. In Serbia, two species have been described: *R. arrhizus* and *R. nigricans*. These parasites have caused significant damage in the country. However in 2003, a higher degree of the disease was noticed. The more intensive attack of the disease appeared due to the high temperatures and dry weather [100]. Fungi of the genus *Rhizopus* are thermophiles, and their importance could increase as a result of global warming.

In the sunflower disease spectrum, the climate change leads to the decreasing intensity of sunflower rust but to an increasing intensity of other pathogenic fungi such as powdery mildew Alternaria blight, Phoma black stem, and Sclerotinia base and stem rot. It also increases the risk of Phomopsis stem canker as a quarantined pest in Australia [101].

5. Changes of Faunal Composition in the Vojvodina Region

A climate change inevitably alters the faunal composition of a given area. It has been observed that in years with extreme climatic conditions, some species become very numerous while others decline in numbers. A climatic change will have the greatest effect on the species that have two generations and develop on annual plants. The reason for this is that the majority of annual plants in such years tend to dry out ahead of their usual time, due to the high temperatures during the spring and early summer. The caterpillars that develop on the plants consequently deteriorate or give rise to fewer viable individuals, leading to a situation where the second generation is either completely absent, is much less numerous, or shows reduced overwintering success. No complete die-off of such a species will ever occur because it is never the case that each and every one of the imagoes of the first generation develops during the year concerned. Nevertheless, these species are coming under increasing threat as a result of the climate change.

Moth species that develop on perennial plants are not as threatened by the climatic changes because such plants are better able to withstand the water deficits. The effects of high spring temperatures were the most pronounced in 2007. That year from April to June, the mean daily air temperature was increased by 2.1 °C. Such high temperatures caused a number of insect species to appear earlier than usual. These included *Lacanobia oleracea* (Bright-line

Brown-eye), *Mamestra brassicae* (Cabbage moth), *Agrotis exclamationis* (Heart and dart), *Agrotis segetum* (Turnip moth) [102], and *Ostrinia nubilalis* (European Corn Borer).

The temperature changes that occur during the spring have a greater effect on insect development than those that occur in the summer. This is explained by the fact that many insects have not only a lower but also an upper threshold of development. Let us assume that an insect has the values of 10 °C and 33 °C for these thresholds. If the spring temperatures rise from 20 °C to 25 °C, the insect will be able to use all five degrees of said increase to speed up its development. However, if the summer temperature increases from 30 to 35 °C, only three degrees will be available for the insect to use. Because of this type of influence in 2007, a large number of moths appeared much earlier than usual in the area studied.

5.1. Changes in the Behavior of the Turnip Moth (*Agrotis segetum*)

The species *A. segetum* (Figure 7) has two generations per year [103]. The first reports of *A. segetum* in the Vojvodina region came from the area of the town of Bečej in 1969 and 1970 [104]. The changes in the occurrence and flight dynamics of this species are twofold. The first aspect of the change in the behavior of the species *A. segetum* consists of a shift in the average date of the peak flight of the moths. Between 1980 and 1993, the peak flight of the first generation of the turnip moth occurred on May 25 on average. From 1994 to 2004, the average date of the peak flight occurrence was May 23, while during the 2005 - 2010 period, the date was May 22. This means that the peak flight of the moth's first generation now occurs three days earlier than before.

In the case of the second generation, the average dates of peak flight occurrence have been as follows: August 3 during 1980 - 1993, July 30 during 1994 - 2004, and July 29 during 2005 - 2010. Therefore, the peak flight of the insect now occurs as many as five days earlier than before on the average. This is a direct result of the increasing mean air temperatures that cause the turnip moth to go through its developmental stages at a faster rate than before.

Significantly, this seemingly small shift in the mean occurrence of the peak flight of the moths is most likely the cause of the second type of behavioral change that has been observed in the species. In earlier years, moths of the second generation usually finished their flight between August 25 and 30. This is no longer the case. Over the last five years, the second generation moths have been observed to end their flying between August 15 and 25. This earlier cessation of flight activity gives the insect enough room for another kind of behavioral change.

By studying the flight dynamics of *A. segetum* moths (Figure 8), it has been determined that the moths appearing after September 1 usually belong to a third generation of these insects [105]. Moreover, this third generation has also been found to have discernible flight dynamics (Figure 8). In some years, the peak flight times could be observed in this generation as well.

A third generation of *A. segetum* is not a new occurrence either. It was reported in Vojvodina as far back as 1950, which was a year with an early and warm spring and a long autumn [106]. The weather data from that year show that 1950 was the year with absolutely the hottest July on record (mean temperature of 24.2 °C) and that May and June were warmer than usual as well.

Figure 7. *Agrotis segetum*.

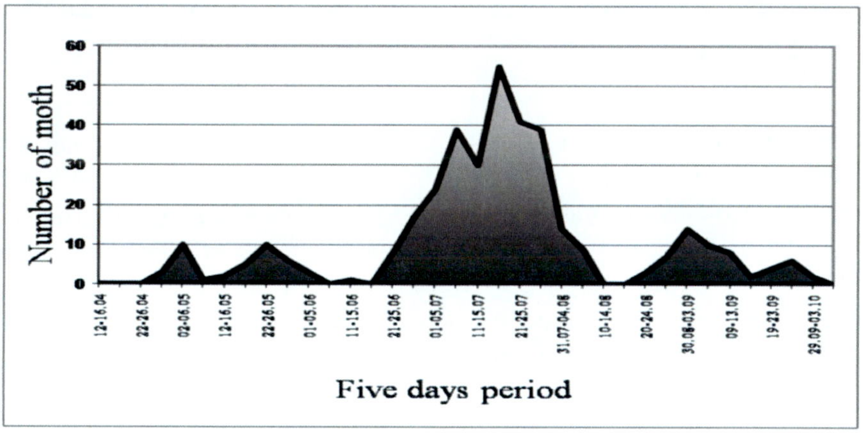

Figure 8. The number of *Agrotis segetum* moths caught in light traps in 2007.

5.2. Mass Occurrence of Warm-Climate Species

The species *Helicoverpa armigera* has been part of the moth fauna of Serbia for a long time. In the past, the abundance of this insect was usually very low because the species typically occurs in the warmer parts of the globe. With the global warming, however, it has become a regular member of the fauna of formerly temperate climates. It is thought that *H. armigera* is capable of overwintering in Central Europe if the winter is mild. According to Vajgand (unpublished data), *H. armigera* specimens in excellent condition can be found as early as the spring, which is when the flight activity of the species begins. This suggests that *H. armigera* can overwinter in small numbers even when the winter is severe. Thanks to its great reproductive capacity, the species can reach high numbers rapidly, depending on the weather conditions.

According Kereši et al. [107], the cotton bollworm (*H. armigera* Hb.) is one of the most harmful pests of cultivated plants worldwide (it feeds on 100 - 250 plant species). This pest, whose natural habitats are tropical and subtropical regions, was found in Serbia only in small numbers prior to 1993. Since then, however, it has spread rapidly and has caused noticeable damage to many cultivated plants (Figure 9a, b and c). In years with mass outbreaks of the cotton bollworm in the north of Vojvodina, 60 to 100 % of maize ears and sunflower heads and 20 to 94 % of string bean pods were reported to be damaged [107].

Figure 9. The cotton bollworm (*Helicoverpa armigera)* causing major damage to a) corn b) pepper and c) tomato.

Between 1981 and 1991 at the Rimski Šančevi site near Novi Sad, not a single *H. armigera* specimen was found caught in light traps [108]. At the Sombor site in 1986, Vajgand (unpublished data) reported catching two specimens of this species using a light trap (August 25 and 28). The records on the species contain reports of its sporadic occurrence in the years 1988, 1989, 1992, and 1993 as well. Reports of the daily presence of the cotton bollworm began appearing only from 1994 onwards, with wide variation in the coloration of the specimens (Figure 10). Since that time, the abundance of the insect has ranged anywhere from 38 to as many as 10 027 moths per year (Figure 11).

Figure 10. Variation of *Helicoverpa armigera* coloration.

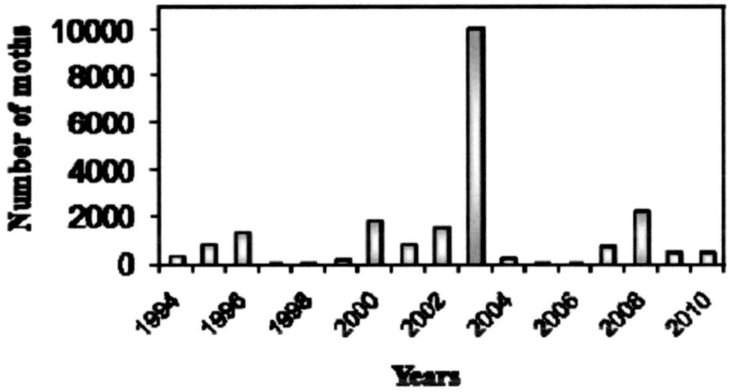

Figure 11. The abundance of *Helicoverpa armigera* in light traps in various years.

Between 1994 and 2003, there was a positive correlation between the mean temperatures during the June-August period and the abundance of *H. armigera*. A negative correlation was found between the amount of rainfall during June-August and the population numbers of the species [99]. From June to August of 2003, the weather was very warm with little precipitation, which promoted the development of *H. armigera* and caused it to appear in record numbers.

Spodoptera exigua (the Beet Armyworm) (Figure 12) is a species that occurs mainly and most numerously in the tropical regions. As a species, the beet armyworm was first recorded at Sombor in 1987, although individual specimens of the insect had been caught in earlier years as well. The first mass outbreak of the species occurred in 2003 [108], a year in which the mean daily temperatures during the summer were much higher than the average. The second mass occurrence was reported in 2006 (Figure 13), when the mean daily temperatures of the air were also above the long-term average.

Figure 12. Spodoptera exigua.

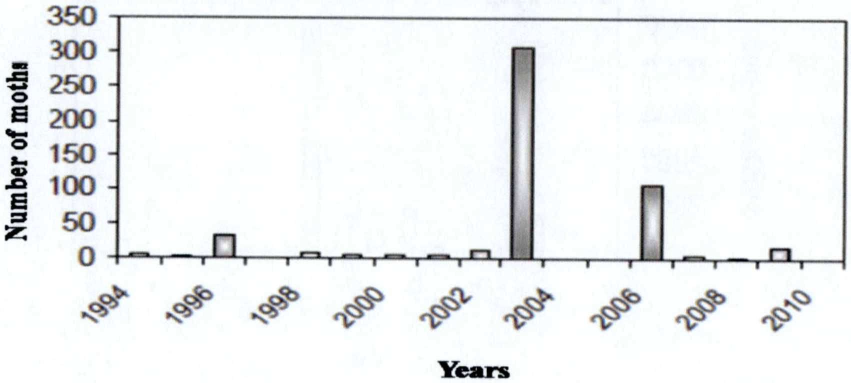

Figure 13. The abundance of *Spodoptera exigua* in light traps.

In both of the above years, the mean daily temperatures in July exceeded 23 °C. In 1994, 1995, 2007, and 2009, the mean July temperature was also above 23 °C, but the abundance of *S. exigua* was not very high. This indicates that other conditions besides temperature must be fulfilled in order for the species to occur in high numbers. One such condition is migrations. *S. exigua* is a migratory species from the Mediterranean to Central Europe. By monitoring this insect, it has been determined that in years in which its abundance is low, individual specimens are usually found for several days in a row.

Such incidence is a result of smaller-scale migrations to the region. According to the data that has been collected, such migrations occur regularly, and when the moths find themselves in favorable conditions, their progeny increase the insect's presence in the region within a short period of time. These favorable conditions have been found to consist of above-average air temperatures during the summer months.

Conclusions

The diseases of small grains that have been occurring frequently in the Vojvodina region in Serbia as a result of climate change have been caused by Biotrophic (powdery mildew and rusts), Necrotrophic (leaf and glume blotch, tan spot) and Grain pathogens (Fusarium Head Blight and Bunts). There are more than 50 sunflower diseases, but only 10 are important worldwide. The following sunflower diseases have occurred frequently in the Vojvodina region: broomrapes, downy and powdery mildew, white rot, stem canker, Alternaria blight, and Phoma black stem.

The climatic change has had the same effect on the occurrence and severity of small-grain and sunflower pathogens as found in the previous seasons, enabling the predictability of their occurrence. An appropriate crop rotation should be applied. In regions where sunflower is cultivated more often (e.g., every second year), more accumulation of the inoculum occurs.

The development of resistant varieties and hybrids, the sowing of healthy, certified seeds that have been properly treated, an appropriate crop rotation, the permanent education of farmers, and the constant monitoring of race compositions are all crucial in preventing the pathogens from spreading to non-infested areas during conditions of climate change.

The cultivation of resistant small-grain varieties and sunflower hybrids can suppress a decreased population of pathogens. The prevention of pathogen spread should also include finding new sources of resistance (in cultivated as well as wild species) that can be used to breed and develop new tolerant varieties and hybrids of small grains and sunflowers.

The race composition of parasites should be monitored continuously. The application of modern techniques is very important in the prompt detection of new aggressive pathotypes and races.

A climate change inevitably alters the faunal composition of a given area. It has been observed that in years with extreme climatic conditions, some species become very numerous while others decline in numbers. In the Vojvodina region, some species, such as *Agrotis segetum* and *Ostrinia nubilaris*, have increased the number of generations they produce. An earlier occurrence of *A. segetum* and *Lacanobia oleraceae* is a direct consequence of the climate change. The species that has definitely become dominant due to the increase in air temperature is the cotton bollworm (*Helicoverpa armigera*).

ACKNOWLEDGEMENTS

This chapter was written as part of the project "Studying climate change and its influence on the environment: impacts, adaptation and mitigation" (No. 43007), which has been financed by the Ministry of Education and Science of the Republic of Serbia within the framework of integrated and interdisciplinary research for the 2011 - 2014 period.

REFERENCES

[51] G.J. Kovics, *Contribuţii Botanice.*, 43, 164-172 (2008).

[52] E.O. Jorgen, M. Bindi, *Eur. J. Agron.,* 16, 239-262 (2002).

[53] N. Fodor, L. Pasztor, *App. Ecol. Environ. Res.,* 8(3), 177-190 (2010).

[54] S. Chakraborty, A.V. Tiedemann, P.S. Teng, *Environ. Pollut.*, 108, 317-326 (2000).

[55] R. Jevtić, B. Lalić, D.T. Mihailović, J. Eitzinger, V. Alexandrov, D. Ventrella, M. Trnka, D.P. Anastasiou, M. Medany, J. Olejnik, M. Nikolaev, *Extended Abstracts of International Symposium Impact of Climate Change and Adaptation in Agriculture,* Vienna, AT, 121-123 (2009).

[56] S. Cook, *Sunflowers and Climate Change* [http://www.hgca.com/document.aspx?fn=loadandmedia_id=5013andpublicationId=3750].

[57] R. Jevtić, M. Telečki, B. Lalić, D.T. Mihailović, M., Malešević, In *Advances in Environmental Modeling and Measurements;* D.T. Mihailovic, B. Lalić, Eds.; Nova Science Publishers Inc: New York, US, 2010; pp. 209-222.

[58] T. Popović, E. Radulović, M. Jovanović (2005). How we are Climate Changing, What Will be our Future Climate? [http://www.sepa.gov.rs/download/5_web.pdf].

[59] D. T. Mihailovic, B. Lalic, I. Arsenic, S. Malinovic, In *Seed Production*; M. Milosevic, M. Malesevic, Eds.; Institute for Field and Vegetable Crops: Novi Sad, RS, 2004; Vol. I, p. 312 (In Serbian).

[60] B. Lalić, D. T. Mihailović (2011). Impact of Climate Change on Food Production in Northern Serbia (Vojvodina) [http://documents.rec.org/topic-areas/Impacts-climage-change-food-production.pdf].
[61] D.T. Mihailovic, B. Lalic (2011). Climate of Serbia [http://polj.uns.ac.rs/ Files/ Meteorologija/Klima%20Srbije1.ppt].
[62] D.T. Mihailović (2007). 1st Project Meeting ADAGIO Climate Change, Vulnerability, and Adaptation in Agriculture - the Situation in Serbia [http://www.adagio-eu.org/documents/1st_Meeting/04_Serbia_Michailovich.pdf].
[63] RHS of Serbia (2011). *Basic Climate Characteristics For The Territory Of Serbia (Standard Normal Period 1961-1990)* [http://www.hidmet.gov.rs/podaci/meteorologija/ Klima_Srbije_eng.pdf].
[64] V. Djurdjevic, B. Rajkovic, *Ann. Geophys.*, 26, 1935-1954 (2008).
[65] D.T. Mihailović, B. Lalić, R. Jevtić, Z. Keserović, Ž. Petrović, S. Jasnić, *Extended Abstracts of International Symposium Impact of Climate Change and Adaptation in Agriculture,* Vienna, AT, 78-80 (2009).
[66] R. Jevtić, M. Telečki, M. Malešević, N. Mladenov, N. Hristov, *Plant Doctor*, 3, 187-191 (2010) (In Serbian).
[67] E. Duveiller, R.P. Singh, J.M. Nicol, *Euphitica*, 157, 417-430 (2007).
[68] L.M. Costamilan, *Fitopatol. bras.*, 30, 420-422 (2007).
[69] R. Jevtić, *Ph.D. Thesis,* Faculty of Agriculture: Novi Sad, RS, 1993 (In Serbian).
[70] A.C. Newton, L. McGurk, *J. Phytopathol.*, 132, 328-338 (1991).
[71] S. Chakraborty, S. Datta, *New Phytol.*, 159, 733-742 (2002).
[72] S.M. Coakley, H. Scherm, S. Chakraborty, *Annu. Rev. Phytopathol.*, 37, 399-426 (1999).
[73] R. Jevtić, M. Telečki, S. Stojanović, M. Staletić, *6th Congress of Plant Protection*, Zlatibor, RS, I, 45 (2009) (In Serbian).
[74] H. Jørgensen, In *Durability of Disease Resistance*; T. Jacobs, J. E. Parlevliet, Eds.; Kluwer Academic Publishers: Dordrecht, NL, 1993; pp. 159-76.
[75] J.T. Kinane, P.W. Jones, *Eur. J. Plant Pathol*, 106, 607-616 (2000).
[76] D.Z. Yu, X.J. Yang, L.J. Yang, M.J. Jeger, J.K.M. Brown, *Plant Breed.*, 120, 279-284 (2001).
[77] A.P. Roelfs, R.P. Singh, E.E. Saari, *Rust Diseases of Wheat: Concepts and Methods of Disease Management*; CIMMYT: Mexico, DF, 1992.
[78] K.A. Garrett, S.P. Dendy, E.E. Frank, M.N. Rouse, S.E. Travers, *Annu. Rev. Phytopathol.*, 44, 489-509 (2006).
[79] A. Legreve, E. Duveiller, In *Climate Change and Crop Production*; M.P. Reynolds Ed.; CAB Ebooks: Wallingford, GB, 2010; pp. 50-70.
[80] E.A. Milus, K. Kristensen, M.S. Hovmøller, *Phytopathol.,* 99, 89-94, (2009).
[81] R. Jevtić, Z. Jerković, S. Denčić, S. Stojanović, *Plant Doctor*, 4, 455-458 (1997) (In Serbian).
[82] HGCA (2003) [http://www.hgca.com].
[83] E.R.L. Jones, B.C. Clifford, *United Kingdom Cereal Pathogen Virulence Survey 1985-2001 Annual Reports* (1986-2002).
[84] E.R.L. Jones, *United Kingdom Cereal Pathogen Virulence Survey 2003-2006 Annual Reports* (2004-2007).

[85] E.R.L. Jones, *United Kingdom Cereal Pathogen Virulence Survey 2002 Annual Reports*, 19-31 (2003).
[86] L.E. Browder, M.G. Eversmeyer, *Phytopathol.,* 76, 1286-1288 (1986).
[87] P.L. Dyck, R. Johnson, *Can. J. Plant Pathol.*, 5, 229-234 (1983).
[88] Z.K. Gerecheter-Amitai, E.L. Sharp, M. Reinhold, *Euphytica*, 33, 665-672 (1984).
[89] J.W. Martens, R.I.H. McKenzie, G.J. Green, *Can. J. Bot.*, 45, 451-58 (1967).
[90] R. Jevtić, Z. Jerković, M. Pribaković, *Plant Doctor*, 1, 42-45 (1995) (In Serbian).
[91] M. Jahn, E. Kluge, S. Enzian, *Asp. Appl. Biol.*, 45, 247-252 (1996).
[92] G.M. Murray, J.P. Brennan (2009). Australian Grains Research and Development Corporation report, 69 [http://www.grdc.com.au].
[93] M. Kalentić, R. Jevtić, M. Telečki, *Abstracts of 8th Counseling on Plant Protection*, Zlatibor, RS, 32-33 (2006) (In Serbian).
[94] Z. Eyal, A.L. Scharen, J.M. Prescott, M. Vanginkel, *The Septoria Diseases of Wheat. Concepts and Methods of Disease Management,* CIMMYT, Mexico, DF (1987).
[95] R. Jevtić, *Plant Doctor*, 5, 320-524 (1997) (In Serbian).
[96] M.V. Nikolaev, In *Advances in Environmental Modeling and Measurements* D.T. Mihailovic, B. Lalić, Eds.; Nova Science Publishers Inc: New York, US, 2010; pp. 209-222.
[97] R. Jevtić, M. Telečki, M. Malešević, N. Mladenov, N. Hristov, *Plant Doctor*, 3, 187-191 (2010) (In Serbian).
[98] M. Telečki, R. Jevtić, *Abstracts of* 8th Counseling on Plant Protection, Zlatibor, RS, 34-35 (2006) (In Serbian).
[99] Maiorano, M. Blandino, A. Reyneri, F. Vanara, *Crop Prot.*, 27, 182-188 (2008).
[100] N. Magan, A. Medina, D. Aldred, *Plant Pathol.*, 60, 150-163, (2011).
[101] Anonymous, Commission Regulation 1881/2006., *Off. J. Eur. Union*, L 364/5 (2006).
[102] I.M. Jajić, R.M., Jevtić, V.B. Jurić, S.Z. Krstović, M.S. Telečki, J.J. Matić, S.M. Đilas, F.B. Abramović, *Proc.Nat. Sci, Matica Srpska Novi Sad*, 120, 19-24 (2011).
[103] V. Rossi, S. Giosue, E. Pattori, F. Spanna, A. Del Vecio, *EPPO Bulletin,* 33, 421-5, (2003).
[104] H.J. van der Fels-Klerx1, M.C. Kandhai1, S. Brynestad, M. Dreyer, T. Börjesson, H.M. Martins, M. Uiterwijk,E. Morrison and C.J.H. Booij, *World Mycotoxin Journal,* 2(2), 119-127 (2009).
[105] R. Jevtić, S. Stamenković, M. Malešević, *Plant Doctor,* 5, 524-526 (1997) (In Serbian).
[106] M. Josifović, *Plant Pathology*; Scientific Book: Belgrade, RS, 1964; p. 573 (In Serbian).
[107] M. Ivanović, D. Ivanović, *Plant Mycoses and Pseudomycoses*, 2nd Edition; P.P. DMV, Belgrade, RS, 2001; p. 553 (In Serbian).
[108] M. Koprivica, *MSc. Thesis*, Faculty of Agriculture: Belgrade-Zemun, RS, 2004 (In Serbian).
[109] V. Župunski, D.I. Ignjatović-Micić, A. Nikolić, S. Stanković, R. Jevtić, J. Lević, D. Ivanović, *Genetika*, 43(1), 183-195 (2011) (In Serbian).
[110] C. Parker, C.R. Riches, *Parasitic Weeds of the World: Biology and Control*; CAB International: Wallingford, GB, 1993; p. 368.
[111] K.I. Mohamed, M. Papes , R. Williams, B.W. Benz, *AMBIO: J. Human Environ.,* 35(6), 281-288 (2006).

[112] J. Sauerborn, J.H. Grenz, *COST Action 849, Parasitic Plant Management in Sustainable Agriculture WG1+4 Workshop on Means for Limiting Orobanche Propagation and Dispersal in Agricultural Fields*, Newe-Yaar Research center, IL, 23 (2005).

[113] S. Maširević, S. Medić-Pap, *2nd National Conference on Climate Change Expected in Vojvodina and their Effects: Climatic Changes and Agriculture Production in Serbia*, Novi Sad, RS, 6-7 (2009a) (In Serbian).

[114] S. Maširević, S. Medić-Pap, *6th Congress of Plant Protection (Book I),* Zlatibor, RS, 117-118 (2009b) (In Serbian).

[115] S. Maširević, *Field Veg. Crop Res.*, 35, 235-241 (2001) (In Serbian).

[116] M. Kojić, S. Maširević, D. Jovanović, *Helia.,* 24(35), 73-92 (2001).

[117] S. Maširević, M. Kojić, *Ratar. Povrt. / Field Veg. Crop Res.*, 36, 161-168 (2002) (In Serbian).

[118] S. Maširević, S. Medić-Pap, *Helia.,* 32(51), 91-100 (2009c).

[119] S. Maširević, T.J. Gulya, *Field Crops Res.*, 30, 271-300 (1992).

[120] G.J Boland, R. Hall, *Plant Pathol.*, 16, 93-100 (1994).

[121] S. Maširević, S. Jasnić, *Plant Doctor.*, 34(4-5), 336-342 (2006a) (In Serbian).

[122] S. Maširević, B. Dedić, *Ratar. Povrt. / Field Veg. Crop Res.*, 42, 87-98 (2006) (In Serbian).

[123] S. Koch, S. Dunker, B. Kleinhenz, M. Rohrig, A. Tiedemann, *Phytopathol.*, 97, 1186-1194 (2006).

[124] B. Šimić, J. Ćosić, I. Liović, M. Krizmanić, J. Poštić, *Proceedings of the 17th International Sunflower Conference*, Cordoba, ES, 261-264 (2008).

[125] H.C. Huang, G.C. Kozub, *Bot. Bull. Academia Sinica.*, 32, 279-286 (1991).

[126] V. Heffer Link, K.B. Johnson (2007). White Mold, *The Plant Health Instructor* [http://www.apsnet.org/edcenter/intropp/lessons/fungi/ascomycetes/Pages/WhiteMold.aspx]

[127] T.J. Gulya, W.E. Sackston, F. Viranyi, S. Maširević, D.Y. Rashid, *Phytopathol.*, 132, 303-311 (1991).

[128] T.J. Gulya, K. Rashid, S. Maširević, In *Sunflower Technology and Production;* A. Schneiter, Ed.; American Society of Agronomy: Madison, Wisconsin, US, 1997; pp. 263-379.

[129] V. Cockerell, In *New and Priority Diseases in Western Europe*; J.D. Hutchins, J.C. Reeves, Eds.; CAB International: Wallingford, GB, 1997; pp. 1-10.

[130] D.E. Zimmer, J.A. Hoes, In *Sunflower Science and Technology;* J.F. Carter, Ed.; American Society of Agronomy: Madison, US, 1978; pp. 225-262.

[131] S.J. Kolte, *Diseases of Annual Edible Oilseed Crops: Peanut Diseases Rapeseed-Mustard and Sesame Diseases*, 1st Edition; CRC Press: Boca Raton, FL, US, p.168 (1985).

[132] F. Virányi, In *European Handbook of Plant Diseases;* I.M. Smith, J. Dunez, D.H. Phillips, R.A. Lelliot, S.A. Archer, Eds.; Blackwell Scientific Publications: Oxford, GB 1988; pp. 228-230 (In Hungarian).

[133] M. L. Molinero-Ruiz, J. Domínguez, J. M. Melero-Vara, *Plant Dis.*, 86, 736-740 (2002).

[134] P. Shindrova, *Helia.,* 33(52), 19-24 (2010).

[135] S. Maširević, *Ratar. Povrt. / Field Veg. Crop Res.*, 20, 405-409 (1992) (In Serbian).

[136] T.J. Gulya (2010). New Virulent Races of Downy Mildew: Distribution, Status of DM-Resistant Hybrids and USDA Sources of Resistance [http://www.sunflowernsa.com/uploads/resources/575/gulya_virulentracesdownymildew.pdf].
[137] A. Marić, S. Maširević, *Plant Prot. Bull.*, 12, 421-423 (1980) (In Serbian).
[138] V.I.Yakutkin (2011). Gray Stem Spot of Sunflower (Phomopsis) - Diaporthe helianthi Munt.-Cvet. et al. (teleomorph), Phomopsis helianthi Munt.-Cvet. et al. (anamorph) (=Diaporthe arctii (Lasch) Nit., =Diaporthe arctii var. achileae (Auers) Wehm. Agroatlas [http://www.agroatlas.ru/en/content/diseases/Olee/Olee_Phomopsis_helianthi/].
[139] P. Debaeke, J. Moinard, *Field Crops Res.*, 15, 50-60 (2010).
[140] A. Marić, S. Maširević, In *Sunflower Diseases and Pests and their Control*; A. Marić, D. Čamprag, S. Maširević, Eds.; Nolit: Belgrade, RS, 1988; pp. 11-121 (In Serbian).
[141] C. Larfeil, G. Dechamp-Guillaume, G. Barrault, *Helia*, 25(36), 153-160 (2002).
[142] A. Marić, S. Maširević, S. Fayzalla, *Plant Prot.*, 32(4), 329-334 (1981) (In Serbian).
[143] S. Maširević, S. Jasnić, *Plant Doctor*, 34(4-5), 326-333 (2006b) (In Serbian).
[144] Z. Loncaric, *Europe Adapts to Climate Change: The Regional Challenges for Biodiversity, Agriculture and Rural Development (BARD)*, Budapest, HU (2007).
[145] J. Šárová, I. Kudlíková, Z. Žalud, K. Veverka, *J. Plant Dis. Prot.,*110, 444-448 (2003).
[146] P. Bokor, *Biologia*, 62(2), 136-138 (2007).
[147] P. Békési, J. Vörös, O.H. Calvert, *Plant Dis. Reptr.*, 54, 286-287 (1970).
[148] J. Vörös, S. Manninger, *Plant Prot.*, 9(5), 193-195 (1973) (In Hungarian).
[149] W.C. Schnathorst, In *Fungal Wilt Disease of Plants;* M.E. Mace, A.A. Bell, C.H. Beckman, Eds.; Academic Press: New York, US, 1981; pp. 81-111.
[150] M. Guenard, S. Maširević, *ISTA Plant Diseases Workshop*, Novi Sad, RS, 30-32 (2004).
[151] S. Thompson (2010). Phomopsis Stem Canker - An Emerging Aisease of Australian Sunflowers [http://www.grdc.com.au/uploads/documents/summer_grains_conference_2010/Sue%20Thompson.pdf].
[152] D. Vajgand, *Plant Doctor*, 2, 121-131 (2008) (In Serbian).
[153] Z. Mészáros, A. Vojnits, Đ. Varga, *Contemporary Agriculture*, XIX (9), 55-66 (1971) (In Serbian).
[154] D. Đukanović, *Climate of Sombor and its Surroundings*, Belgrade, RS, 1970; p.134 (In Serbian).
[155] D. Vajgand, *Handbook about the Owlet Moth (Noctuidae, Lepidoptera) on Light Trap*, Garden print: Sombor, RS (2011) (In Serbian).
[156] A. Petrik, M. Jovanić, *Proc. Nat. Sci, Matica Srpska Novi Sad*, 3, 119-132 (1952) (In Serbian).
[157] T. Kereši, D. Čamprag, R. Sekulić, D. Vajgand, G. Forgić, K. Radonić, *Abstracts of 5th Congress of Plant Protection with International Participation*, Zlatibor, RS, 20-23 (2004) (In Serbian).
[158] T. Kereši, R. Almaši, *Acta Ent. Serbica*, 14(2), 147-162 (2009).

In: Essays on Fundamental and Applied Environmental Topics ISBN 978-1-61942-522-4
Editor: Dragutin Mihailovic

Chapter 14

Climate Change Impact on Water River Flow: A Case Study for the Sarajevo Valley (Bosnia and Herzegovina)

E. Hadžić[1] *and N. Drešković*[2]*
[1]Department of Water Resources, Faculty of Civil Engineering,
University of Sarajevo, Sarajevo, Bosnia and Herzegovina
[2]Department of Geography, Faculty of Sciences,
University of Sarajevo, Sarajevo, Bosnia and Herzegovina

Abstract

It is widely accepted that global warming has had a significant impact on water resources. This impact is reflected in different ways and is dependent on the impact size of the main climatic elements. Thus, like the tiles of a complex mosaic, any research conducted to analyze the characteristics of specific places contributes to the global recognition of the impact that climate change has on water resources. For the Sarajevo valley, which is a mountainous, morphological area that is geographically well-isolated from the direct impact of the large, urban centers of the European continent, the impact of air temperature and precipitation fluctuation on the flow of rivers has been analyzed in the Miljacka River, as well as the Bosna River, during the hydrological instrumental period (1926 - 2010), which is referred to as the HI period in the text. Climate variability is considered for the meteorological instrumental period (1888 - 2010), which is referred to as the MI period in the text, using meteorological data on an annual basis from the meteorological station at Bjelave, which is representative for the Sarajevo Valley. We concluded that although an increase in the air temperature and a slight increase in the amount of precipitation exist, there is still evident stagnation in the considered river flow. Sarajevo Valley is truly representative of the general climate conditions of the mountainous part of southern Europe. Therefore, this research is a useful contribution to completing the picture on the impact of climate change on the water flow of rivers in Europe as a whole.

---

* Corresponding author: E-mail address: nusret2109@gmail.com; Phone: +387 61 109 203. Fax: +387 33 649 359.

1. Introduction

Water, whether in liquid, gas or solid form, is the basis for all life on Earth. In economic terms, water holds a central place in the development and sustainability of a society. Water makes it possible to perform agricultural and energy production. It is the most basic need in the development of industry and for mining, tourism and other economic sectors, as well as for supplying people with water. Due to all of these factors, it is very important to know the amount of water available and if it is available for use. Of course, the issue of water quality cannot be separated from its quantity. However, in this chapter, we will not address this issue, which certainly does not mean that it is a less important question; rather, this question deserves special attention itself. Due to its natural ability for self renewal and self cleaning during its cycling, it was once believed that the amount of water is inexhaustible. However, when you take into account estimates of global water distribution, in one cycle of water circulation (Table 1), it is easy to understand the basic illusion of humanity with regards to water's endlessness. Due to this reason, people have historically been careless in their use of water, which implies minimal or no investment in cleaning waste water or any consideration toward the problem of protecting water. The growth of the human population, urbanization and a concentrated population in urban environments, the increase in pollution and the need for the increased production of food and energy have led to an exponential growth for water requirements for both drinking and industrial purposes. Thus, we can easily claim that the problem of water scarcity will represent one of the most important questions for the future preservation of not only humanity, but also all known life forms on Earth.

Until the end of the twentieth century, the main problem that could occur in relation to water and its use was considered to be pollution and thus, the problems that could occur due to its quality. Shortly thereafter, larger challenges emerged, for example, water scarcity and polluted water. With regards to water scarcity and increased pollution, the reasons for this concern have been recognized. It is unclear what actions can solve this crisis with the least amount of consequences for later generations. In recognition of the problems and their causes, in this case, the solution cannot be simple or one-sided. The reason for this point of view is the fact that the water cycle is a complex and stochastic process and is very difficult to predict. Insofar as these uncertainties and problems add an as yet unknown impact on the climate changes in the water cycle and water resources, it is certain that the future of the 'blue planet' becomes questionable. A large amount of contemporary scientific research warns that climate change will lead to intensive changes in the water cycle with regards to quantity disorders in the mean amount of water inside of a circulation cycle, which in turn will lead to short but very intense precipitation periods followed by severe thunderstorms and long-term droughts [1].

Apart from quantity, climatically caused disorders of a water cycle will certainly negatively reflect on water quality because a drop in the amount of freshwater will increase water intake and all other negative forms of anthropogenic pressures [2]. Thus, the conclusion can be made that studying and monitoring the function of recent climate systems is something invaluable for understanding the impact of climate change on a river's regime. One of the basic scientific methods for monitoring climate change is studying the fluctuation in climate elements during monitoring on meteorological instruments, especially at the level of fluctuations in air temperature and the amount of precipitation [3]. Although systematically measuring and monitoring the time-space dynamics of climate elements is relatively new, it is already possible to find certain estimates about the feedback effect that climate change has had on the water cycle.

With respect to these listed facts, this chapter analyzes the impact of basic climate elements on river flow in the area of the Sarajevo valley, which consists of the mountain rim and Sarajevo Field with the alluvion plateau of the Miljacka River. In Section 2, we give a short description of the prevalence of water on Earth and the main numerical values of the water cycle. Subsection 2.1 comprises the general climatic characteristics of Bosnia and Herzegovina, which are a prerequisite for understanding the climate fluctuations in different parts of this area. Subsection 2.2 is devoted to the description of the water balance in Bosnia and Herzegovina and includes basic indicators of the hydrological potential for surface and groundwater.

Section 3 consists of three subsections. In subsection 3.1, we describe the main geographical features of the Sarajevo Valley, which is the research subject for the climate change impact on river flow.

In subsection 3.2, there is a short summary on the present climate of the Sarajevo Valley. An analysis of the fluctuations in air temperature and precipitation are given in subsections 3.3 and 3.4, respectively. Recent climatic characteristics of the Sarajevo Valley are given in detail through the analysis of the air temperature and the amount precipitation during the period from 1961 - 1990. Analyses of the fluctuations in the aforementioned elements were performed using data for the period from 1888 - 2010 (the meteorological instrumental period, MI period in the further text), which was observed from the meteorological station (MS) Bjelave-Sarajevo.

A statistical analysis of the fluctuations in the mean annual air temperature and precipitation was performed by linear trend using a five-year moving average. Section 4 deals with an analysis of the hydrological conditions and the fluctuation in the river flow in the Sarajevo Valley through the following: (i) the current hydrological state in the Sarajevo Field and alluvial plateau of the Miljacka River (subsection 4.1); (ii) an analysis of the fluctuations in river flow for the Bosna River (subsection 4.2) and (iii) fluctuations in river flow for the Miljacka River. For the period from 1926 - 2010 (the hydrological instrumental period, HI period in the further text), the fluctuations in river flow for both rivers were analyzed using data from both the Bosna-Reljevo and Miljacka-Sarajevo (subsection 4.3) hydrological stations (HS). Finally, Section 5 provides concluding remarks and a quantitative analysis between (i) air temperature and river flow and (ii) the amount of precipitation and river flow for the HI period in HS Bosna-Reljevo is given.

2. Prevalence of Water and the Water Cycle

The total amount of water on our planet is estimated at approximately 1386 million km^3, of which approximately 2.5 % is freshwater. The bulk (approximately 96.5 %) of this water is contained in saltwater seas, which largely changes the picture in terms of the available amount of water and about ostensible water abundance. Of the total freshwater, 68.7 % is water in the form of icebergs and snow, approximately 30 % is underground water and the remaining 0.3 % consists of available surface water, out of which 2 % is river flow.

Table 1. Estimation of the global prevalence of water in one cycle of circulating water [4]

Water reserve	Volume of water (in cubic kilometers)	Percent of freshwater	Percent of total water
	km^3	%	%
Oceans, seas and bays	1 338 000 000	--	96.50
Ice caps, glaciers	24 064 000	68.70	1.74
Underground water	23 400 000	--	1.70
Freshwater	10 530 000	30.10	0.76
Saltwater	12 870 000	--	0.94
Underground ice and permafrost	300 000	0.86	0.022
Soil moisture	16 500	0.05	0.001
Lakes	176 400	--	0.013
Freshwater	91 000	0.26	0.0070
Saltwater	85 400	--	0.0060
Atmosphere	12 900	0.04	0.001
Swamp water	11 470	0.03	0.0008
Rivers	2120	0.01	0.0002
Other	357 480	1.02	0.0258
Total	1 386 000 000	100	100

While taking the form of all three aggregate states, water circulates from oceans to land and back again. The process of water circulation has interested people throughout the history of civilization. Even in the period Before Christ, Thales and Plato were observing the circulation cycles of moving water and writing that everything is born out of water and that everything is coming back to the oceans. After the sixteenth century, when hydrological measurements began, the existence of water cycles was confirmed. The French physicists Pierre Perrault (1608 - 1680) and Eduard Mariotte (1620 - 1684) were separately performing rough measurements of the precipitation in the catchment area of the Seine River in France. They performed a correlation of these measurements with the flow of the Seine River to illustrate the relationship between the amount of water from precipitation and river flow. It was understood that the water cycle represents the process of constant circulation, updating and an apparent loss of water on Earth due to the activity of solar energy and gravitational forces (Figure 1).

Water constantly evaporates from the surface of the ocean, the sea and other land and water surfaces and then rises and condenses in the atmosphere. At the end of this process, it then returns back to the surface of the Earth in the form of precipitation. Thus, circulating water repeats this process 42 times a year. When entering into the ground through surface or underground channels, water drains toward a collection zone, which could be a river or lake.

The majority of surface water or groundwater (approximately 80 %) returns to the atmosphere by evaporation. A part of the precipitating water infiltrates into the ground and can be temporarily held in the form of soil moisture or penetrate into deeper zones where, in the form of underground water, it can remain for a long period of time.

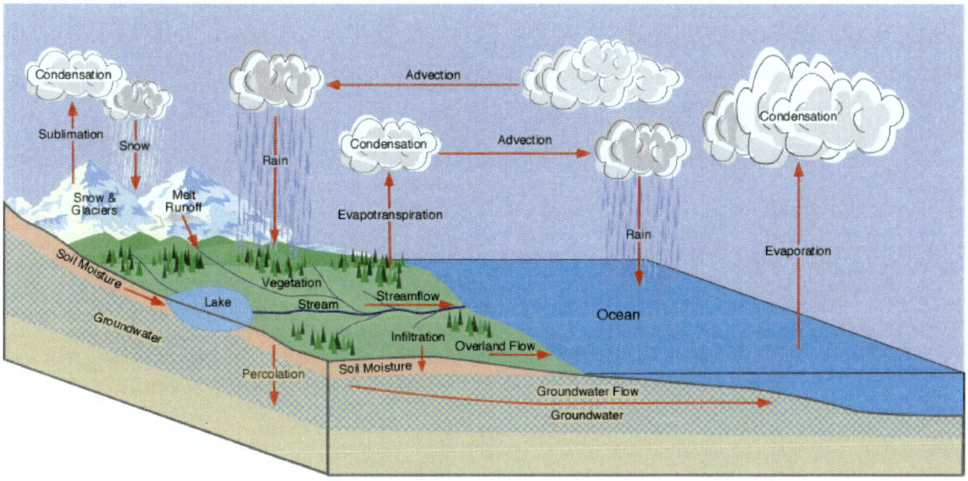

Figure 1. Schematic diagram of the water cycle [5].

However, regardless of its retention time, all of this water again arrives at waterways or lakes and then returns to the oceans for the process to begin again.

To visualize the water resources of an area within the required allotment of time, it is necessary to quantify them. A quantitative measurement of water cycles is performed using a water balance equation, which is commonly used distinguish the amount of water that enters, leaves or is maintained in a defined area for a given time period. The input component of a water balance equation is precipitation, while the output components are evaporation and runoff. For an overview of the water resources of an area, it is important to understand the climatic characteristics from which the components of the water balance are directly dependent.

2.1. Climatic Characteristics of Bosnia and Herzegovina

For Bosnia and Herzegovina, there is a wide zone of transition between two climatic belts of the eastern Balkan Peninsula, the continent and the Mediterranean. Through the wide spectrum of climatic types and different factors, the most important are the general circulation of the atmosphere, dynamic relief, the directions in which the mountain range pull and the vicinity of the Adriatic Sea.

The result is a mild continental climate, which is mostly represented by the north and central parts of the Bosnian and Herzegovinian submountain and mountain types (above 1000 m), the Adriatic (Mediterranean) and changed Adriatic climate type, which represent the Bosnian and Herzegovinian coast and hinterland.

In the Peri-Pannonian region of the country, the mean monthly air temperature generally varies between (i) - 1 °C and - 2 °C in January and (ii) 18 °C and 24 °C in July. At an altitude of 1000 m above sea level, the mean monthly temperature varies from - 4 °C to - 7 °C in January and 9 °C to 14 °C in July [7]. On the Adriatic coast and mainland hinterland, the mean monthly air temperature varies between 3 °C and 7 °C in January and from 22 °C and 25 °C in July.

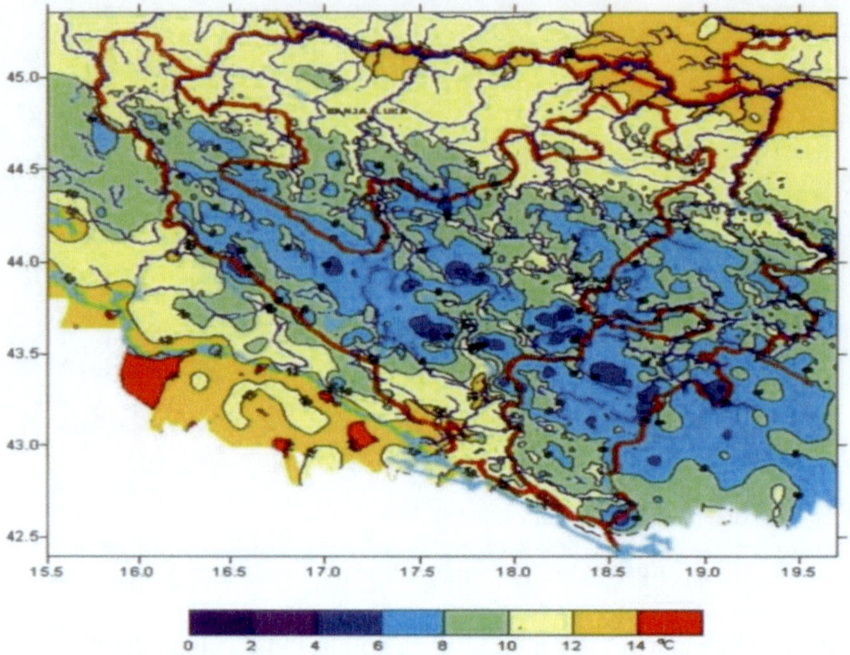

Figure 2. Spatial distribution of the mean annual air temperatures in Bosnia and Herzegovina [6]. The red line indicates the administrative borders between the state entities.

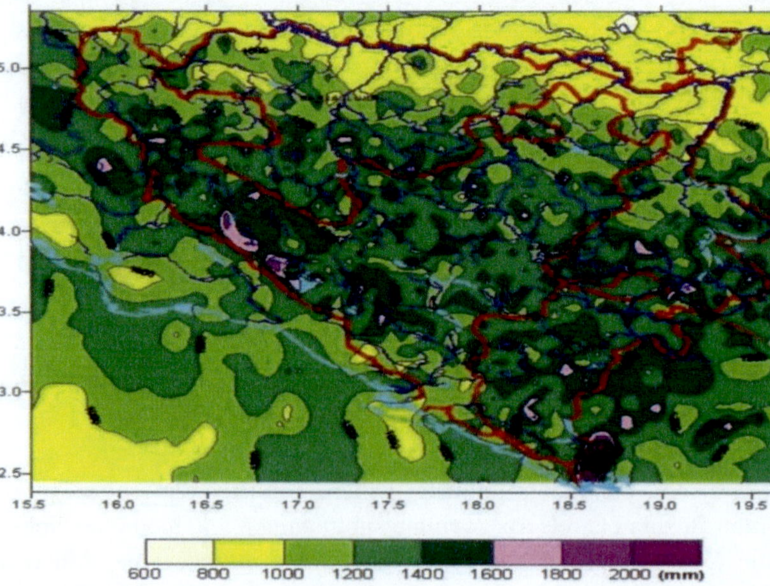

Figure 3. Spatial distribution of the annual amount of precipitation in Bosnia and Herzegovina [6]. The red line indicates the administrative borders between the state entities.

Extreme temperatures have been recorded in the broad range from - 41.9 °C (lowest) to 42.2 °C (highest). The lowland areas of northern Bosnia and Herzegovina have a range of mean annual air temperatures between 10 °C and 12 °C while, in the areas above 400 m, this temperature is below 10 °C. The mean annual air temperatures of the coastal areas are in the interval between 12 °C and 17 °C (Figure 2). The annual amounts of precipitation vary from

800 mm in the north along the Sava River to 2000 mm in the central and southeastern mountainous regions of the country (Figure 3). In the central part of Bosnia and Herzegovina, which belongs to the catchment area of the Sava River, the main portion of the annual amounts of precipitation appears in the warmer half of the year, which reaches its maximum in June. The central and southern parts of the country have numerous mountains and narrow coastal areas and are characterized by a Mediterranean pluviometric regime under the influence of the Mediterranean Sea. In these areas, the monthly maximum amount of precipitation reaches late into autumn and the beginning of winter, mostly in November and December. The amounts of precipitation, which depend on the part of Bosnia and Herzegovina, have displayed minimal changes in the last 100 years (with variation from - 5 to + 5 %). In the middle mountain belt, a trend of increases in the annual amount of precipitation exists.

In the southwestern, northern and northeastern parts of the country, a decreasing trend in the amount of precipitation is observed and has different values depending on the climatological seasons. Figure 4 illustrates the spatial distribution of surplus or deficient amounts of precipitation over a ten-year period (1999 - 2008) in comparison with a reference period (1961 - 1990) in Bosnia and Herzegovina. A large part of Bosnia and Herzegovina display a negative trend during the spring and summer while an increase in precipitation is observed during the winter half of the year [6]. Annually, the areas of Bosnia and Herzegovina with surplus precipitation are located in the mountain and sub mountain regions, especially around the Sokolac-Glasinac plateau and a wider area around the central part of the Bosna River Valley. In contrast, areas with precipitation deficits are located around the Neretva River Valley and the northwestern part of Bosnia and Herzegovina, specifically, a wider area around the Sava River.

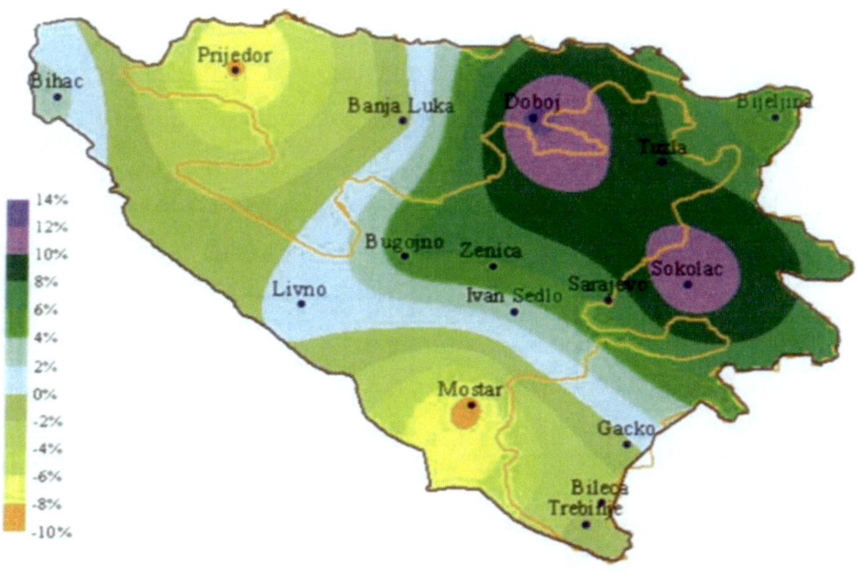

Figure 4. Spatial distribution of the relative surplus and deficit of the annual amount of precipitation in Bosnia and Herzegovina [6]. The yellow line indicates the administrative borders between the state entities.

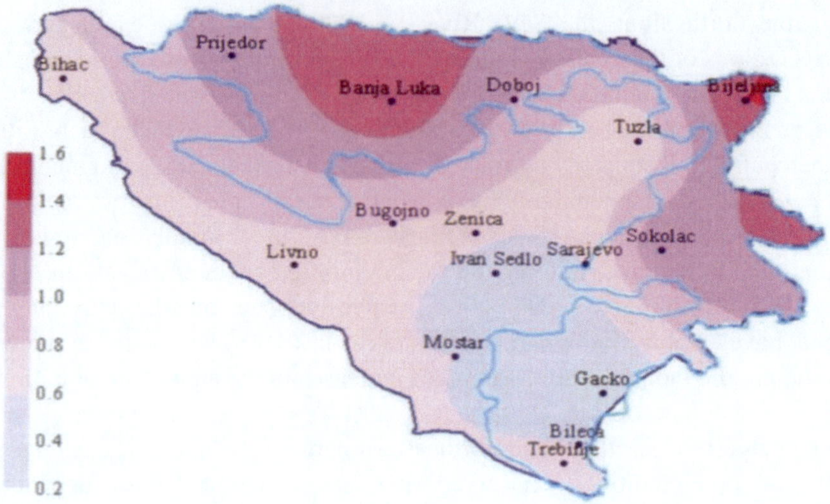

Figure 5. Increase in the mean annual air temperature over a ten-year period (1990 - 2000) in comparison with the reference period (1961 - 1990) in Bosnia and Herzegovina [6]. The blue line indicates the administrative borders between the state entities.

Recently, there has been an increase in the mean annual air temperature over a ten-year period (1990 - 2000) in comparison with the reference period (1961 - 1990) in Bosnia and Herzegovina, which is depicted in Figure 5.

The greatest increase of mean annual air temperatures, more than 1.5 °C, was in the central Peri-Pannonian region of the country and the wider area of the Drina River Valley downstream of the town of Zvornik.

2.2. Water Balance of Bosnia and Herzegovina

Although the summary of the annual amount of precipitation that is produced over the land part of Bosnia and Herzegovina is relatively high (about 1250 mm), on the surface, the runoff accounts for only a small fraction of more than half of this precipitation. The coefficient of surface runoff is approximately 0.57, or a mean annual runoff of 1200 $m^3\ s^{-1}$. From this value, surface runoffs in the catchment area of the Black Sea and the catchment area of Danube River have waste amounts of about 722 $m^3\ s^{-1}$ or about 433 $m^3\ s^{-1}$, respectively, in the catchment area of the Adriatic Sea (Figure 6). The total length of the surface river network in Bosnia and Herzegovina is 20 919 km and the average for the entire national territory is 0.41 km km^{-2}. Using advanced GIS technology, we have created a picture of the basic catchment areas of the water surface network in Bosnia and Herzegovina. Let us note that this is the first time that data for lengths of river network have been comprehensively elaborated for Bosnia and Herzegovina (given in Table 2).

The natural hydrographical system of Bosnia and Herzegovina has developed in accordance with geological structures and climatic characteristics and appears in the forms of surface and underground types of runoff.

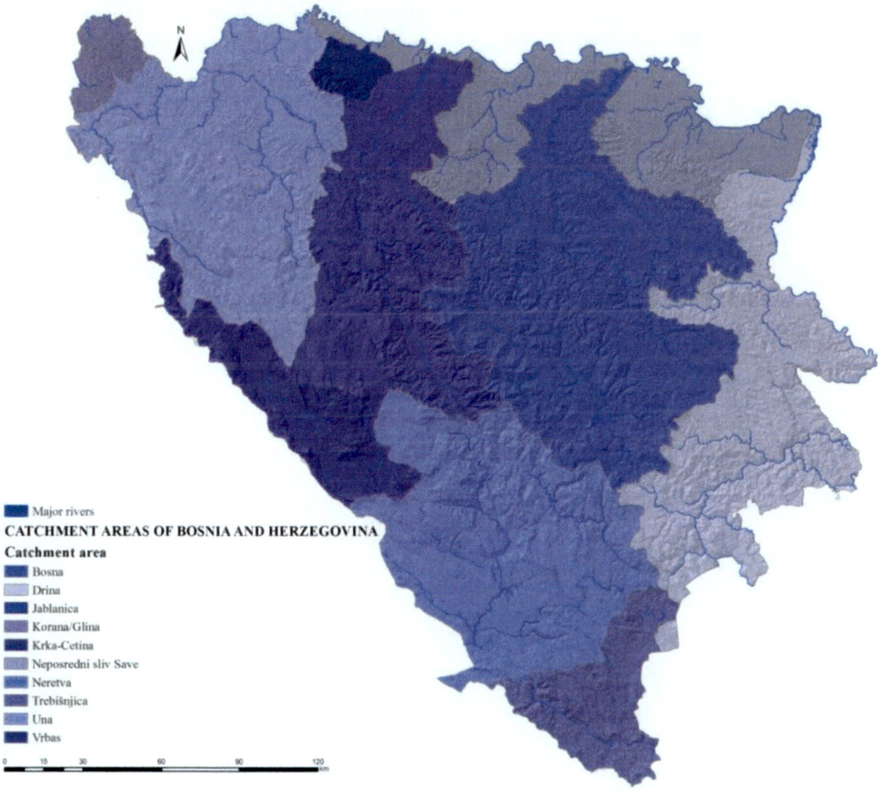

Figure 6. Catchment areas in Bosnia and Herzegovina (Designed by the authors).

More specifically, the largest part of the hydrological system is composed of freshwater, which appears in surface and underground forms and the surface system has developed from its spatial aspects as well as from aspects of the stock of water that circulates in it. Analogous to the geological composition of the field and general physical geographical conditions in Bosnia and Herzegovina and in accordance with its porosity structure, we can categorize different parts of aquifers: (i) fields with an aquifer sinter granular porosity, (ii) fields with aquifers with an intergranular and cracking porosity, (iii) fields with aquifers with a cavernous cracking porosity, (iv) fields with aquifers with a cracking porosity and (v) fields practically without aquifers. The estimated values for reserve aquifers with a sinter granular porosity for the Black Sea and Adriatic Sea catchment areas are approximately 3.93 $m^3 s^{-1}$ and 1.05 $m^3 s^{-1}$, respectively. The reserve water in the zone of karst cracking-type porosity in the catchment areas of the Black Sea and Adriatic are approximately 28.26 $m^3 s^{-1}$ and 35.94 $m^3 s^{-1}$, respectively. A significant amount of water that participates in the freshwater water balance is represented by lakes in Bosnia and Herzegovina.

Based on the existing data, surface natural lakes account for approximately 169.45 km^2, while approximately 11.6 km^2 is drawn from artificial lakes. Sea surface water is located in the Neum-Klek bay, which is the only freshwater area that belongs to Bosnia and Herzegovina. This water is composed of the Neum bay and half from Mali Ston canal and is approximately 12.2 km^2, which represents the territorial waters of Bosnia and Herzegovina [8].

Table 2. The basic catchment areas of the water surface network in Bosnia and Herzegovina. The meaning of the symbols are: L is the total length of the river network in the catchment, while Q is the surface runoff in the catchment

No.	Catchment areas	Area(km^2)	Area (%)	L (km)	Q (ls^{-1} km^{-2})
1	Immediate catchment area of Sava	5323.10	10.40	3197.3	11.4
2	Korana/Glina	705.57	1.38	500.22	24.6
3	Una	7962.10	15.56	2747.71	26.1
4	Vrbas	6288.59	12.29	2654.14	19.9
5	Bosna	10 758.99	21.02	5910.79	16.1
6	Drina	7185.10	14.04	3167.67	21.2
	Catchment area of Black Sea in Bosnia and Herzegovina:	38 223.45	74.69	18 177.83	
1	Neretva	7947.51	15.53	1961.79	38.1
2	Trebišnjica	2254.94	4.41	298.79	49.4
3	Krka-Cetina	2753.10	5.38	480.27	34.6
	Catchment area of the Adriatic Sea in Bosnia and Herzegovina:	12 955.54	25.31	2740.85	
	Total in Bosnia and Herzegovina:	51 179.00	100.00	20 918.68	

3. Analysis of Climate Conditions in Sarajevo

In the wider space of the Sarajevo Valley, the Sarajevo agglomeration is formed, which has recently been reported to contain more than 500 000 citizens and, thus, represents the densest populated area in Bosnia and Herzegovina with approximately 6650 inhabitants per km^2. Apart from demographic overcrowding, the Sarajevo Valley is characterized by its economic development and infrastructure capacities, which additionally impact the anthropedic load of the area.

3.1. Geographic Features of the Sarajevo Valley

The Sarajevo Valley is a space between mountain depressions and is between the massive Bjelasnica and Igman on the southwest as well as the low mountains and middle mountains on the northeast. The valley generally stretches in the NW-SE direction and there are low mountains and middle mountain areas on the southeastern slopes of Trebevic and on the northwestern slopes between valley peaks (Figure 7). The mean altitude of the bottom of the valley is approximately 515 m. The lowest height above sea level is found in the north of the valley (in Reljeva, 493 m) and points of slightly higher heights are in the catchment area and enlargements of the Željeznica, Kasindolka, Tilava, Miljacka and Zujevina Rivers.

The mathematical geographical position of the valley is defined with the following coordinates: (i) northernmost point: 43°57'05" N; 18°38'20" E; (ii) southernmost point: 43°43'05"N; 18°29'25" E; (iii) westernmost point: 43°50'10" N; 18°08'50" E; and (iv) easternmost point: 43°55'55" N; 18°41'55" E. The surface of the valley is 75.25 km^2, while the valley bottom has a surface area of approximately 58 km^2 with the center of the area being Ilidza and Plandista [9]. The valley is a hydrological input for the source area of the Bosna River, in which the following are the main tributaries: Zujevina with Rakovica, Željeznica with Kasindolka, Dobrinja with Tilava and Miljacka. On the basis of its geological composition and geomorphological set, the Sarajevo Valley can be differentiated into two morphological units, i.e., (i) the mountainous rim and (ii) the Sarajevo field, which contains the valley bottom at the Miljacka River (see Figure 7). The area of the mountain rim has two special features, which have different hydrogeological features conditioned by different structures that display composition structure and surface composition. They are the highest mountain peak, mountain areas of middle height and the belt of the low rim. The highest mountain peaks are approximately 1000 m and higher and their structures are dominated by karstified limestone and dolomite. These include Bjelašnica, Igman, Trebević, Jahorina, Crepoljsko and similar mountain areas of middle heights (700 to 1000 m) and are morphologically more complex. Here, limestone masses are represented in mainly smaller zones by isolated tables or blocks and, in between, widespread narrow and wider zones of verven sediments. The bulk of the lower rim includes areas from 520 to 700 m and is morphologically significantly different from the previous two zones. Here, next to the limestone and dolomite of the early Triassic, much broader zones occur and, in the higher parts, tertiary freshwater remains. The lowest part is composed of complex tertiary sediments and a quaternary plateau is formed from these sediments from different ages.

The Sarajevo Field together with the alluvial plateau of the Miljacka River composes the second largest morphological location within which freshwater Triassic complexes cover the deposits of gravel-sand sediments of different widths. Sarajevo Field, in the narrow sense, can be differentiated into two parts. First, the youngest part extends directly from the foot of Igman (south from Vojkovica to Vrela Bosne) and further toward the east, approximately to the Blažuj-Ilidža-Kotorac Valley (where we find the first barrier to underground accumulation). On the eastern edge of this area are thermal and sorrel waters, which furthermore continue to the northwest. Based on the results of a series of bores, which have been created at more localities, we can conclude that the thickness and facial marks illustrating quaternary sediments are different (Figure 8). The thinnest are around Vojkovica and south of Plandista. However, as Stojčevac-Bačevo is approached, their thickness grows. Quaternary deposits of mud and clay lay relatively deep (approximately 40 m), which indicates upon positive oscillation of the bottom that only finely graded material was deposited and a large grade is held further south at the old mouth of the river. Here, the existence of an upper and lower water bearing has been demonstrated, which have substantially different hydraulic properties. The second part of the field is located north and northeast from the Blažuj-Ilidža-Kotorac line. It differs structurally from the young quaternary depression in that the Triassic deposits here have sunken significantly deeper as a freshwater petrographic complex. The quarter is part of the field and is relatively shallow from a few to approximately ten meters (the area around Lukavice, Nedžarića, Rajlovca and others).

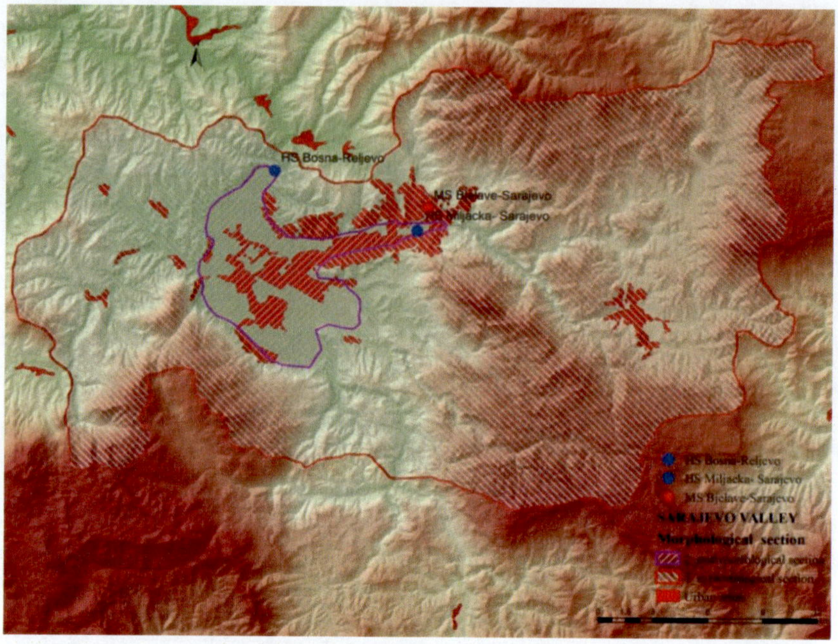

Figure 7. Topological location of the Sarajevo Valley and the locations of meteorological and hydrological stations used in this study (Designed by the authors).

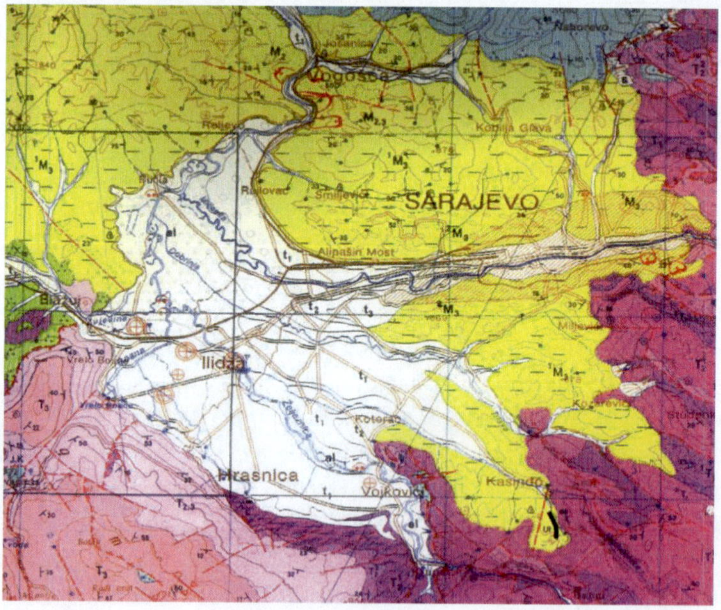

Figure 8. Geological characteristics of the Sarajevo Valley [10]. (T_1) Lower Triassic: Quartz-micasandstones, shales, marl and limestone; ($^1T_1$) Lower Triassic: Quartz-mica sandstones; ($^2T_1$) Lower Triassic: Sandstones, marls, shales; (T^1_1) Lower Triassic: Sandstones, shales, marls; (T^1_2) Middle Triassic: Limestones with crinoids, brachiopods and ammonites; (T^2_2) Middle Triassic: Limestones, sandstones, shales, chert, tuff; ($T_{2,3}$) Middle, Upper Triassic: Dolomite and limestone; (T_3) Upper Triassic: Limestone and dolomite with megalodons; (K_2) Middle Cretaceous: Flysch - breccia limestones, sandstones, conglomerates; ($M_{2,3}$) Middle, Upper Miocene: Conglomerates, sandstones, marls; ($^1M_3$) Upper Miocene: Clay, marl, sand; (Al) Alluvium; (t_1,t_2,t_3) The lowest, middle and highest river terrace.

The eastern and southeastern parts of the Sarajevo Valley are predominantly composed of Triassic limestone, while on the north and partially on the west, is dominated by deposits of Jurassic-Cretaceous in terms of cretaceous flysch. The northwestern and partial western rim of the valley is composed of Miocene clastic sediment and the base of the valley is composed of submolas sediments, i.e., loose clastics in which powdery deposits dominate less Pelitic with intergranular waters of a higher power. The genesis of the field is related to the subsidence of the field view and the escalation of sulfur compounds of a thermomineral source at Ilidza and synchronized so as to cause a depression in the alluvial-proluvial deposits in the aforementioned river flows.

3.2. The Current State of the Climate in the Sarajevo Valley

The Sarajevo Valley has specific climate characteristics that are strongly influenced by the surrounding high mountains Bjelašnica, Jahorina and Treskavica. These mountains are barriers to the influence of a maritime climate. This barrier is particularly reflected in the great differences in the annual amounts of precipitation. For example, the Bjelašnica and Jahorina receive approximately 1800 mm and the neighboring Romanija, which is located at the same altitude, receives approximately 1000 mm [11]. Based on the level of humidity, the summer months can be classified from subhumid to very humid and the air temperature can range from 1.5 °C in January to 18.9 °C (July) in the warm months. The valley is characterized by its rainfall and the snow, with the snow representing up to 20 % of the precipitation, while in the Igman - Bjelašnica mountains, snowfall can represents up to 50 % of the precipitation. The Sarajevo Field, which contains the alluvial valleys of the Miljacka River, is characterized by a fairly stable average monthly precipitation, which provides a modified Mediterranean pluviometric regime. Precipitation is significantly higher in the autumn compared to the rest of the year. The annual amount of precipitation in the Sarajevo Field is approximately 879 mm. The absolute air temperature amplitude is very high (around 70 °C), which is characteristic of the continental climate. The mean annual air temperature in the Sarajevo Field is about 9.7 °C with an average annual value for the vertical air temperature gradient of 6.5°C / 1 km.

3.3. Analysis of the Fluctuations in the Air Temperature in the Sarajevo Valley

Analyses of climate fluctuations in the instrumental period are based on measurements of the air temperature and the amount of precipitation, i.e., two the most frequently used climate elements for this type of analysis. The meteorological station at Bjelave has taken a homogeneous series of observations and measurements since 1888. This analysis is very important because the Sarajevo Valley is a good representative area of a mountainous region in southeastern Europe. Therefore, this study will contribute to our knowledge about the climate of this region and to the current climate situation in Europe as a whole.

A commonly used method for the analysis of air temperature fluctuations is a method involving moving or sliding in the mean air temperature. It aims to remove the shortest period of oscillation related to one-year intervals but without pronouncing the prevailing fluctuation trends over extensive periods. Based on the results of many studies, the moving average in a time series has been shown to be the most efficient method for climate analysis [12]. It is defined as shown below

$$X_{n+1} = \frac{1}{6} X_{n+1} (X_{6+n} - X_n)$$

$$X_n = \frac{1}{6} \sum_{j=n}^{n+5} X_j \tag{2}$$

where X is the quantity of interest (air temperature, precipitation or river flow) and n is the volume 0, 1, 2, 3, 4, ... m, where m is the last number in the time series that is being considered. Based on the above formula, a series is calculated to provide the relevant averages, which are presented in Figure 9. The annual air temperature averaged over the MI period (in the remainder of the text, this will be referred to as the normal air temperature) was 9.55 °C, but there were very pronounced interannual variations, which range in air temperatures between the limits 7.71 °C (1893) and 11.57 °C (1946). In the MI period, we determined a total of 60 years, where the mean annual air temperatures were above the normal air temperature, while for 62 years, the mean annual air temperatures were lower than the specified norm. In the analyzed series, a total of 25 actual periods of different lengths were extracted (2 to 10 years), which are characterized by an increase or decrease in the mean air temperatures. More specifically, the existence of 12 shorter or longer time intervals when the mean annual air temperatures were higher than the normal ones were characterized, while during 13 intervals, the mean annual air temperature was lower than normal. Also, it should be stressed that the longer homogenous periods of growth or decline in air temperature (approximately 14 times) were quite often separated by one-year intervals of opposite air temperature dependencies. The trend of air temperature changes for the period considered is calculated using the last mean square method. We found that this trend is best described by a linear function as follows: $Y = aX + b$, where a = 0.0075 and b = 9.1368. Thus, a certain trend in the mean annual air temperature is positive with an increase of 0.82 °C in the MI period.

Based on the analysis of the air temperature characteristics, it is possible to select four dominant periods during the MI period. The first such characteristic period was recorded from 1888 to 1925 and displayed low air temperatures with an average of 9.20 °C, which is 0.35 °C less than the normal air temperature. Although there are several (mostly) one-year breaks with a lower mean annual air temperature, the predominant increase in the mean annual air temperature is a basic characteristic feature of the second period (1926 - 1961). The period's average has a value of 9.82 °C, which is 0.27 °C above the normal temperature. The next period (1962 - 1986) is again characterized by a decline in the prevailing mean annual air temperatures and the average is 9.45 °C, which is 0.10 °C lower than the normal air temperature and indicates a less pronounced cooling compared to the first period. During the last quarter period, which runs from 1987 to 2010, the most prominent characteristic is an increase in the mean annual air temperature and indicates an average of 10.02 °C, which is 0.47 °C above the normal air temperature.

In establishing the relationship between the actual river flow and the mean annual air temperature, a five-year moving average is a powerful tool because it significantly removes oscillations in air temperature values and provides an easier way to determine time intervals with air temperature tendencies. The first of these was registered at the beginning of the MI period (1888 - 1895) with low air temperatures, which has otherwise been a characteristic of the air temperature regime of the entire European continent since the beginning of the 20th century.

The average value was 8.95 °C, which is 0.61 °C below the normal air temperature. Over the next ten years (1896 - 1905), the air temperature increased slightly so that the average was 9.35 °C. Colder weather was again observed and characterized the following (third) eight-year period (1906 - 1913) with a mean air temperature value of 8.93 °C, which is close to the value of the first period.

A new moderate warming period occurred from 1914 -1924 and the mean air temperature was 9.40 °C. This period was followed by a six-year period (1925 - 1930) with an increase in the mean air temperature to 9.91 °C, which is 0.35 °C above the normal air temperature. During the period from 1931 - 1943, there were very pronounced interannual air temperature variations in terms of the existence of a one- to two-year interval during which the air temperatures were lower than normal.

The mean air temperature for this period was 9.47 °C, i.e., 0.11 °C below the normal air temperature calculated for the MI period. During the period from 1944 - 1953, the mean air temperature reached a value of 10.27 °C. In 1946, the highest mean annual air temperature, 11.57 °C, was observed (Figure 9). However, it is of note that this time interval is known as one when extremely high air temperatures were recorded in Europe. In comparison to the previous period, during the eight-year period from 1954 - 1961, there was a decrease in the mean air temperature to 9.63 °C. The tendency toward a further decrease in the mean air temperature is a basic characteristic of the period from 1962 - 1974 and thus, the average for this period was 9.50 °C.

Although the interannual air temperature variations were more pronounced, similar air temperature characteristics were observed for the next period (1975 - 1986). For this period, the mean air temperature was 9.47 °C and there were one- to three-year intervals during which increases and decreases in the air temperature were observed. Considering that all of the mean annual air temperatures, except for two years, were above the observed average, the last selected period (1987 - 2010) had a distinct tendency toward an increase in the air temperature. For this period, the mean air temperature was approximately 10.0 °C, which indicates that this was one of the periods with the most pronounced warming trends in the 20th century.

Based on the data from the five-year moving average, it can be concluded that there are 11 fluctuating periods of different lengths ranging in the interval of 6 - 14 years, which indicates a non-uniformity in the oscillations of air temperature. A general feature of the MI period is that the end of the 19th and the beginning of the 20th century were characterized by remarkably colder weather during the first half of last century with air temperatures oscillating more or less around the normal temperature. The second half of the 20th century can be identified as a period of warming within which there are two ten-year periods of an intense increase in air temperature.

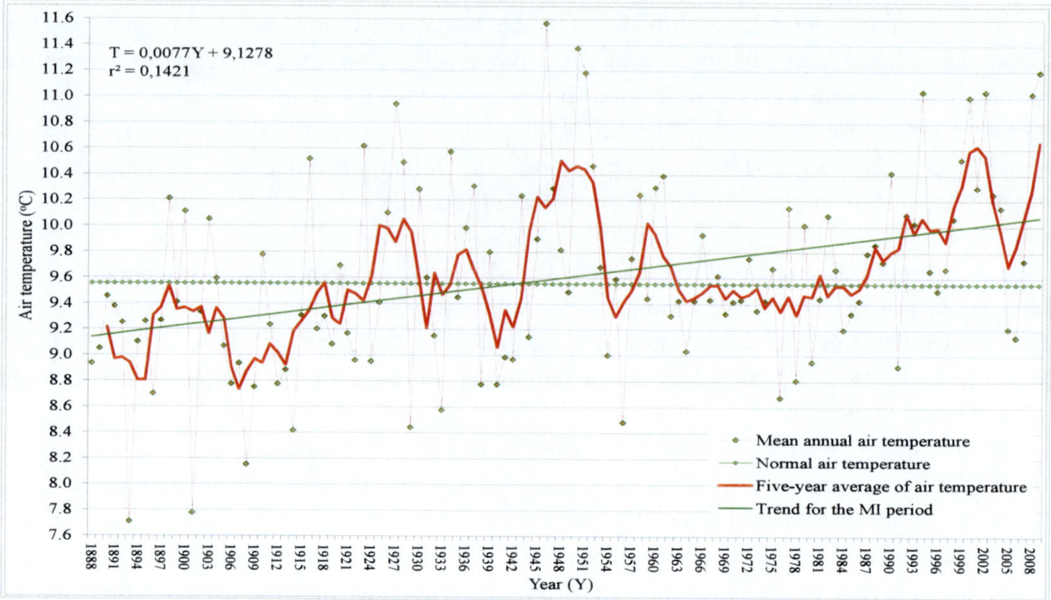

Figure 9. Fluctuations in the mean annual air temperatures during the MI period on the MS Sarajevo-Bjelave.

Based the data presented here, we can emphasize a few general conclusions about the air temperature regime of the Sarajevo Valley for the MI period: (i) The basic characteristic of the air temperature regime is warming when considering that the linear trend of the air temperature is positive with an increase of 0.82 °C for the MI period. (ii) Oscillations in the mean annual air temperatures are evident, but it is not possible to determine a regularity in their occurrence during the MI period. (iii) Based on polynomial fitting of the order of the sixth degree, cyclic trends showed that in comparison with the normal air temperature, we can sort through the four time periods with increases or decreases in air temperatures. However, the lengths of these periods are quite different and change from 18 to 40 years, which does not allow us to make more specific conclusions about the origin of these fluctuations. (iv) An analysis of the five-year average indicates that there are 11 periods with time lengths of 6 - 14 years and mainly different air temperature tendencies. The lengths of these periods are as follows: 8 (at the beginning of the MI period), 12, 12, 10, 9, 13, 9, 14, 8, 11 and 7 years (at the end of the instrumental period).

3.4. Analysis of the Fluctuations in Precipitation in the Sarajevo Valley

In addition to studying air temperature fluctuations, we considered fluctuations in the amount of precipitation as one of the basic indicators of climate variability research. The methodology applied to the calculation was the same as used in the analysis of air temperature fluctuations [12].

The annual amount of precipitation averaged over the MI period in further text will be referred to as the normal amount of precipitation. That amount for the MS Sarajevo-Bjelave is 918.8 mm (Figure 10). However, from this figure, a very pronounced oscillation in the individual annual sum amount of precipitation in the instrumental average is observed. The lowest annual sum of precipitation was registered in 1926 (625.0 mm), while the highest was measured in 1915 (1534.0 mm). If we look at the MI period, we can see that for 58 years, the annual sum was over the normal amount precipitation, while for 64 years, the annual sum of precipitation was below the normal value.

The analysis of the linear trend leads us to the conclusion that the annual amount of precipitation during the instrumental period slightly increased. More specifically, for this 122-year period, the total increase in the annual amount of precipitation was only approximately 14 mm. Thus, there was a barely noticeable humidification of air inside the Sarajevo Valley. By the five-year moving average analysis, it was determined that eight cycles of different time lengths represent periods with amounts of precipitation above and below the normal amount of precipitation (red line in Figure 10). The first of these cycles is the instrumental period until 1896, within which there are two four-year minicycles with approximately identical values. The next cycle is the time interval from 1897 to 1905, which had an average value of 954.8 mm. The third round precipitation cycle was one of the longest since the last one from 1906 to 1925. In comparison to a normal cycle, it had a slightly reduced average value of 913.4 mm. The main characteristic of the next cycle (1926 - 1947) is the pronounced differences in the time duration of the period with precipitation trends of opposite tendencies. The average amount of precipitation during this period was 906.2 mm, which was lower than the corresponding value in the previous cycle. For the fifth cycle (1948 - 1972), the average was 921.2 mm and the sixth period (1973 - 1983) had a five-year average precipitation amount higher than the fifth cycle (933.7 mm).

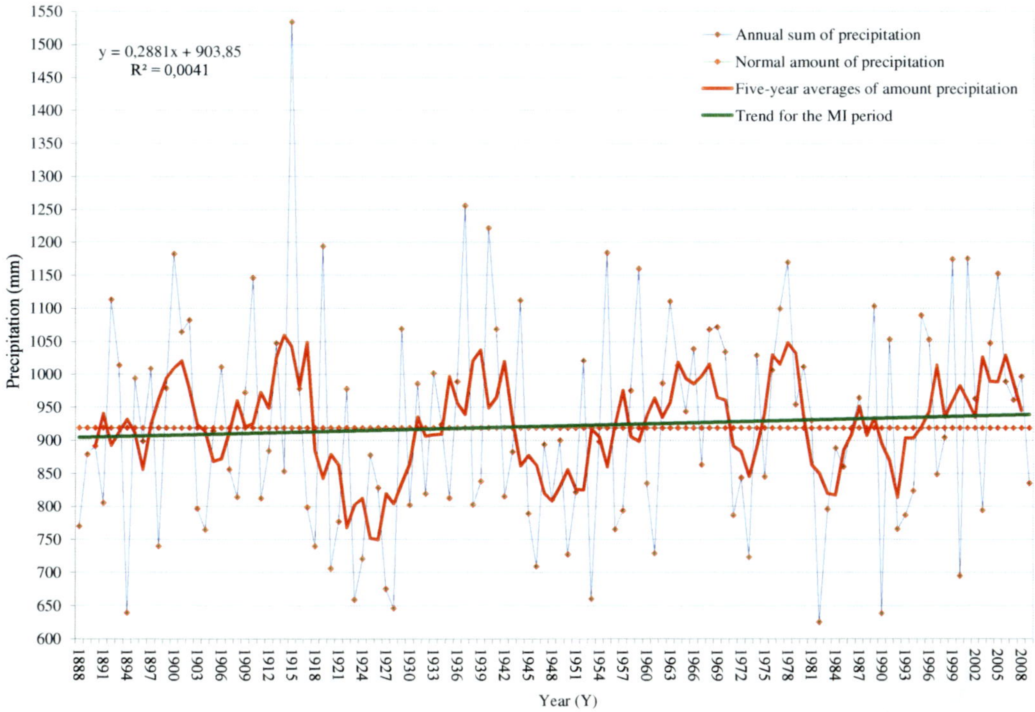

Figure 10. Fluctuations in the annual amount of precipitation for the MI period on MS Sarajevo-Bjelave.

The seventh cycle was a very short period (1984 - 1991) with a precipitation value of 859.9 mm, which was the lowest amount of precipitation in the MI period. This period can be considered a time of a pronounced arid climate. Finally, the last period (1992 - 2010) can be identified as one part of a new cycle that has not yet been completed. It indicates a moderate humid climate with regard to its amount of precipitation (930.5 mm).

4. ANALYSIS OF HYDROGRAPHICAL CONDITIONS IN THE SARAJEVO VALLEY

The Sarajevo Valley has a very developed network of surface and underground waterways. The studied area includes the Miljacka, Željeznica, Dobrinja and Zujevina Rivers and the region upstream of the Bosna River to the Reljevo VS. The total length of this river network is approximately 120 km. Other specific characteristics of these rivers are given in Table 3.

The internal depression of the Field area is ruled by different hydrogeological relations in its parts and is completely different from its mountain rim. The structural diversity of the morphogenetic differences along with the hydrogeological features of the rocks in a litofacial sense has a direct effect on all of the hydrological and hydrogeological phenomena, processes, dynamics and regime of the underground waters, as well as the formation of the issued type, especially larger reservoirs, how they recharge and discharge and their hydraulic characteristics. Coarse clastic sediments, such as sand, gravel and sand stone, are primarily porous rocks and the magnitude of their porosity depends mainly on separated grains. As a rule, groundwater in these porous media flow at low speeds or its flow can be regarded as laminar. Measurements of the effective speed of the field show that the speeds range from 20 to 50 m day^{-1} and that the direction of movement corresponds to the movement of the water on the surface of the terrain and on the topography of the terrain.

To determine the water balance of the exploration area and to understand the impact of climate change on the hydrological regime, the Miljacka and Bosnia Rivers have been determined to be the most relevant. Due to some specific features related to the abstraction of water immediately after the source zone of the river and the infiltration of river water into the upper granular accumulation that exists in the Sarajevo Field, the Bosna River undoubtedly has significance for the entire Sarajevo Valley.

Table 3. The main rivers of the Sarajevo Valley. The meaning of the symbols are: L is the total length of the river network, while C.A. is the catchment area

River	L (km)	C.A. (km^2)
Zujevina	23.50	171.20
Dobrinja	9.20	57.50
Bosna	8.30	35.80
Zeljeznica	40.0	481.80
Miljacka	38.0	387.60
Total:	122.80	1133.90

On the other side, the Miljacka River is the backbone of urban development for the Sarajevo city agglomeration and is significantly burdened by different types of anthropogenic pressures, which is certainly reflected in its hydrological regime.

4.1. The current Hydrological State in the Sarajevo Field and the Alluvial Plateau of the Miljacka River

Part of the Sarajevo Field, which is situated inside the second morphological section (Figure 11), is an area surrounded by the edge of the Igman Mountain, the Željeznica River and the Bosna River, which is an area of approximately 58 km^2.

In morphological terms, the plateau gradually decreases from the Igman rim to the output profile of the river with its highest peak at approximately 525 m above sea level and its lowest elevation at approximately 493 m above sea level.

The underground reservoirs are naturally recharged and renewed through the infiltration of water from the Željeznica, Bosnia, Večerica and Stojčevac streams; the flow of ground water from Igman-Bjelašnica through the contact zone; and the vertical infiltration of precipitation into the field [13]. Groundwater coming from the Igman-Bjelašnica mountains, which in part forms a spill way, appears from Grlice to Vrelo Bosna forming the waterways Stojčevac and Vecerica and partly as a subsurface water supply in the form of an underground reservoir. Groundwater recharge in Igman-Bjelašnica is performed exclusively through the infiltration of atmospheric water, which occurs mainly in the sinkhole zone and system of fissures.

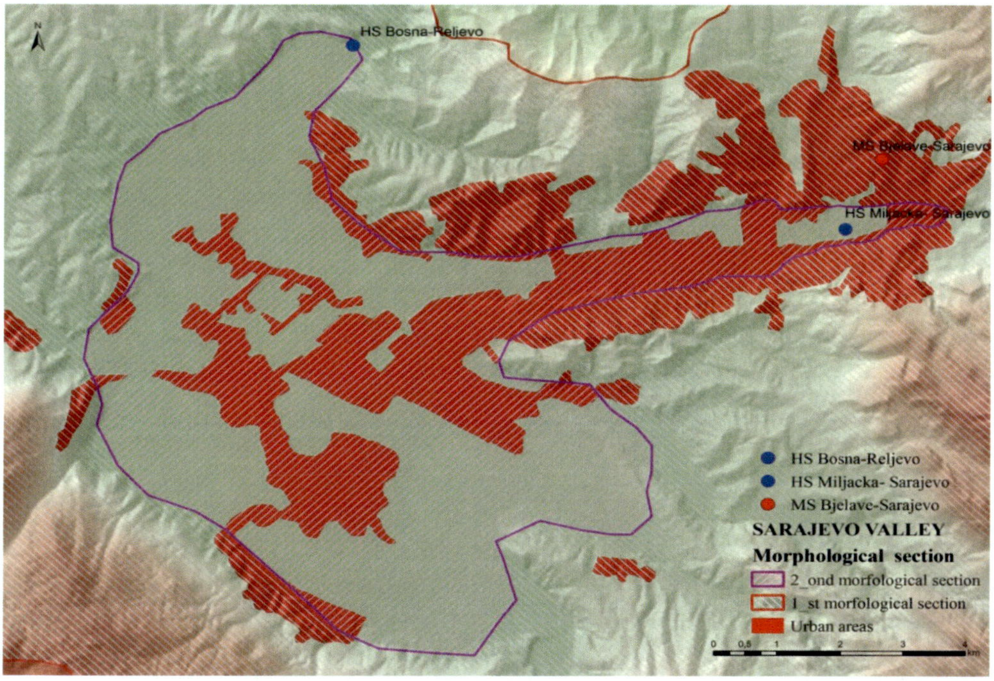

Figure 11. Sarajevo Field and the alluvial plateau of the Miljacka River (Designed by the authors).

Let us note that according to the results from the water balance of the Sarajevo Field from 1984, 1985, 1986 and 1987, the value for the vertical infiltration from precipitation (0.271 $m^3 s^{-1}$) and evaporation (0.297 $m^3 s^{-1}$) are similar and are neglected in the calculation for the balance. The basic features of the groundwater regime are as follows. (i) The regime for the underground water is of rain-snow quality and is similar to the water regime for open watercourses. Due to the large supply of snow in the catchment area, regardless of the precipitation in the spring, the groundwater is high in May and even in June. (ii) The system reacts almost instantaneously to effective precipitation. The accumulation point of the water level for the ground water during higher precipitation delays the accumulation counts in open streams by only two hours. (iii) The direction of ground water flow is approximately parallel to the Igman rim and the Željeznica River. (iv) According to some studies, the natural renewable quantity of ground water ranges from 1.25 to 800 $m^3 s^{-1}$ [14].

For studies of hydrological fluctuations, analyses were performed using the mean annual river flow at selected sections of the Bosna and Miljacka Rivers. On the Bosna River, there are several measurement locations: Bosnia-Vrelo, Bosna-Plandište and Bosna-Reljevo. Among these locations, Bosnia-Reljevo was chosen because it is located on the outskirts of the Sarajevo Field, i.e., it is representative of all of the collected tributary waters. Also, the specified location has a representative and reliable instrument for hydrological monitoring from 1926. From t he Miljacka River, there are also multiple measurement locations both in the upstream part of the alluvial section and the downstream part of the following measurement locations where measurements for three reference profiles has been established. To study the fluctuations in river flow, a reference measurement location was established on the Skenderija bridge, which also has a reliable instrument for hydrological monitoring.

4.2. Analysis of Flow Fluctuations for the Bosna River

The Bosna River is made up of about 50 karst springs formed in contact with the limestone sections of Igman and intergranular Sarajevo plain deposits, which have a geological structure as described in subsection 3.1. At the main source of the Bosna River, the mouth point, the mean annual river flow is about 8 $m^3 s^{-1}$, except during the precipitation season when this value increases to approximately 24 $m^3 s^{-1}$. It is very characteristic that the source zone does not dry even during periods of little water as a result of, among other things, the recharge of the underground karst [15]. On the part of the source zone, until the HS Reljevo Bosnia, the Bosna River receives water from the Željeznica, Zujevina and Dobrinja Rivers and Miljacka.

HS Bosnia-Reljevo is located on the pedestrian bridge in Reljevo village, which is approximately 11.6 km away from the source zone. The water meter is continually performing regular hydrological measurements, except for a discontinuity that occurred during the war and immediate post-war periods (1990 - 2002). For the above mentioned periods, the correlation was determined that there was neither a specific hydro-climate nor any other (anthropogenic) disturbance that could affect the quantitative value of the trend or the general appearance of fluctuations in river flow charts. In other words, the data used can be taken as completely representative of the river flow research that is correlated with climatic fluctuations. The methodology used in the study of fluctuations in river flow is identical to the methodology described in Section 3. More specifically, in this study, we used data from five-year moving averages of mean annual river flow.

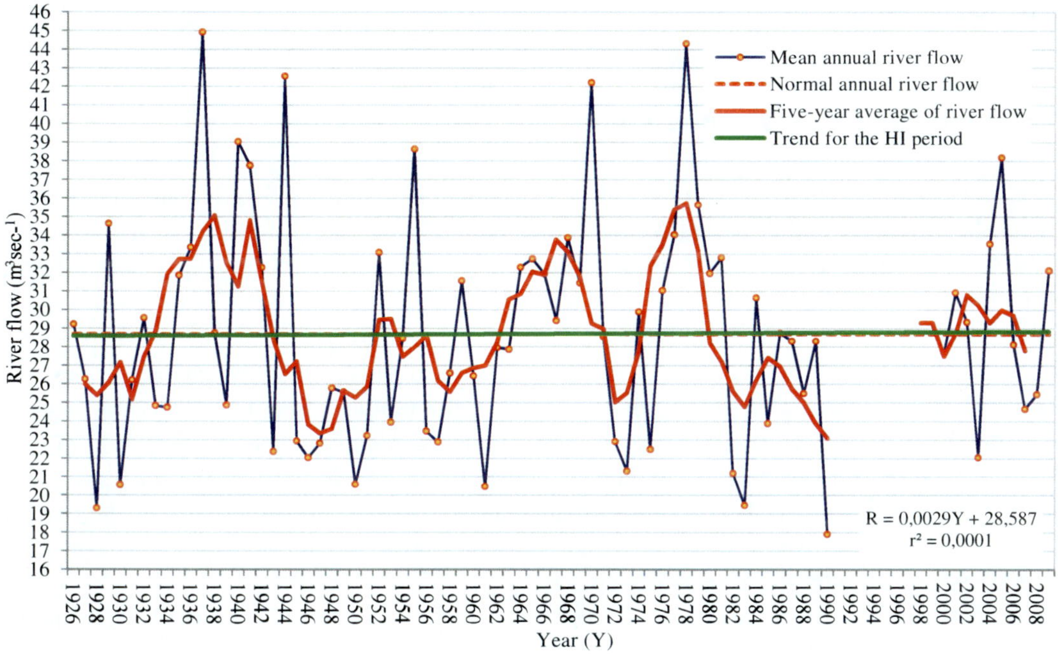

Figure 12. Fluctuations in the mean annual river flow for the HI period on the HS Bosna-Reljevo.

The normal river flow during the HI period is 28.70 $m^3 s^{-1}$ a significant variation from this average is characteristic of this river flow (Figure 12). More specifically, the largest mean annual river flow during the analyzed period was registered in 1937 and amounted to 44.93 $m^3 s^{-1}$. Another extreme analog, a minimum of the mean annual river flow, was recorded in 1990 and had a value of 17.92 $m^3 s^{-1}$. A fluctuation in the first period, which registered from 1926 to 1932, is characterized by a reduced value given that this period's average was 26.0 $m^3 s^{-1}$. After this performance period, which is compared to the normal and is characterized with high values for the river flow, i.e., the period from 1933 - 1940, the average river flow was 31.9 $m^3 s^{-1}$. The next fluctuating period, which lasted from 1941 to 1948, developed a slow growth trend and the average river flow amounted to 28.4 $m^3 s^{-1}$. The following fluctuating period lasted 12 years (1949 - 1960) with an average close to the normal period (26.8 $m^3 s^{-1}$). From 1961 to 1973, which is the fifth period of fluctuations, the average river flow was 30.0 $m^3 s^{-1}$, which indicates a further slight increase in river flow.

For the period from 1974 - 1983, subsequent fluctuations occurred with similar characteristics to the previous three and developed the trend of slow growth with a periodic average river flow of 30.5 $m^3 s^{-1}$. The fluctuations during the next seven-year period (1984 - 1990) were characterized by a decrease in river flow below the normal value of 25.4 $m^3 s^{-1}$. Although the period from 1990 - 2000 formed an instrumental break in the hydrological monitoring, a continuity of declining river flow can be discerned given that from 2000 to 2003, the registered value was 28.5 $m^3 s^{-1}$. The last fluctuating period ending in 2010 and has a slightly increased periodic average as compared to the normal value of 29.5 $m^3 s^{-1}$. Based on the data presented, it can be concluded that during the period from 1926 - 2010, there were a total of 9 fluctuating periods whose durations varied within the range of 7 - 12 years. The general characteristic of the quantitative dynamics of f low in the HS Bosna Reljevo in the HI period, which determined nearly stagnant trend changes or the value of the trend, is positive at only approximately 0.2 $m^3 s^{-1}$.

4.3. Analysis of Flow Fluctuation for the Miljacka River

In hydrological terms, the Miljacka River results in the confluence of its two source arms, Paljanska Miljacka and Mokranjska Miljacka, at the wider area at the site of the Goat's Bridge. The longer tributary and, in this sense, the source stream is the Paljanska Miljacka River, which arises from the plentiful waters of the karst springs at the west-north western edge of Mount Jahorina at an altitude of 1010 m. The entire river system upstream of the expansion Miljacka alluvial plains (near the Babic Basta village) has the characteristics of a mountain stream with a very steep and un harmonized longitudinal profile. Downstream from this site, the Miljacka River flows through the alluvial plateau with a very small drop of only 3 % to 5 %. In this part of the Miljacka River, the Sarajevo Valley agglomeration was developed, which has over 400 000 inhabitants. HS Miljacka-Sarajevo was established on the bridge at Skenderija, which has representative hydrological monitoring and was suspended for the reasons already given for HS Bosnia-Reljevo.

Analogous to the previous methodology applied, the normal river flow throughout the HI period was calculated from 5.46 $m^3 s^{-1}$ with a characteristic and significant deviation from this average (Figure 13). More specifically, the largest mean annual river flow in the analyzed period was registered in 1937 and had value of 9.09 $m^3 s^{-1}$. Another extreme analog minimum of the mean annual river flow was recorded in 1990 with a threefold lower value of 3.03 $m^3 s^{-1}$. The first fluctuating period was registered from 1926 to 1933 and had a registered periodic average river flow of 5.3 $m^3 s^{-1}$. The next fluctuating period lasted 7 years (1934 - 1940) and had an increased periodic average river flow of 6.0 $m^3 s^{-1}$. From 1941 to 1947 (the third period), there was a lower fluctuation in the average river flow (5.0 $m^3 s^{-1}$). The fourth fluctuation period lasted 11 years (1948 - 1959) and was characterized by a lower average (5.2 $m^3 s^{-1}$) than the normal or average periodic values. The next seven-year period had a less pronounced oscillation with a higher average compared to normal, i.e., 5.6 $m^3 s^{-1}$. The trend of a slight increase in river flow was continued during the sixth fluctuation (1966 - 1973) and the average was 6.2 $m^3 s^{-1}$. In the period from 1974 - 1983, the trend registered a slight decline in the river flow value. Although still above normal, the value was 5.8 $m^3 s^{-1}$. During the eighth fluctuating period (1984 - 1990), a further downward trend in river flow was registered when considering the periodic average of 4.7 $m^3 s^{-1}$. During the downtime period (1990 - 2002), similar conclusions can be derived as in HS Bosnia-Reljevo or the empirically determined period of reduced river flow , which was continued for the last fluctuating period (2003 - 2010) during which the period's average was below normal (5.2 $m^3 s^{-1}$).

Based on the presented data, the existence of 9 fluctuating periods was also determined and the durations ranged in time from 7 - 11 years. Characteristics of the prevailing trend of stagnation also have characteristics that may have determined the trend of a slight decrease of only 0.1 $m^3 s^{-1}$.

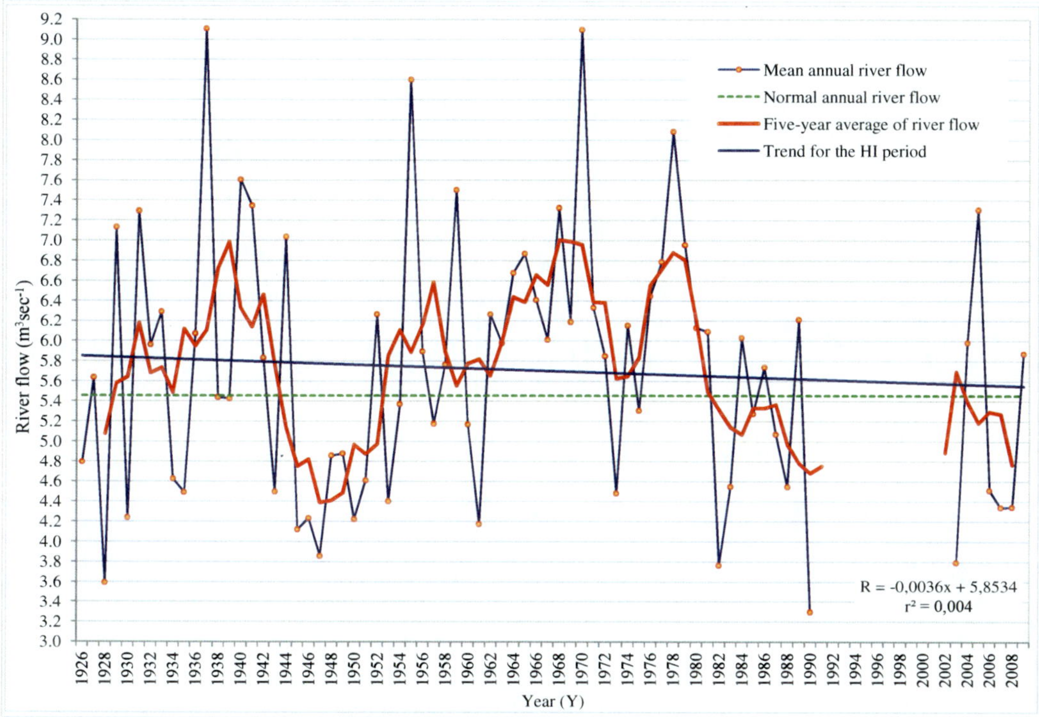

Figure 13. Fluctuations in the mean annual river flow for the HI period on the HS Miljacka-Sarajevo.

CONCLUSIONS

Based on data presented in Sections 3 and 4, the climate elements (air temperature and amount of precipitation) and a hydrological element (river flow) were analyzed in terms of defining their magnitude and duration of fluctuation. To remove the oscillations in the annual values, the analyzed elements were calculated on a five-year moving average, which were the basis for the study of fluctuations. The results of the study indicate that during the monitoring of data during the MI period, all investigated climate elements can be separated into time periods of different sizes (7 - 12 years) within which a full fluctuation can be observed, i.e., the time period within which the climate element analyzed exhibits all quantitative characteristics describes its perennial value.

The first and a particularly interesting conclusion to be drawn from these studies, is the high correlation between the quantitative values in the annual courses of the amount of precipitation and the mean annual river flow.

For an annual course, the air temperature was found to mainly have an opposite trend due to the mentioned two climate elements. This finding is fully in agreement with the spatial and temporal dynamics of the elements analyzed. These conclusions are derived from the graphical representation of the mean annual river flow and air temperature in April during the HI period. April was chosen as a representative example because, in this month, the highest monthly river flow was registered.

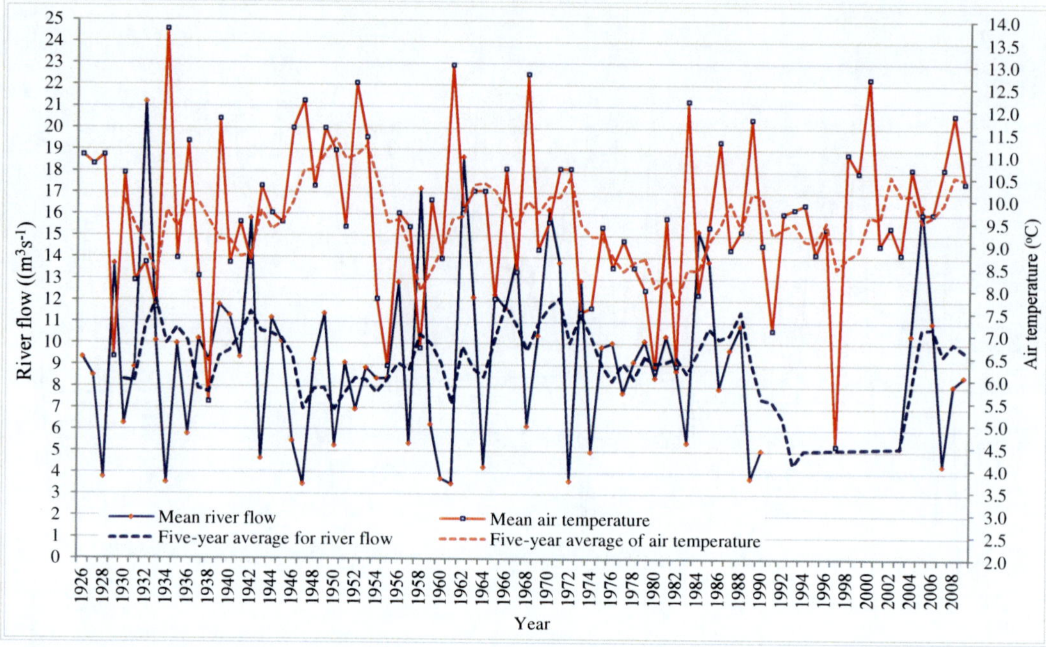

Figure 14. Comparative presentation of the fluctuations in the mean monthly river flow and air temperature in April during the HI period on the HS Bosnia-Reljevo.

Specifically, this research was conducted only at HS Bosnia-Reljevo and takes into consideration that the previous analyses indicated that the river flow during the HI period for both the hydrological stations highly correlated with each other and that it is sufficient to consider one of them or HS Bosnia-Reljevo.

The results of the analysis are presented in Figs. 14 and 15. In Figure 14, a comparative review is given of the mean monthly air temperature and mean river flow in April for the HI period. The main conclusion that can be obtained on the basis of the presented data is that to be almost completely expressed, the air temperature regime is inversely related to river flow. In any case, the present small deviations, which are related to slight time delay changes in river flow in relation to air temperature effects, are fully consistent with the mechanism of heating the land area and contact watercourses.

Figure 15 presents the results of an analysis of the mean river flow and amount of precipitation in April during the HI period. The main conclusion to be drawn from the analyzed data is that there is a high correlation between the mean river flow and amount of precipitation, i.e., the mean river flow is directly proportional to the amount of precipitation. Similar to the previous analysis, it can also be determined here that there are periods with small time delays between the appearance of high water in relation to the maximum amount of precipitation. The main reason for these delays can be found in the specific geological structure of the ground contact mountain units, which are characterized by a certain retardation due to the specific characteristics of the karst terrain. The completed analyses of air temperature and amount of precipitation have shown that fluctuations with average time durations of 7 - 11 years can be extracted from the MI period. Through the analysis of additional data, it was also determined that within the specified time interval, they exhibit all of the quantitative and qualitative characteristics of their spatial and temporal dynamics.

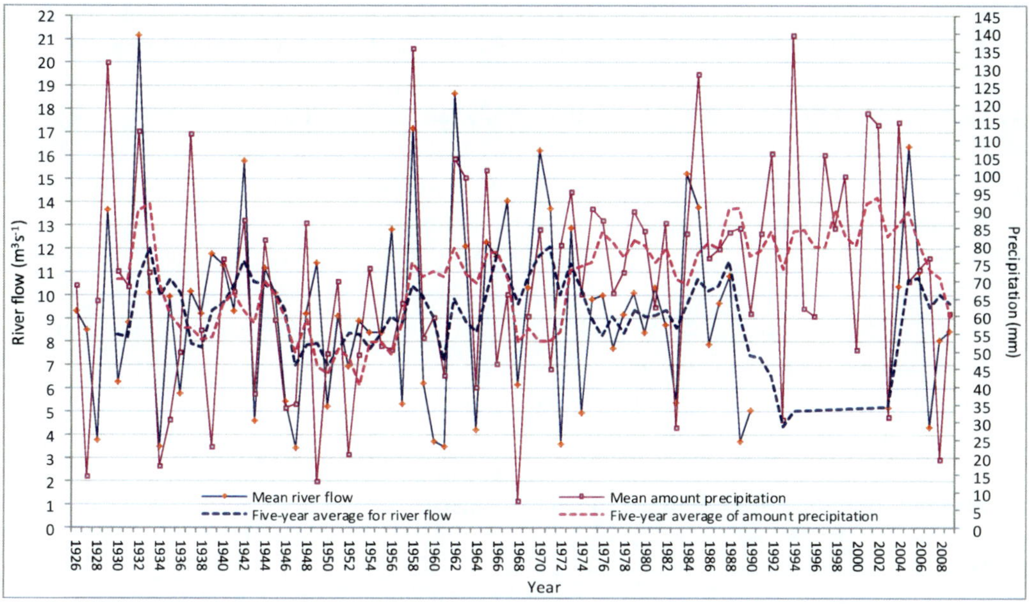

Figure 15. Comparative presentation of the fluctuations in the mean monthly river flow and the sum of precipitation in April during the HI period on the HS Bosnia-Reljevo.

The main conclusion, which was obtained by analysis of the climate data, is that in the Sarajevo Valley there is a positive trend in mean annual air temperature of 0.82 °C. The trend of the annual amount of precipitation shows an increase of 14 mm. From analyses of the river flow during the monitoring of the HI period on HS Bosna-Reljevo and HS Miljacka-Sarajevo, we can conclude that trends in river flow characteristics are almost stagnant with respect to values of 0.2 $m^3 s^{-1}$ for HS Bosna-Reljevo and 0.1 $m^3 s^{-1}$ for HS Miljacka-Sarajevo.

Correlating the basic elements of climate, precipitation and air temperature with the river flow in the Miljacka and Bosna Rivers, the conclusion was obtained that there is a high degree of quantitative compliance in their mean annual river flows. These results indicate that climate change in the Sarajevo Valley, in relation to the northern hemisphere, is still less pronounced in terms of the impact on the water regime. The above conclusions can be explained by the fact that the Sarajevo Valley is in the zone of a mountain depression for a geographic region that is well isolated from orographic- expressed anthropogenic influences, which come mainly from large urban agglomerations of the northern hemisphere.

REFERENCES

[1] F. Ludwig, P. Kabat, H. van Schaik, M. van der Valk, Eds.; *Climate Change Adaptation in the Water Sector*; Sterling: London, GB, 2009; pp. 35-44.
[2] C. Pahl-Wostl, P. Kabat, J. Moltgen, Eds.; *Adaptive and Integrated Water Management,* Springer: Berlin, DE, 2008; p. 440.
[3] F. Rubel, M. Kottek, *Meteorol. Z.*, 19(2), 135-141 (2010).
[4] Internet Source [http://ga.water.usgs.gov/edu/watercyclesummary.html].
[5] Internet Source [http://www.physicalgeography.net/fundamentals/8b.html].

[6] *Initial National Communication (INC) Of Bosnia and Herzegovina under the United Nations Framework Conventionon Climate Change (UNFCCC)*, Banja Luka, BA, 2009; p. 195.

[7] *Climate Atlas of Yugoslavie*, Federal Hidrometeorological Institute, Belgrade, RS, 1981 (In Bosnian).

[8] *Framework Water Management Plan of Bosnia and Herzegovina*, Institute for hydraulicity: Sarajevo, BA, 1994; p. 186 (In Bosnian).

[9] N. Drešković, *M.Sc. Thesis*, Faculty of Sciences, University of Sarajevo: Sarajevo, BA, 2004, p. 208 (In Bosnian).

[10] *Basic Geological Map*, Scale: 1:100 000, Sheet of Sarajevo, Federal Geological Institute, Belgrade, RS, 1979 (In Bosnian).

[11] N. Drešković, *Ph.D. Thesis*, Faculty of Science, University of Sarajevo: Sarajevo, BA, 2011, p. 224 (In Bosnian), (In proc).

[12] Đ. Radinović, *Weather and Climate of Yugoslavia*; Federal Hidrometeorological Institute, Belgrade, RS, 1974; p. 245 (In Bosnian).

[13] E. Hadžić, *M.Sc. Thesis*, University of Sarajevo: Sarajevo, BA, 2001; p. 214 (In Bosnian).

[14] E. Hadžić, *Ph.D. Thesis*, University of Sarajevo: Sarajevo, BA, 2006; p. 265 (In Bosnian).

[15] *Research Works for the Water Supply Company of Sarajevo, 1983-1986*; Sarajevo Institute for hydraulicity, Sarajevo, BA, 1989; p. 250 (In Bosnian).

In: Essays on Fundamental and Applied Environmental Topics ISBN 978-1-61942-522-4
Editor: Dragutin Mihailovic

Chapter 15

NONGOVERNMENTAL ORGANIZATIONS AND THE LOCAL ENVIRONMENTAL POLICY: A CASE STUDY FOR THE CITY OF NOVI SAD (SERBIA)

J. Čikić[*] *and Ž. Petrović*
Department for Agricultural Economics and Rural Sociology, Faculty of Agriculture, University of Novi Sad, Novi Sad, Serbia

ABSTRACT

In the era of bottom-up development planning, social networking and capacity building, one of the key issues in governing local communities is how to enable citizen participation in policy making and practice. This especially refers to environmental issues as the environment is considered a public good. The ecological modernization as a contemporary theoretical mainstream in researching and planning environmental policy argues that citizen participation is one of the crucial elements in establishing an environmentally sensitive institutional framework for policy making. In exploring the possible patterns of citizen participation in environmental policy, it can be concluded that some type of connection among citizens often resulting in the formation of nongovernmental organizations (NGOs) can be an efficient way to influence local policy making. Such organizations, although frequently balancing between formal and informal 'ways of doing things', are more likely to succeed in achieving their environmental interests and goals than are individuals because the organizations have greater access to resources such as financial support, information availability, social capital, human resources etc.

In such context, the role and status of NGOs related to environmental issues is being analyzed. As a subject of the research, the NGOs are especially interesting at the local community level because they seem to be more closely connected to the citizens. In the post-socialist and transitional societies, such as contemporary Serbia, the role of environmental NGOs is regarded not only from the standpoint of environmental protection but also as a pattern of political decentralization and a manner of modernizing (local) governance.

[*] Corresponding author: E-mail address: cikicj@polj.uns.ac.rs; Phone: +381 21 485 3270. Fax: +381 21 635 0822.

The role and status of NGOs in local environmental policy is analyzed according to the data collected in the survey that covered all environmental NGOs registered at the city of Novi Sad. Five major issues are being analyzed: human resources, available equipment and facilities, areas of interest and activities of NGOs, cooperation and information and funding. Those issues are addressed not only with the respect to the potential negotiating power of the environmental NGOs in the policy actors' network but also with the respect to their ability to influence development of a community's environmental awareness.

1. Introduction

It has been almost 40 years since scientists gathered around the Club of Rome and published their first report on the demographic, economical and social aspects of global environmental issues [1]. After their, some might say, terrifying predictions based on the world system model, this report raised public attention and launched environmental problems to the very top of the list of global concerns, along with issues such as world peace, hunger, poverty elimination and nuclear weapons. Those who are more environmentally sound are probably going to say that environmental problems are the most important issue of all as they address the very base of human survival on this planet.

Since the 1980s, environmental issues have been incorporated into development policies concerning not only environmental protection but also economic, demographic, spatial, agricultural and rural development. Since the publication of another important report of the Brundtland Commission [2], the sustainable development comprised of economic, environmental, sociopolitical and cultural sustainability has been actively deployed in many societies as a dominant development strategy. According to the concept of sustainable development, contemporary societies can achieve long term wellbeing and welfare only if they significantly change their resource management. However, the effects of the changes in production and consumption patterns are hardly to be achieved if there is no basic sociopolitical and cultural sustainability. According to the cultural dimension of sustainability, the principle of responsibility, both individual and social, is one of the crucial elements in establishing political legitimization of the concept itself. Such political legitimization requires decentralization of power in decision making, so enabling and strengthening participation of different social actors is essential in order to accomplish sustainability.

Contemporary environmental policies (from the global to the national and the local levels) strive to broaden and enhance the network of social actors involved in policy planning, implementation, monitoring and evaluation. This means that environmental policies today (attempt to) engage, to as great a degree as possible, social actors that differ in aspects such as organization, power, social capital and financial resources. The important actors are the nongovernmental organizations (NGOs) that serve as a bridge between the citizens and the institutions. Similar, environmental policy is understood as a network of associated social actors or stakeholders, different in levels of organization, power and influence that mutually work, applying available measures, to eliminate and prevent environmental problems and improve environmental quality.

This chapter is focused on examining the role and position of NGOs in environmental policies in specific social conditions of post-socialist transition. Such context enables analysis of the two crucial questions regarding environmental protection in transitional societies: (i) how environmental policies are established and implemented in such societies and (ii) what is the role of NGOs in establishing environmental policy and facilitating democratization of such transitional societies.

This analysis is built upon a few general hypotheses. The first hypothesis is that urban areas are characterized by remarkable environmental pollution. This assumption is based upon structural characteristics of urban areas, both spatial and social. Namely, urban areas have always been the locus of spatial and temporal concentration of people and social functions. Therefore, urban areas have been the locus of social actions and relations that generate changes in environment. The second hypothesis includes the relation between physical and social space. Accordingly, urban areas are considered to be social communities, specific for their social structure and dynamic of social processes. A social community is a social construction made by locals for purposes of their social identification and sense of belonging as well as a social arena for individual and group need fulfillment. The third hypothesis indicates that the social community, as a methodological framework, is a fruitful ground for the analysis of environmental policies because it is an intersection of general social conditions and local specifics. Brought together, those circumstances create a network of factors that stimulates or has restrictive impact on the environmental policy.

In addition, this analysis requires some specific assumptions regarding an efficient environmental policy. The first specific assumption explains that the characteristics of environmental awareness have the most significant influence on the establishing and functioning of the environmental policy as an organized social action. Namely, environmental awareness is nothing but a way in which the social community recognizes its interest in environmental protection. Secondly, it is presumed that the efficiency of the environmental policy is most influenced by the characteristics of the network of social actors engaged in the system. And, thirdly, the characteristics of environmental awareness and the characteristics of the network of social actors are interdependent – the higher the environmental awareness (especially its manifest dimension), the more the social actors are engaged in the functioning of the environmental policy. And *vice versa*, the more complex is the social actors' network, the more developed the environmental awareness is.

Therefore, in this chapter, we discuss the role and the status of NGOs in the environmental policy stakeholders' network. In Section 2, we argue for the role of NGOs within the broader context of participation issues. Participation of the stakeholders in the environmental policy network is regarded from two perspectives: (i) the ecological modernization theory and (ii) community capacity building. Also, the issue of the role of environmental NGOs in the post-socialist and transitional societies, such as contemporary Serbia, is regarded not only from the standpoint of environmental protection but also as a pattern of political decentralization and a manner of modernizing (local) governance. In Section 3, we analyze the role and status of NGOs in local environmental policy based on the data collected in the survey that covered all environmental NGOs registered at the city of Novi Sad. Five major issues are analyzed: human resources, available equipment and facilities, areas of interest and activities of NGOs, cooperation and information and funding. These issues are addressed not only with the respect to the potential negotiating power of environmental NGOs in the policy actors' network but also with respect to their ability to influence the development of a community's environmental awareness.

2. The Environmental Protection and Stakeholders' Network – Participation as a Necessity or Trouble?

With respect to environmental policy, one of the most important questions is who can participate in policy planning, decision making and practice and to what extent can they participate. The issue of participation has become a crucial factor in the evaluation of an environmental policy's effects and a compulsory element in policy planning. Due to the complexity of its manifestations, however, there is no unique definition of participation. The most common mistake, when arguing participation, is to regard it only as an engagement in decision making. Such miscomprehension derives from the most obvious and, according to some authors, most important goal of participation in every political process, which is to practice power by influencing what kind of decisions are going to be made.

Exploring the participation of the American citizen in public affairs in the late 1960s, Arnstein [3] emphasized that the true meaning of citizen participation is practicing power. According to the author, participation can be viewed through the eight degree typology. The first two levels, manipulation and therapy, can be regarded as nonparticipation because their "*objective is not to enable people to participate in planning or conducting programs, but to enable powerholders to 'educate' or 'cure' the participants*" [3]. The next three levels, informing, consultation and placation, the author describes as "tokenism". These forms of participation suggest that participants' opinions can be heard, but they do not have "*power to insure that their views will be heeded by the powerful*". At the very top of the participation ladder, according to Arnstein [3], are three true forms of participation: partnership, delegated power and citizen control. The author argues that those forms are real demonstrations of power that change the balance between actors in decision making. Such participation forms illustrate actual public interests and willingness to be proactive in achieving goals that correspond with the new social values.

Similar to [3], Pimbert and Pretty [4] defined their own typology of participation consisting of passive participation, participation in information giving, participation by consultation, participation for material incentives, functional participation, interactive participation and self-mobilization. According to [4], interactive participation and self-mobilization are true forms of citizen involvement in environmental policy as they led to significant changes in the environmental management system.

Participation is often regarded in the context of the citizens' broader active engagement in public affairs. The idea itself can be observed as a response to the crisis of representative democracy. In other words, it is an expression of citizen discontent with the governmental (in)efficiency, (ir)responsibility, (un)knowing of the real circumstances and the (in)sufficiency of an offered solutions. At the same time, participation can be viewed as an example of passing the "hot" issue from the government to the citizens. Through participation, citizens can experience the "bitterness of power and decision making" while government makes scapegoats if something goes wrong.

The virus of participation is global and impacts environmental policy, such as many other spheres of public interest. In this case, the significance of participation for efficient environmental policy is pointed out because of the very nature of the policy's subject, which suggests apprehension of the environment as a public and a general good.

Not everyone is advocating for citizen participation, however. According to Cooke and Kothari [5], participation can also be regarded as a new tyranny. Namely, authors argue that participatory development facilitates the "*illegitimate and/or unjust exercise of power*". In addition, we point out that the tyranny of participation is one of the major characteristics of contemporary developed societies. Contemporary strategic documents, normative acts, development plans or scientific studies, among other things, label necessity and the right of the citizenry to participate in the processes of decision making, realization, monitoring and the evaluation of political decisions important for community development. Inevitably, the issue of participation became, along with the final effect, the measure of success in every organized social activity. The authors further argue that tyranny can be understood as a rejection of critiques that can be fatal for the development of the participation as a theoretical and practical concept.

As a theoretical idea, citizen participation in environmental policy has its background in the concept of sustainable development as well as in the ecological modernization theory. According to Mol and Sonnenfeld [6], ecological modernization as a set of social and institutional changes implies political modernization in the form of decentralization, strengthening the bottom-up approach and providing more opportunities for non-state actors. Namely, in the time of establishing the ecological modernization theory (early 1980s), authors such as Martin Jänicke [7] introduced a critical review on the role of the bureaucratic state in environmental reform. This analysis is in line with the review of a market in the environmental reform dynamics. Such an argument clearly emphasizes one of the primal postulates of the ecological modernization theory, which regards the necessity of bringing together environmental protection and economic growth and development.

Jänicke [8], like Huber [9] (one of the founders of the ecological modernization theory), argues that the main goal of ecological modernization is the upgrading of the level of eco-efficiency. Unlike Huber [9], however, Jänicke [10] focuses his analysis on environmental policy (i.e., policy of ecological modernization), its planning and stakeholders. Thus, Jänicke [10] defines the policy of ecological modernization as a totality of management activities towards stimulating diffusion of ecological innovations. According to the author [11], ecological modernization represents the process of modernization of the developed market economies that overtake the role of pioneers in the ecological modernization policy implementation and diffusion of policy innovations.

Among the authors that argue necessity of ecological modernization, there are two streams regarding the question of the main initiator of changes. The first stream includes those authors who are more orientated towards technological changes and the second stream is called 'sociocratic' because the authors are more inclined to look at the ecological modernization as a process of institutional reforms and reforms in behavioral patterns.

According to the ecological modernization theory, environmental problems are seen as a challenge, not an unconquerable obstacle [12]. In addition to this, supermodernization, as a set of gradual changes in contemporary societies, involves (i) changes in production, (ii) macroeconomic changes, (iii) changes in social institutions and (iv) changes in environmental policy. The last two are especially important with respect to participation in environmental policy. Changes in social institutions imply establishing an adequate base for environmental policy formation. Very important is the role of new social movements or NGOs, particularly when it comes to their participation in decision making. Capacities for such participation are associated with the changes in environmental policy itself in terms of embracing new ideologies and approaches. Such modifications are primary transformations in a dominant value system that results in new environmental policy goals and measures, as well as changes in behavioral patterns.

If we stand on the point that citizen participation is an important aspect of an efficient environmental policy, the central question is what is, "What is the optimal form for citizens to participate?" In defining such a form, we must begin at the very base, which implies factors that determine participation itself. Thus, we can speak of at least three factors. These are (i) the characteristics of a social structure, (ii) the characteristics of environmental awareness and (iii) the way an environmental policy works.

Another factor that should be taken into account is the social level of participation. Thus, we can speak about participation at the global, national, regional and local levels. Citizen participation is more likely to be linked with environmental policy at the local community level because the local community is a spatial and social locus of individual and group identification. Therefore, the local community can be considered as a basic social locus for achievement of individual and group interests.

Citizen participation within the local environmental policy is a fruitful ground for testing the idea of community capacity building. Being the "*local solutions to local problems*" [13], the concept of community capacity building implies relying on internal resources such as (i) human capital, (ii) social and cultural capital (the willingness of local populations, groups and organizations to cooperate to solve problems), (iii) physical capital (available infrastructure and other utilities that enable community development) and (iv) financial capital (e.g., structure of local community budget and available income sources). Therefore, community capacity building can be defined as an improvement in individual, group, institutional and organizational capacities to provide adequate support for community development.

The essence of community capacity building is to strengthen the participation of all interested stakeholders in the planning and implementation of activities relevant to the local community. Within this, particular emphasis is given to the active role of citizens. Namely, as Cuthill and Fien [14] point out, representative democracy, typical of most modern societies, carries with it certain disadvantages. These deficiencies derive from the fact that community development planning is a very complex process, burdened with a number of specific problems of which locally elected representatives may not have sufficient knowledge. Hence, the need to involve in planning and implementation activities as many different social actors possible arises. Such a common network allows social actors to connect and exchange their knowledge and experiences, as well as their individual and specific interests, within a unified plan of action to achieve certain goals.

Defining community capacity as "*a process that increases the assets and attributes which a community is able to draw upon in order to improve their lives*", Gibbon et al. [15] argue that community capacity building can be improved through relationships between government and NGOs.

One of the most frequently discussed issues today is the possibility of direct citizen participation in environmental policy. Opinions range from those who argue that participation of all stakeholders (thus, the citizens) is the only proper way to run environmental policy to those who posit that participation must be rejected as an unrealistic concept, full of restraints and difficulties.

Nevertheless, when discussing forms for citizen participations, NGOs seem to be often the appropriate mode as they are considered to be a mediator between citizens and governmental institutions. Such organizations can be defined as a group of citizens gathered around specific goals and brought together by shared specific values who organize specific activities aimed at achieving their mission. Furthermore, NGOs often have a not-for-profit orientation and they tend to influence and change public opinion as well as governmental activity toward similar or contrary objectives.

Environmental NGOs are not exclusive in contemporary societies. In fact, the first environmental movements were founded in the late 19th century in Great Britain. They were engaged in the preservation and protection of natural landscapes and/or plant and animal species. However, the first modern environmental NGOs were founded in the late 1960s and the beginning of the 1970s, concurrent with the recognition of environmental problems as social risks and with the process of democratization of modern societies.

According to Despotović [16], these organizations, as a specific form of new social movements, function as political innovators because their basic values and their political behaviors represent "*a challenge to existing institutional political system*". This political system due to interactions with those new movements adopts the concerns and interests of the NGOs and implements the existing political mechanisms to address them.

When speaking of new social movements, such as NGOs, in contemporary societies and politics, we argue, like Nesh [17], that their two crucial characteristics are (i) an orientation towards civil society and a changing public state of mind and (ii) a high dependency on the mass media.

3. A Case Study for the City of Novi Sad (Serbia)

It was mentioned before that the role and position of environmental NGOs can be understood the best when analyzed at the local community level. The local community, as an analytical framework, assists in an adequate understanding of participation as well as an understanding of the coordinated stakeholders' actions as one of the crucial factors for efficient environmental policy.

This analysis is based on the results from the empirical survey of all NGOs registered at the city of Novi Sad and engaged in local environmental protection issues. The city of Novi Sad is selected as a case study because its spatial and social structures have influenced changes in environmental quality, especially during the last two decades.

The city of Novi Sad is a major settlement in the municipality of Novi Sad which covers an area of 700 km^2 and has more than 380 000 inhabitants. The city of Novi Sad is also a regional center of the Autonomous Province of Vojvodina with approximately 262 000 inhabitants, which is 69 % of the total municipality population and 13 % of the total population of Vojvodina. Population density is very high. At the municipal level up, it reaches up to 430 inhabitants per km^2 and at the city level it reaches almost 1 600 inhabitants per km^2.

The development of the community can be continuously observed over the last three centuries [18]. The first mention of the city of Novi Sad dates back to the late 17th and early 18th centuries as a commercial and crafts center. The most significant changes in urban landscape and surroundings occurred in the middle 19th century, determined by the increase of the population and the heavy industrial and traffic development. Additionally, in the late 19th century, environmental problems such as the disposal of household waste and waste water from the households, craft workshops and industry were first noted. Moreover, there were a problem of air pollution as a result of the combustion of coal in households and workshops. Due to the lack of technical and technological solutions for wastewater treatment and the prevention of air pollution, the main causes for the environmental quality degradation was the lack of spatial and urban planning, including utility infrastructure. In addition, a constant increase in the city's population caused the spreading of arable land and an intensification of agricultural production, which resulted in changed environmental conditions in the periurban and rural areas.

Industrial development, mainly metal, textile, food and chemical industry, brought intensive rural to urban migration and a rapid increase in population. The city area expanded rapidly, changing the natural landscape. The city has experienced the most intensive development in the second half of the 20th century. The urban population has tripled over the last 50 years. The last decade of the 20th century was a period of dramatic changes in both physical and social urban space. These changes were reflected in the degradation of environmental quality. Most of those changes can be attributed to the general social conditions that the Serbian society has been facing during the 1990s. Along with the economic and political crises, these social conditions include the non-functioning of basic social institutions and organizations with respect to the urban management, a lack of financial resources for the implementation of urban planning measures, the actual nonfunctioning of local autonomy and the emergence of the interest groups of illegal nature. Combined, these factors resulted in urban disorganization. Hence, the environmental quality has also been significantly negatively impacted during the bombardment in the 1999. Perhaps the only positive result from these unfortunate events was the awakening of an environmental awareness.

The city of Novi Sad has an interesting history of environmental policy. The first steps in environmental protection at the local community level were made in the early 1970s with the implementation of quality monitoring of the air and drinking water. Since 1990, the Department of City Administration, which is responsible for addressing environmental protection issues, publishes a monthly bulletin *Ekobilten* that disseminates information on environmental quality. In 1992, the city of Novi Sad was the first city in Serbia to enact a declaration *Novi Sad – towards an ecological city* that guarantee the citizen the right to be informed on local environmental issues. In 1995, the city of Novi Sad was the first Serbian city to dedicate a part of the city budget solely for the purpose of environmental protection. From 1990 until 2000, the city of Novi Sad was the only Serbian city to have a representative of the environmental movement who would participate in the city assembly [19].

As previously stated, environmental NGOs vary in terms of their structure, organization, number of active participants, areas of activities and funding. In this chapter, those characteristics are regarded as guidelines that shape their capacity to take an active role in local environmental policy. Because of its specific environmental policy history and the present state of its environmental quality, the city of Novi Sad is used as an analytical framework for examining the role and position of environmental NGOs in local environmental policy. Data on local environmental NGOs are collected by standardized questionnaires consisting of fifty questions classified into six groups: (i) nine general questions on organization, (ii) eight questions on human resources, (iii) seven questions on equipment and facilities, (iv) ten questions on the organization's areas of activity, (v) eleven questions on cooperation and information issues and (vi) five questions on financial resources.

The research results show that there is a significant difference between the number of registered and currently active environmental NGOs. To be precise, there are twenty six registered environmental NGOs in the city of Novi Sad, but only half of them are currently active. Another half are most likely fictive organizations founded for unknown purposes and they have no intention of being environmentally active.

As was expected, most of the organizations (five) were established after 2000, as a consequence of the changes in general social circumstances and in the national environmental policy. Three organizations were established in the early 1970s and they are considered to be a response to the first steps in establishing a national environmental policy. By the late 1950s, systematic national activities geared toward environmental issues were primarily concerned with the ratification of international conventions dealing with the protection of nature. In addition, during this early stage of environmental policy development, there were certain activities that focused on the legal protection of natural landscapes such as nature parks, natural monuments and unique natural areas. Nevertheless, there was no defined national environmental policy or precise and coordinated network of policy social actors.

After the first United Nations Conference on Environment and Development, Serbia as one of the former Yugoslav republics initiated the first steps in establishing a national and federal environmental policy. This milestone was the legal foundation of the environmental rights in the constitution in 1974 [20]. Another important step was the constitutional establishment of the environmental policy's social actors network. That is, according to the constitution, this network will consist of "*working people and citizens, organizations of joint work, political organizations, local communities and other self-governed organizations*" [20].

In this second stage of the national environmental policy's development, some of the major environmental laws were declared. Some of these laws were directed towards activities in environmental protection and others dealt with these issues from the perspective of establishing legal regulation. As the ratification of international environmental conventions and normative acts is continued, until the end of the 1970s, former Yugoslavia had ratified 19 international protocols on environmental protection [21].

Despite the first steps taken in the development of a national environmental policy, we question whether this policy was only formally implemented and whether the social circumstances contributed to the lack of full recognition of the significance of the environmental issues in the everyday lives of the citizens.

Three other environmental NGOs were founded in the late 1980s and in the beginning of the 1990s. These organizations can be regarded as a product of the significant changes in the social, economic and, foremost, political scene. By the middle of the 1980s Yugoslav/Serbian society faced the end of the self-management system. Serbian society was also introduced to the origins of the plurality in economic relations, as well as in politics. These changes reflected on the environmental stakeholders' network. An economic crisis which began in the middle of the 1980s, conditioned transformation in the environmental policy stakeholders' network, that is, the leading actors in decision making became state and governmental institutions. This stage in environmental policy development was characterized by the foundation of the institutions authorized in environmental protection. The role of state as social actors in environmental policy became dominant. This is reflected in the fact that autonomous provinces, such as Vojvodina (with the city of Novi Sad as a capital city), although a legal part of the Republic of Serbia, had no jurisdiction in the area of environmental protection.[1] The most important change in this phase was the creation of basis for the plurality of social actors engaged in environmental policy. This served as fruitful ground for founding NGOs. First NGOs in Serbia were focused on human rights issues and on environmental protection.

As the process of the foundation, development and functioning of environmental NGOs is influenced by the social context in which they occur, Despotović [16] notes their specific role and position in the post-socialist societies. In his opinion, these movements and organizations, except for their primary environment objectives, have had the task of "*breaking the monopolistically organized political life*". In this context, the foundation and functioning of environmental NGOs should be seen as a significant contribution to the promotion of the role of citizens as legitimate social (political) actors and, hence, as assistants in the creation and strengthening of civil society. Furthermore, environmental NGOs should also be viewed in the broader context, as a, more or less, organized expression of chronic social unrest and dissatisfaction with certain social problems.

After this stage, the 1990s were a period of economic crisis, political totalitarianism and social isolation. Like many other spheres of organized social activities, environmental policy was entirely dysfunctional. Activities of institutions and organizations were minimized due to issues such as the lack of financial resources, to structural problems and to the lack of functional coordination. It is obvious that general social conditions at the time discouraged systematic actions in environmental protection. Therefore, the 1990s can be described as a period of frozen activity in environmental policy.

After 1999, however, environmental issues began to rise as one of the major social problems/consequences of the bombardment of Serbia. According to the *Report on environmental consequences of bombing in Yugoslavia*, the quality of the environment in the city of Novi Sad was seriously threatened [22]. These threats mostly concerned water, air and land pollution and damage caused to protected natural landscapes.

Only one year after the bombardment, Serbian society faced democratic changes. After the year 2000, Serbian society focused its attentions on economic renewal, political democratization and stabilization. EU accession and integration also became an important political issue. Social changes drove national environmental policy towards a process of re-activation. A focus was placed on complying with national environmental goals, standards and activities and aligning national environmental goals with the regional and international goals. The major difference compared to previous periods in national environmental policy development was the existence of social interest in environmental protection. This increased interest in environmental protection was a consequence of the global efforts in environmental protection, but and Serbian efforts to become an integral part of the international (foremost, European) community.

One of the first steps in the modernization of the national environmental policy was the renewal of the environmental legislation and its harmonization with EU legislation. In 2004, four major environmental laws were enacted. These laws were the *Environmental Protection*

---

[1] The Autonomous Province of Vojvodina, with the capital city of Novi Sad, gained legal option for establishing its own principal institution in environmental protection no sooner than 2002 after the law on establishing certain responsibilities of Autonomous Province which reinforce Vojvodinian authorities.

Act, the *Strategic Environmental Impact Assessment Act*, the *Environmental Impact Assessment Act* and the *Integrated Pollution Prevention and Control Act*. Each of these acts states decentralization in environmental policy to enable broader participation of citizenry as well as any other social actor interested in environmental issues. The *National environmental protection program* argues that one of the key factors in modernizing the national environmental policy is a redistribution of power in decision making [21]. This is primarily aimed to intensify the capacities of local communities (citizens and NGOs) for the preparation and realization of environmental action plans as a base for efficient local environmental policy.

According to the results from the research on citizens' environmental awareness,[2] the citizens of Novi Sad think that they, along with the NGOs, must have a large degree of involvement in local environmental policy [23]. Such results are quite promising as they reflect the willingness of citizens to participate. The same research suggests that citizens think of environmental NGOs as the most convenient form of their involvement in local environmental policy. On the other hand, political parties are seen as an inappropriate form of citizen engagement because of the parties' tendency to exploit environmental issues as well as the commitment of the citizens. In such a context, the research of the environmental NGOs attempts to answer the following question "Do those organizations have a capacity to answer to the citizens expectations?" This question implies the analysis of present potentials but it also points to the problems in environmental NGOs and their possible solutions. From at total of eleven environmental NGOs registered and active at the city of Novi Sad, the research included an equal number of local, regional and national organizations. Furthermore, it is worth noting that some organizations implemented their activities at all of the levels, depending on the specifics of the current projects.

According to the survey, the main areas of interest and involvement of currently active environmental NGOs are (i) the protection of natural landscapes, (ii) the protection of water, air and soil, (iii) the preservation and enhancement of biodiversity and (iv) the animal welfare. These areas can be regarded as traditional or common areas of work for the environmental NGOs. One aspect of the currently active environmental NGOs (five out of eleven) deals with issues of environmental legislation, for example, the monitoring and analysis of the legal measures, the monitoring and analysis of the implementation of proposed legislative solutions and the cooperation with relevant institutions dealing with new legal regulations in this area. Recent developments in environmental protection drove activities of these NGOs toward directions such as projects related to sustainable rural development, the promotion of organic agriculture, waste management and sustainable tourism.

Within the identified areas of interest, the environmental NGOs fulfill their goals through preparing and implementation a series of activities. The most common forms of these activities consist of (i) cooperation with the media, (ii) organization and implementation of concrete actions to protect and improve the environment (e.g., protests and campaigns), (iii) organization of events, conferences, seminars, panel discussions, workshops, among

[2] The research was conducted on a sample of the total population from the four settlements in the municipality of Novi Sad (the city of Novi Sad and three more settlements with tight spatial, structural and functional relations to Novi Sad). The survey was based on a questionnaire containing fifty one questions divided into five groups: (i) demographic data on the respondent, (ii) environmental quality assessment in the residential block, (iii) environmental quality assessment at the local community level, (iv) participation in environmental policy and (v) environmental information.

others, (iv) cooperation with educational institutions (primary and secondary schools), (v) publications such as magazines, brochures, manuals and books and (vi) research. Working with media, as the most common activity, is of great importance because the media can promote the goals, roles and positions of the environmental NGOs to the other social actors in the environmental stakeholders' network. Furthermore, the dissemination of environmental information is vital for raising public awareness.

The research on the environmental awareness and participation of the citizens of Novi Sad in environmental policy indicates that there is a disparity between the latent and the manifest dimensions of environmental awareness. This means that citizens have significant knowledge of the environmental issues upon which they form their opinions and attitudes; however, they do not practice or implement environmental policy measures in their everyday lives. In such context, the organization of different activities towards environmental protection, such as campaigns, protests, seminars and workshops, has a major impact on strengthening the motivation of the citizens to participate in environmental policy. Such activities are not only the way to introduce environmental issues to the public but also an opportunity to raise awareness for all stakeholders to participate and to facilitate networking.

The majority of the environmental NGOs (eight of eleven) are satisfied with the effects of their activities. As main reasons for their contentment, they cite (i) good cooperation with the media, (ii) trust of the citizens, (iii) realization of projects goals, (iv) successful experience exchange and (v) innovative approaches in project realization.

It was previously mentioned that cooperation between government and NGOs is an essential for building community capacity. Therefore, potentials for the improvement of environmental policy lay in establishing close links between these two sectors. The media, however, should also be included as an essential link due to their role in raising environmental awareness. The results illustrate that the environmental NGOs that have participated in the research are not active enough on issues regarding the application of proposals to the environmental government institutions. The results further indicate that the NGOs need to become more active in the process of planning environmental protection at the local, regional and/or national levels. The relatively small number of environmental NGOs involved in such activities indicates that there is a problem establishing connections between the governmental and the nongovernmental sector. This is a common problem in post-socialist societies; however, usually during the transitional period, these two sectors learn how to encourage and engage in civil discourse.

It is also important to emphasize that the problem with the insufficient participation of environmental NGOs, and NGOs in general, in policy planning has two sides. Namely, governmental institutions and organizations often lack the capacities to cooperate with the NGO sector. In transitional societies, governmental organizations are often burdened with massive, inflexible, bureaucratic internal structures. Such structures represent important obstacles when it comes to both internal and external functional coordination. The most common problems cited regarding capacity for cooperation among governmental institutions are a lack of qualified human resources and the insufficient procedures for participation of the NGO sector. On the other hand, it is important to note that environmental NGOs also have internal difficulties that restrain them from participating in policy planning and decision making. Such difficulties include environmental NGO understaffing, a lack (in some cases) of information on the available modes of participation and even a dogmatic adherence to the goals and interests of the NGO that is unwillingness to compromise. All previous factors make participation more complex than it should be which is very unfortunate for the citizens, especially because the citizens perceive environmental NGOs as the most appropriate model for their inclusion in environmental policy issues.

Most of the environmental NGOs are satisfied with their contribution to the environmental protection at the local community level. They also consider themselves to be recognizable by the public, especially by the environmental stakeholders.

Previously, we stated that one of the purposes of this chapter is to point out the environmental NGOs' potentials for taking an active role in local environmental policy. That implies that we address some of the characteristics of the NGOs that can be seen as problems in their everyday functions and can impact goal achievement. According to the research results (Figure 1), we have identified the four main problems of environmental NGOs: (i) a lack of financial resources, (ii) problems in cooperating with other stakeholders in local environmental policy network, (iii) a lack of equipment and facilities and (iv) problems in the internal workflow.

A lack of financial resources seems to be the dominant problem characteristic of all local environmental NGOs. This problem implies two main obstacles: (i) low incomes and (ii) irregularity of incomes. According to the research, the principal strategy to diminish these problems is diversification of income sources. As the most common income sources, local environmental NGOs declare donations, membership fees and project contracts and some organizations fill their budget by providing services.

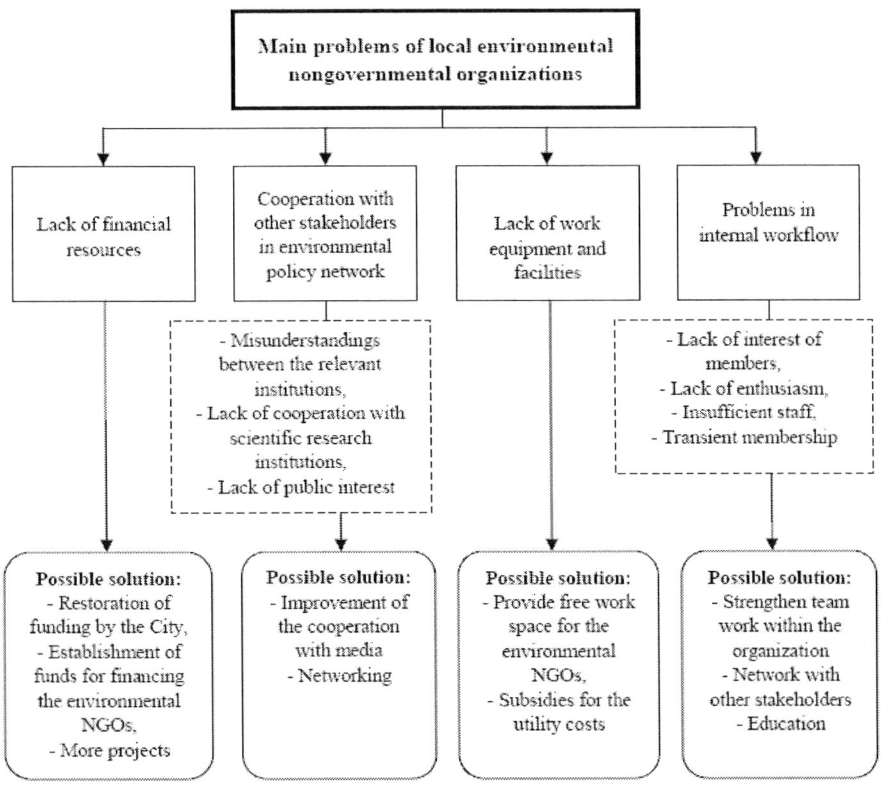

Figure 1. Main problems of local environmental nongovernmental organizations and possible solutions.

Donations are the most common income source and they include donations from various sources. The most frequent are the donations from national, provincial and local government institutions and business corporations. Membership fees are also a frequent source of income, but it is questionable what the share of this type of income is in overall budget. The third most usual type of income source is project contracts, particularly with national, provincial and local government institutions. For example, local environmental NGOs from the city of Novi Sad have an opportunity to participate in tenders announced by the Provincial Secretary for Environmental Protection and Sustainable Development, the Ministry of Environment, Mining and Spatial Planning of the Republic of Serbia and other similar institutions. The most frequent project contracts, however, are those with the Department for Environmental Protection of the city of Novi Sad. For instance, every year the Department for Environmental Protection of the city of Novi Sad announces a public tender for financing projects related to environmental protection and sustainable development. In 2011, twenty three projects were contracted with environmental NGOs. It is worth noting that the total value of these projects is 29 % of the total projects' budget [24].

The research further indicates that the most significant source of income, according to the share in the organization budget, is project contracts with national, provincial and local government institutions. This type of income is not regular, however, and the project budget addresses project activities. Therefore, there are no sufficient finances to redirect to parts of the organizational budget intended for other purposes, such as utility costs and staff costs.

Most of the local environmental NGOs are partially satisfied with their current financial state. As a way to improve their budgets, organizations emphasize the need for more project contracts. More project contracts not only increase income flow, but they also promote the organizational goals.

Another important problem is inadequate or insufficient equipment and facilities. This problem is closely connected to the previous one and can be regarded as an internal organizational problem. Most of the local environmental NGOs have their own work facilities. However, because of the low and irregular incomes, the main problem regarding work facilities is their rent and utility costs. According to the research results, those costs are the highest expenses for environmental NGOs.

The second half of the problem is a shortage in necessary equipment, such as computers and vehicles. Donations are the most common source of funding for the acquisition of required equipment and most of the organizations are satisfied with the quality of their existing equipment.

The data further show that older environmental NGOs, those founded before 2000, are more satisfied with their work facilities, but not with the quality of their equipment. As a solution for this problem, local environmental NGOs propose to be provided with free work facilities as a part of the city project to improve the NGO sector and, thus, improve citizen participation. Local environmental NGOs argue that subsidies for the utility costs would be of great assistance in lowering their total expenses and would, consequently, increase the part of the NGOs budget that is projected for other activities.

The third problem regarding internal issues of the local environmental NGOs is a problem with workflow and delegation. Such problems are due to (i) a fluctuating membership, (ii) a lack of interest among members to participate in organizational activities, (iii) an insufficient staff and (iv) a lack of enthusiasm.

Work structure is an inherent issue in every organization. In the case of local environmental NGOs, the issue of work structure carries specific weight because most of these organizations have no permanent employees. Therefore, the organizational activities that are designed to help the NGOs meet their goals are based on the work of volunteers. In such a context, it is not possible to define precise work structure. Also, because we are analyzing local environmental NGOs, it is expected that they have a minimal number of members (potential volunteers) and a fluctuating/transient membership. The majority of the organizations are currently not satisfied with the flexible work structure and they are planning some organizational changes, which implies a reorganization of existing human resources in terms of redistribution of their engagement.

Problems in internal workflow are mainly human resource issue. Under conditions of scarce human resources, local environmental NGOs often emphasize the need of increase members interest and willingness to participate in organizational activities. A lack of interest and enthusiasm among members of these organizations is partly caused by the state of the post-socialist transition.

With respect to improving human resources in local environmental NGOs, it is important to note their (the NGOs) willingness to invest in education and training of staff or volunteers. According to the research results, all environmental NGOs reported that their employees or volunteers attended certain level of organized professional education and/or training, such as conferences, seminars and workshops. Professional education is not only a model for improving human resources, but also a channel for establishing connections and, thereby, strengthening environmental NGOs as a specific environmental policy stakeholder.

One of the answers for the deficiency in human resources is to strength team work. Mutual activities toward achieving projected goals are often found beneficial in raising enthusiasm and skills development, a crucial factor when dealing with staff shortage.

Depending on the specific problem, one possible solution is to make valuable connections outside the organization. Actually, networking with other NGOs, as well as other stakeholders in environmental policy, is of fundamental significance for efficiency in policy planning and in the implementation of policy measures. This is closely connected with the problem of cooperation among other stakeholders within the environmental policy network. This problem includes issues such as misunderstandings with relevant governmental institutions, a lack of cooperation with scientific research institutions and a lack of public interest.

The capacity for networking is based on the role and position of each environmental NGO in the stakeholders' network. The analysis of these capacities identifies on two levels. The first level is in regards to the cooperation of an environmental NGO with other environmental NGOs and with NGOs in general. The second level is based on the characteristics of the cooperation between environmental NGOs with other stakeholders within the environmental policy network.

The first case implies the analysis of the internal characteristics of environmental NGOs as a homogenous group of social actors. Research shows that environmental NGOs registered in the city of Novi Sad are primarily orientated towards cooperation with other (Serbian) environmental NGOs and other NGOs related to nonenvironmental issues. Most of the environmental NGOs also cooperate with foreign environmental NGOs. The analysis of the internal characteristics of environmental NGOs as a relatively homogenous stakeholders group implies not only the frequency of contacts but also the quality of cooperation. The research results illustrate an interesting feature. Namely, environmental NGOs are more positive about cooperation with foreign NGOs than they are about their cooperation with Serbian NGOs. The average evaluation score of such cooperation (with foreign NGOs) is 3.9 (on the scale of 1 to 5, where 5 is the best), while the cooperation with Serbian environmental NGOs is evaluated at 3.4. All environmental NGOs argue the need to improve cooperation by communication, engaging in mutual projects, arranging meetings and other similar activities.

It was previously stated that networking plays an important part in the improvement of the role and position of environmental NGOs. In general, networks, as a model for organizing mutual activities, imply reaching consensus on mutual priorities and facilitate more determined approach in goal achievement. The research suggests that half of all environmental NGOs are members of, mostly international, environmental NGO networks. Those organizations think highly of the benefits received from the networking (average evaluation score is 3.9, on the scale of 1 to 5). They stress that the most common forms of cooperation and activities within the network are: (i) the exchange of experience, knowledge and information, (ii) professional education and (iii) mutual projects.

The second level of the analysis provides insight into the characteristics of cooperation of the environmental NGOs with other stakeholders in the environmental policy network. The research indicates that local environmental policy networks can be defined as heterogeneous, according to the present types of stakeholders involved. The majority of the environmental NGOs cooperate with the (mostly local) media, with scientific and educational institutions and with the citizens. When considering the quality of cooperation, environmental NGOs felt most positively about their relationships with the scientific and educational institutions (average evaluation score of 4.2, on the scale of 1 to 5) and the citizens (average evaluation score of 4.1, on the scale of 1 to 5). The high scores for cooperation with scientific and educational institutions is very important from the aspect of the role that education plays in raising environmental awareness. It is also a key factor in the process of the diffusion of knowledge and information regarding adequate environmental decision making. Similarly, high evaluation scores for cooperation with citizens are a great incentive and motivating factor for facilitating more active engagement in local environmental policy.

When we previously discussed citizen participation, we found that the research confirms our conclusion that environmental NGOs are, in opinion of the citizens, the most convenient way to become involved in local environmental policy. In fact, 48.1 % of the citizens who believe they should participate in environmental decision making state that NGOs are the best way to so. As expected, there is a positive correlation between the choice of ways to participate and the level of information on NGOs in that the: higher the level of information about the NGOs, the more frequent they are selected as the preferred form of participation.

The results show that 73.3 % of the citizens of Novi Sad who have participated in the research have heard of at least one of the environmental NGOs. However, the concern is that only one-third of the citizens are familiar with the NGOs' activities with respect to environmental protection. Citizens get their information about environmental NGOs mainly through the media (68.9 % of citizens identified the media as their primary information source). Informal sources, such as friends and relatives, are the second most common information source.

The majority of the citizens of Novi Sad who participated in the research (95 %) have never been a member of any NGO, even though most of them insist that environmental NGOs are the most appropriate way for them to participate in shaping environmental policy. Only

1.7 % of the citizens of Novi Sad who participated in the research are currently active members of environmental NGOs. Because of that, environmental NGOs argue that cooperation with citizens, as well as other groups of stakeholders should be facilitated and supported by improving the flow of information, organizing joint activities and offering educational opportunities.

In stark contrast to the cooperation with citizens, environmental NGOs think that the lowest quality of cooperation is characteristic of their relations with the local government institutions. Low evaluation scores can be interpreted as a limitation factor for the functioning and overall efficiency of the local environmental policy as local government institutions have mandatory rights and obligations to coordinate the environmental stakeholder network.[3]

Improving the NGOs' cooperation with the media is one of the possible solutions for strengthening the environmental policy network. The role of the media is particularly important when there is no functional or active public relation service for the local government institutions. Such a role should be overtaken by the Department for the Environmental Protection of the city of Novi Sad but there must be more structural changes within the institution to efficiently enact a completely new approach in environmental policy planning and practice.

The majority of the environmental NGOs argue that information on environmental issues is not adequately represented in the local media, which correspond to the citizens' opinion on that matter. When it comes to regarding media as a channel for promoting the environmental NGOs, the results suggest that most of the NGOs are not satisfied with the quality of information that is being disseminated about their work and activities. To improve cooperation with the local media, environmental NGOs posit that it is important to develop new presentational approaches, both within the media and within the NGOs themselves. This implies improved self-presentation and public relations by the environmental NGOs through more public-friendly web-sites, current information, more frequent contacts with the media and a more professional and responsible relationship with the media.

The success of each organization's activities depends on, among other factors, the scope and quality of the information presented. Available internal and external information grouped in the form of an information system that represents the basis for planning, monitoring and evaluating of organizational activities would be a logical answer. According to the value of information about the functioning of environmental NGOs, an important segment of the analysis was centered on public accessibility of the information. Environmental NGOs concur that information on environmental protection and related issues is partially public and partially accessible. This means that, according to the NGOs themselves, there is selectivity in the dissemination of environmental information, especially when speaking of accessibility. This is due, in large part, to bureaucratic procedures. This concern is additionally burdened with the fact that there is no unified system of environmental information that would enable quick and systematic collecting, processing and exchanging of information[4] at either the local or the national level.

---

[3] Citizens gave similar evaluation score of cooperation with local governmental institutions which supports the previous conclusion on necessary internal reform of the local environmental governmental sector.

[4] In January 2008, EU declared an initiative for establishing SEIS (Shared Environmental Information System). Such a system enables environmental quality monitoring and represents the basis for decision making from the local level to the level of the EU.

Conclusions

In this chapter, we suggested that environmental NGOs play a significant role in the development and implementation of environmental policy, especially at the local community level. In addition, from the research and from the theoretical background, three main conclusions can be underlined. The first conclusion regards the role of NGOs as an environmental policy stakeholder. We found that these organizations are vital in facilitating citizen participation. This is not only with respect to participation in the decision making but also in practicing and implementing the environmental policy measures.

The second conclusion concentrates on the role of environmental NGOs in practicing democracy. Namely, the functioning of the environmental NGOs (or any other type of NGO) facilitates the process of democratization in post-socialist societies, in terms of breaking the monopoly in policy making and practice. Even though Serbian society, as a transitional society, has embraced plurality in the political system over 20 years ago, it is still coping with civil engagement. Furthermore, the conditions of transition do not promote citizen participation as there are no cultural traditions of political participation and there are conflicts of individual interests in process of participation. In such context, NGOs are a useful solution for making the necessary connection between citizens and governmental institutions.

Our final conclusion regards the position of environmental NGOs. We argue that their position within the environmental stakeholder network is determined not only by the relations of NGOs within the homogenous, but also within the heterogeneous group of stakeholders. The level of cooperation among environmental NGOs within these two groups of stakeholders determines the NGOs' financial status and their level of power and influence, which ultimately, determines the overall functioning of the environmental policy network.

The research indicates that the environmental NGOs' ability to take an active role in environmental policy is constrained due to four main types of problems: (i) a lack of financial resources, (ii) the problems with cooperation among stakeholders in the local environmental policy network, (iii) an insufficiency in equipment and facilities and (iv) the problems within the internal workflow.

We argue that two major solutions should be imposed to address the identified problems. The first solution regards networking as a basic principle in environmental stakeholders' activities. Common project activities, the exchange of experiences and information and the mutual education of members and staff of NGOs are all practical elements of stakeholder networking that correspond with the contemporary theoretical issues explained in the theory of resource mobilization and the theory of social capital.

The second solution concerns the information flow. The necessary development of the environmental information system is closely linked with networking. The necessity of a functional system of environmental information implies that information is public and accessible. Furthermore, such a system must operate and distribute at least three types of information: (i) information on environmental quality, (ii) information on environmental policy goals and measures and (iii) information on environmental policy stakeholders. A complex and updated system of required information can contribute not only to increasing environmental awareness. It will also help to mediate prejudice among stakeholders thus resulting in an increased efficiency in implementing environmental policy measures.

ACKNOWLEDGEMENTS

This chapter was realized as a part of the project "Sustainable agriculture and rural development in achieving strategic goals of the Republic of Serbia within Danube Region" (No. 46006), which is financed by the Ministry of Education and Science of the Republic of Serbia within the framework of integrated and interdisciplinary for research over the period 2011 - 2014.

REFERENCES

[1] D. Meadows, D. Meadows, J. Randers, W. Behrens III, *Limits of Growth*; Stvarnost: Zagreb, HR, 1974; p. 182 (In Croatian).
[2] G. Bruntland et al., *Our Common Future;* United Nations: Washington, US, 1989.
[3] S. Arnstein, *JAIP*, 35(4), 216-224 (1969).
[4] M.P. Pimbert, J.N. Pretty, In *Social Change and Conservation: Environmental Politics and Impacts of National Parks and Protected Areas;* K.B. Ghimire, M.P. Pimbert, Eds.; UNRISD and Earthscan: London, GB, 1997; pp. 297-330.
[5] B. Cooke, U. Kothari, *Participation: The New Tyranny?;* Zed Books: London, GB, 2001; p. 211.
[6] A.P.J. Mol, D.A. Sonnenfeld, *Ecological Modernization Around the World;* Frank Cass Publishers: London, GB, 2000; p. 300.
[7] M. Jänicke, *State Failure;* Pennsylvania State University Press: Pennsylvania State University, US, 1990; p. 175.
[8] M. Jänicke, *J. Clean. Prod.*, 16, 557-565 (2008).
[9] J. Huber, *New Technologies and Environmental Innovation;* Edward Elgar Publishing Ltd.: Cheltenham – Northampton, GB, 2004; p. 384.
[10] M. Jänicke, *Ecological Modernization: Innovation and Diffusion of Policy and Technology (FFU-report 00-08);* Forschungsstelle für Umweltpolitik: Berlin, DE, 2000.
[11] M. Jänicke, *Eur. Env.*, 15, 129-142 (2005).
[12] A. Blowers, *Urban Stud.*, 34(5-6), 845-871 (1997).
[13] R. Atkinson, P. Willis, *Community Capacity Building: A Practical Guide;* UTAS: Hobart, IN, 2006; p. 15.
[14] M. Cuthill, J. Fien, *Aust. J. Pub. Admin.*, 64(4), 63-80 (2005).
[15] M. Gibbon, R. Labotne, G. Laverack, *Health Soc. Care Comm.,* 10(6), 485-491 (2002).
[16] Lj. Despotović, *Ecological Paradigm – Contributions in Establishing the Political Ecology;* Stilos; Novi Sad, RS, 2002; p. 152 (In Serbian).
[17] K. Nesh, *Contemporary Political Sociology: Globalization, Politics and Power;* Blackwell Publishing: Malden - Oxford, GB, 2000; p. 313.
[18] Đ. Srbulović, *Short History of Novi Sad;* Prometej: Novi Sad, RS, 2000; p. 311 (In Serbian).
[19] N. Aleksić, *20 years of Ecological Movement of the City of Novi Sad*; Ecological Movement of Novi Sad: Novi Sad, RS, 2010; p. 280 (In Serbian).

[20] Constitution of Socialistic Federal Republic of Yugoslavia; Official Gazette of SFRY: Belgrade, RS, 1974 (In Serbian).

[21] National Environmental Protection Program – Proposition; Government of the Republic of Serbia: Belgrade, RS, 2007 (In Serbian).

[22] The Environmental Consequences of NATO Bombing on SR Yugoslavia; Federal Ministry of Development, Science and Environment: Belgrade, RS, 2000; p. 87 (In Serbian).

[23] J. Čikić, Ž. Petrović, In *Environmental, Health and Humanity Issues in the Down Danubian Region - Multidisciplinary Approaches; Proceedings of the 9th International Symposium on Interdisciplinary Regional Research, University of Novi Sad 21 - 22 June 2007;* D. Mihailović, M. Vojinović Miloradov, Eds.; World Scientific: Singapore, SG, 2008; pp. 309-316.

[24] The Conclusion of Public Tender for Allocation of Funds for Co-financing the Research Programs, Projects and Activities to Strengthen Environmental Awareness at the Territory of the City of Novi Sad in 2011; City Department for Environmental Protection: Novi Sad, RS, 2011 (In Serbian).

INDEX

B

C

D

E

F

G

H

I

P

Q

R

S

U

V

W

Y

Z